This report contains the collective views of an international group of experts and does not necessarily represent the decisions or the stated policy of the United Nations Environment Programme, the International Labour Organisation, or the World Health Organization.

Environmental Health Criteria 108

NICKEL

Published under the joint sponsorship of the United Nations Environment Programme, the International Labour Organisation, and the World Health Organization

First draft prepared by Dr R.F. Hertel,
Dr T. Maass and Ms V.R. Müller,
Fraunhofer Institute of Toxicology and Aerosol Research, Germany

World Health Organization
Geneva, 1991

The International Programme on Chemical Safety (IPCS) is a joint venture of the United Nations Environment Programme, the International Labour Organisation, and the World Health Organization. The main objective of the IPCS is to carry out and disseminate evaluations of the effects of chemicals on human health and the quality of the environment. Supporting activities include the development of epidemiological, experimental laboratory, and risk-assessment methods that could produce internationally comparable results, and the development of manpower in the field of toxicology. Other activities carried out by the IPCS include the development of know-how for coping with chemical accidents, coordination of laboratory testing and epidemiological studies, and promotion of research on the mechanisms of the biological action of chemicals.

WHO Library Cataloguing in Publication Data

Nickel

(Environmental health criteria ; 108)

1.Nickel – adverse effects 2.Nickel – toxicity 3. Environmental exposure I.Series

ISBN 92 4 157108 X   (NLM Classification: QV 290)
ISSN 0250-863X

© World Health Organization 1991

Publications of the World Health Organization enjoy copyright protection in accordance with the provisions of Protocol 2 of the Universal Copyright Convention. For rights of reproduction or translation of WHO publications, in part or *in toto*, application should be made to the Office of Publications, World Health Organization, Geneva, Switzerland. The World Health Organization welcomes such applications.

The designations employed and the presentation of the material in this publication do not imply the expression of any opinion whatsoever on the part of the Secretariat of the World Health Organization concerning the legal status of any country, territory, city or area or of its authorities, or concerning the delimitation of its frontiers or boundaries.

The mention of specific companies or of certain manufacturers' products does not imply that they are endorsed or recommended by the World Health Organization in preference to others of a similar nature that are not mentioned. Errors and omissions excepted, the names of proprietary products are distinguished by initial capital letters.

Computer typesetting by HEADS, Oxford OX7 2NY, England

PRINTED IN FINLAND
Vammalan Kirjapaino Oy
91/8809 — VAMMALA — 5200

CONTENTS

ENVIRONMENTAL HEALTH CRITERIA FOR NICKEL 11

1. SUMMARY AND CONCLUSIONS 13
   1.1 Identity, physical and chemical properties,
       and analytical methods 13
   1.2 Sources of human and environmental exposure 13
   1.3 Environmental transport, distribution, and
       transformation 14
   1.4 Environmental levels and human exposure 15
   1.5 Kinetics and metabolism in human beings and animals . . . 17
   1.6 Effects on organisms in the environment 18
   1.7 Effects on experimental animals and *in vitro* test systems . 18
   1.8 Effects on human beings 21

2. IDENTITY, PHYSICAL AND CHEMICAL
   PROPERTIES, ANALYTICAL METHODS 25
   2.1 Identity, and physical and chemical properties
       of nickel and nickel compounds 25
       2.1.1 Nickel carbonate hydroxide 25
       2.1.2 Nickel carbonyl 29
       2.1.3 Nickel chloride and nickel chloride hexahydrate . . 29
       2.1.4 Nickel hydroxide 29
       2.1.5 Nickel nitrate 29
       2.1.6 Nickel oxide 29
       2.1.7 Nickel sulfate 30
       2.1.8 Nickel sulfide 30
       2.1.9 Nickel subsulfide 30
   2.2 Analytical methods 30
       2.2.1 Determination of trace amounts 43
       2.2.2 Sample collection 43
       2.2.3 Sample pretreatment 44
       2.2.4 Analytical methods 45

3. SOURCES OF HUMAN AND ENVIRONMENTAL
   EXPOSURE 48
   3.1 Natural occurrence 48
       3.1.1 Rocks 48
       3.1.2 Soils 49
       3.1.3 Water 50
       3.1.4 Fossil fuels 50
       3.1.5 Air 51
   3.2 Man-made sources 51
       3.2.1 Production, use, and disposal 51
             3.2.1.1 Primary production 51
             3.2.1.2 Intermediate products and end-use 55
             3.2.1.3 World production levels and trends 57
             3.2.1.4 Emissions from the primary nickel
                     industry 60
             3.2.1.5 Emissions from the intermediate
                     nickel industry 62
             3.2.1.6 Emissions from the combustion of
                     fossil fuels 62
             3.2.1.7 Emissions from sewage sludge and
                     waste incineration 65
             3.2.1.8 Miscellaneous emission sources 66
             3.2.1.9 Waste disposal 66

4. ENVIRONMENTAL TRANSPORT, DISTRIBUTION,
   AND TRANSFORMATION 68
   4.1 Transport and distribution between media 68
       4.1.1 Air 68
       4.1.2 Water 70
       4.1.3 Rocks and soil 71
       4.1.4 Vegetation and wildlife 72
   4.2 Uptake and bioaccumulation 73
       4.2.1 Terrestrial organisms 73
       4.2.2 Aquatic organisms 76
   4.3 Biomagnification 79

5. ENVIRONMENTAL LEVELS AND HUMAN
   EXPOSURE 80
   5.1 Environmental levels 80
       5.1.1 Air 80

	5.1.2	Drinking-water	81

         5.1.2   Drinking-water 81
         5.1.3   Food 82
         5.1.4   Terrestrial and aquatic organisms 88
    5.2 General population exposure 88
         5.2.1   Oral 88
         5.2.2   Inhalation 92
         5.2.3   Dermal 92
    5.3 Iatrogenic exposure 94
    5.4 Occupational exposure 95

6. KINETICS AND METABOLISM 101
    6.1 Absorption 101
         6.1.1   Absorption via the respiratory tract 103
                 6.1.1.1   Particulate nickel 104
                 6.1.1.2   Nickel carbonyl 108
         6.1.2   Absorption via the gastrointestinal tract 109
                 6.1.2.1   Experimental animals 110
                 6.1.2.2   Human beings 111
                 6.1.2.3   Factors influencing gastro-
                           intestinal absorption 111
         6.1.3   Absorption through the skin 113
                 6.1.3.1   Experimental animals 114
                 6.1.3.2   Human beings 115
         6.1.4   Other routes of absorption 116
                 6.1.4.1   Experimental animals 117
                 6.1.4.2   Human beings 117
         6.1.5   Transplacental transfer 117
                 6.1.5.1   Experimental animals 118
                 6.1.5.2   Human beings 120
         6.1.6   Nickel carbonyl 120
    6.2 Distribution, retention, and elimination 121
         6.2.1   Transport 121
         6.2.2   Tissue distribution 122
                 6.2.2.1   Experimental animals 123
                 6.2.2.2   Kinetics of metabolism 133
                 6.2.2.3   Nickel carbonyl 140
                 6.2.2.4   Nickel levels in human beings 140
                 6.2.2.5   Pathological states influencing
                           nickel levels 154

        6.3 Elimination and excretion 155
             6.3.1 Experimental animals 155
             6.3.2 Human beings 156

7.   EFFECTS ON ORGANISMS IN THE ENVIRONMENT ... 158
     7.1 Microorganisms 158
     7.2 Aquatic algae and plants 159
     7.3 Aquatic invertebrates 161
     7.4 Fish 164
     7.5 Terrestrial organisms 167
          7.5.1 Plants 167
          7.5.2 Animals 169
          7.5.3 Essentiality of nickel for bacteria and plants 169
     7.6 Population and ecosystem effects 170

8.   EFFECTS ON EXPERIMENTAL ANIMALS AND
     IN VITRO AND OTHER TEST SYSTEMS 171
     8.1 Animals 171
          8.1.1 Essentiality 171
                 8.1.1.1 Nickel deficiency symptoms 171
          8.1.2 Acute exposures 175
                 8.1.2.1 Nickel carbonyl 175
                 8.1.2.2 Other nickel compounds 180
                 8.1.2.3 Possible mechanisms of acute
                          nickel toxicity 182
          8.1.3 Short- and long-term exposures 183
                 8.1.3.1 Effects on the respiratory tract 183
          8.1.4 Relationship of nickel toxicity and mixed
                 metal exposure 189
          8.1.5 Endocrine effects 189
          8.1.6 Cardiovascular effects 190
          8.1.7 Effects on the immune system 192
          8.1.8 Skin and eye irritation and contact
                 hypersensitivity 193
                 8.1.8.1 Skin and eye irritation 193
                 8.1.8.2 Contact hypersensitivity 194
          8.1.9 Reproduction, embryotoxicity, and
                 teratogenicity 195
                 8.1.9.1 Effects on the male reproductive
                          system 195

		8.1.9.2	Effects on the female reproductive system	196

       8.1.10 Embryotoxicity and teratogenicity 196
  8.2 Mutagenicity and related end-points 199
       8.2.1 Mutagenesis in bacteria and mammalian cells 199
       8.2.2 Chromosomal aberration and sister
              chromatid exchange (SCE) 222
       8.2.3 Mammalian cell transformation 223
  8.3 Other test systems 225
  8.4 Carcinogenicity 225
       8.4.1 Inhalation 225
       8.4.2 Oral 235
       8.4.3 Other routes 235
       8.4.4 Interactions with known carcinogens 252
       8.4.5 Possible mechanisms of nickel carcinogenesis 254
       8.4.6 Factors influencing nickel carcinogenesis 256

9.   EFFECTS ON HUMAN BEINGS 257
  9.1 Systemic effects 257
       9.1.1 Acute toxicity - poisoning incidents 257
              9.1.1.1 Nickel carbonyl 257
              9.1.1.2 Other nickel compounds 258
       9.1.2 Short- and long-term exposure 259
              9.1.2.1 Respiratory effects 259
              9.1.2.2 Renal effects 261
              9.1.2.3 Cardiovascular effects 262
              9.1.2.4 Other effects 262
  9.2 Skin and eye irritation and contact hypersensitivity 263
       9.2.1 Skin and eye irritancy 263
       9.2.2 Contact hypersensitivity 263
  9.3 Reproduction, embryotoxicity, and teratogenicity 266
  9.4 Genetic effects in exposed workers 266
  9.5 Carcinogenicity 266
       9.5.1 Epidemiological studies 267
              9.5.1.1 Nickel refining industry 267
              9.5.1.2 Nickel alloy manufacturing 277
              9.5.1.3 Nickel plating industry 278
              9.5.1.4 Welding 279
              9.5.1.5 Nickel powder 280
              9.5.1.6 Nickel-cadmium battery manufacturing . . 280

                9.5.1.7   Case-control studies 281
        9.5.2   Carcinogenicity of metal alloys in
                orthopaedic prostheses 281
10.   EVALUATION OF HUMAN HEALTH RISKS AND
      EFFECTS ON THE ENVIRONMENT 285
      10.1 Exposure 285
      10.2 Human health effects 286
      10.3 Environmental effects 287

11.   RECOMMENDATIONS 288

12.   PREVIOUS EVALUATIONS BY INTERNATIONAL
      BODIES 289

REFERENCES 291

RESUME ET CONCLUSIONS 357

RESUMEN Y CONCLUSIONES 371

WHO TASK GROUP ON NICKEL

Members

Professor D.A. Calamari, Institute of Agricultural Entomology, University of Milan, Milan, Italy

Dr R.F. Hertel, Fraunhofer Institute of Toxicology and Aerosol Research (ITA), Hanover, Germany (*Rapporteur*)

Professor S.M. Hopfer, University of Connecticut School of Medicine, Farmington, Connecticut, USA

Professor B.A. Katsnelson, Occupational Health Research Institute, Sverdlovsk, USSR

Professor Yasushi Kodama, Department of Environmental Health, School of Medicine, University of Occupational and Environmental Health, Kitakyushu City, Japan

Professor V. Yu. Kogan, Occupational Health Research Institute, Erevan, USSR (*Vice-Chairman*)

Ms V.R. Müller, Fraunhofer Institute of Toxicology and Aerosol Research (ITA), Hanover, Germany

Dr G.D. Nielsen, Department of Environmental Medicine, Odense University, Odense, Denmark

Professor T. Norseth, National Institute of Occupational Health, Oslo, Norway (*Chairman*)

Dr J. Pastuszka, Institute of Environmental Protection, Katowice, Poland

Professor J. Peto, Section of Epidemiology, Institute of Cancer Research, Belmont, Surrey, United Kingdom

Dr E.A. Soyombo, Environmental and Occupational Health Division, Federal Ministry of Health, Lagos, Nigeria

Dr S.H.H. Swierenga, Genetic Toxicology Section, Bureau of Drug Research, Health Protection Branch, Health and Welfare Canada, Tunneys Pasture, Ottawa, Ontario, Canada

Dr A.P. Tossavainen, Institute of Occupational Health, Helsinki, Finland

Representatives of nongovernmental organizations

Professor N. Izmerov, Institute of Industrial Hygiene and Occupational Diseases, Moscow, USSR, representing the International Commission on Occupational Health (ICOH)

Observers

Professor A. Horie, Department of Environmental Health, School of Medicine, University of Occupational and Environmental Health, Kitakyushu City, Japan

Dr J. Ishmael, Central Toxicology Laboratory, ICI plc, Macclesfield, Cheshire, United Kingdom

Professor M.I. Mikheev, Institute for Advanced Medical Studies, Leningrad, USSR

Dr L.G. Morgan, INCO Europe Limited, Swansea, United Kingdom

Dr M. Richold, Unilever Research, Colworth Laboratory, Bedford, United Kingdom

Professor A.V. Roscin, Central Institute for Advanced Medical Studies, Moscow, USSR

Secretariat

Dr A. Aitio, International Agency for Research on Cancer, Lyon, France

Dr E. Smith, International Programme on Chemical Safety, Division of Environmental Health, World Health Organization, Geneva, Switzerland

NOTE TO READERS OF THE CRITERIA DOCUMENTS

Every effort has been made to present information in the criteria documents as accurately as possible without unduly delaying their publication. In the interest of all users of the environmental health criteria documents, readers are kindly requested to communicate any errors that may have occurred to the Manager of the International Programme on Chemical Safety, World Health Organization, Geneva, Switzerland, in order that they may be included in corrigenda, which will appear in subsequent volumes.

* * *

A detailed data profile and a legal file can be obtained from the International Register of Potentially Toxic Chemicals, Palais des Nations, 1211 Geneva 10, Switzerland (Telephone no. 7988400/7985850).

ENVIRONMENTAL HEALTH CRITERIA FOR NICKEL

A WHO Task Group on Environmental Health Criteria for Nickel met at the Leningradskaya Hotel, Moscow, USSR, from 17 to 21 April 1989, under the auspices of the USSR State Committee for Environmental Protection, Centre for International Projects. Dr S.N. Morozov welcomed the participants on behalf of the host institution and Dr E. Smith opened the meeting on behalf of the three cooperating organizations of the IPCS (ILO/UNEP/WHO). The Task Group reviewed and revised the draft criteria document and made an evaluation of the health risks of exposure to nickel.

The first draft of this document was prepared by Dr R.F. Hertel, Dr J. Maass, and Ms V. Müller, Fraunhofer Institute of Toxicology and Aerosol Research, Hanover, Germany. This draft was reviewed in the light of international comments by a Working Group comprising Dr V. Bencko, Prague, Czechoslovakia, Dr M. Piscator, Stockholm, Sweden, and Dr F.W. Sunderman, Farmington, Connecticut, USA, with the assistance of Dr R.F. Hertel, Ms V. Müller and Dr G. Rosner. The revised draft resulting from this Working Group was submitted for the Task Group review. Dr E. Smith, IPCS Central Unit, was responsible for the overall scientific content of the document and for the organization of the meetings, and Mrs M.O. Head of Oxford, England, was responsible for the editing.

The efforts of all who helped in the preparation and finalization of the document are gratefully acknowledged.

* * *

Financial support for the Task Group was provided by the United Nations Environment Programme, through the USSR Commission for UNEP. Partial financial support for the publication of this criteria document was kindly provided by the United States Department of Health and Human Services, through a contract from the National Institute of Environmental Health Sciences, Research Triangle Park, North Carolina, USA, a WHO Collaborating Centre for Environmental Health Effects.

1. SUMMARY AND CONCLUSIONS

1.1 Identity, physical and chemical properties, and analytical methods

Nickel is a metallic element belonging to group VIIIb of the periodic table. It is resistant to alkalis, but generally dissolves in dilute oxidizing acids. Nickel carbonate, nickel sulfide, and nickel oxide are insoluble in water, whereas nickel chloride, nickel sulfate, and nickel nitrate are water soluble. Nickel carbonyl is a volatile colourless liquid that decomposes at temperatures above 50 °C. The prevalent ionic form is nickel (II). In biological systems, dissolved nickel may form complex components with various ligands and bind to organic material.

The most commonly used methods for the analysis of biological and environmental materials are atomic absorption spectroscopy and voltammetry. In order to obtain reliable results, especially in the ultratrace range, specific procedures have to be followed to minimize the risk of contamination during sample collection, storage, processing, and analysis. Depending on sample pretreatment, extraction and enrichment procedures, detection limits of 1–100 ng/litre can be achieved in biological materials and water.

1.2 Sources of human and environmental exposure

Nickel is a ubiquitous trace metal and occurs in soil, water, air, and in the biosphere. The average content in the earth's crust is about 0.008%. Farm soils contain between 3 and 1000 mg nickel/kg. Levels in natural waters have been found to range from 2 to 10 μg/litre (fresh water) and from 0.2 to 0.7 μg/litre (marine). Atmospheric nickel concentrations in remote areas range from <0.1 to 3 ng/m^3.

Nickel ore deposits are accumulations of nickel sulfide minerals (mostly pentlandite) and laterites. Nickel is extracted from the mined ore by pyro- and hydro-metallurgical refining processes. Most of the nickel is used for the production of stainless steel and other nickel alloys with high corrosion and temperature resistance. Nickel alloys and nickel platings are used in vehicles, processing machinery, armaments, tools, electrical equipment, household appliances, and coinage. Nickel compounds are also used as catalysts, pigments, and in batteries. Global mining production of nickel was approximately 67 million kg in 1985. The primary sources of nickel emissions into the ambient air are the combustion of coal and oil for heat or power generation, the incineration of waste and sewage sludge, nickel mining and primary production, steel manufacture, electroplating, and miscellaneous sources, such as cement manufacturing. In polluted air, the predominant nickel compounds appear to be nickel sulfate, oxides, and sulfides, and to a lesser extent, metallic nickel.

Nickel from various industrial processes and other sources finally reaches waste water. Residues from waste-water treatment are disposed of by deep well injection, ocean dumping, land treatment, and incineration. Effluents from waste-water treatment plants have been reported to contain up to 0.2 mg nickel/litre.

1.3 Environmental transport, distribution, and transformation

Nickel, which is emitted into the environment from both natural and man-made sources, is circulated throughout all environmental compartments by means of chemical and physical processes, and is biologically transported by living organisms.

Atmospheric nickel is considered to exist mainly in the form of particulate aerosols containing different concentrations of nickel, depending on the source. The highest nickel concentrations in ambient air are usually found in the smallest particles. Nickel carbonyl is unstable in air and decomposes to form nickel oxide.

The transport and distribution of nickel particles to, or between, different environmental compartments is strongly influenced by particle size and meteorological conditions. Particle size distribution is primarily a function of the emitting sources. In general, particles from man-made sources are smaller than natural dust particles.

Nickel is introduced into the hydrosphere by removal from the atmosphere, by surface run-off, by discharge of industrial and municipal waste, and also following natural erosion of soils and rocks. In rivers, nickel is mainly transported in the form of a precipitated coating on particles and in association with organic matter; in lakes, it is transported in the ionic form, also mainly in association with organic matter. Nickel may also be absorbed on to clay particles and via uptake by biota. Absorption processes may be reversed leading to release of nickel from the sediment. Part of the nickel is transported via rivers and streams to the ocean. Riverine suspended particulate input is estimated to be 135×10^7 kg/year.

Depending on the soil type, nickel may exhibit a high mobility within the soil profile finally reaching ground water and, thus, rivers and lakes. Acid rain has a pronounced tendency to mobilize nickel from the soil. Terrestrial plants take up nickel from soil primarily via the roots. The amount of nickel uptake from soil depends on various geochemical and physical parameters including the type of soil, the soil pH and humidity, the organic matter content of the soil, and the concentration of extractable nickel. The best known example of nickel accumulation is the increased nickel levels, in excess of 1 mg/kg dry weight, found in a number of plant species ("hyperaccumulators") growing on relatively infertile serpentine soils. Nickel levels above 50 mg/kg dry weight are toxic for most plants. Accumulation and toxic effects have been observed in vegetables grown on sewage sludge-treated soils and in vegetation close to nickel-emitting sources. High concentration factors have been found in aquatic plants. Laboratory studies showed that nickel had little capacity for accumulation in all the fish studied. In uncontaminated waters, the range of concentrations reported in whole fish (on a wet-weight basis) ranged from 0.02 to 2 mg/kg. These values could be up to 10 times higher in fish from contaminated waters. In wildlife, nickel is found in many organs and tissues, due to dietary uptake by herbivorous animals and their carnivorous predators. However, there is no evidence for the biomagnification of nickel in the food chain.

1.4 Environmental levels and human exposure

Nickel levels in terrestrial and aquatic organisms can vary over several orders of magnitude. Typical atmospheric nickel levels for human exposure range from about 5 to 35 ng/m^3 at rural and urban sites, leading to a nickel uptake via inhalation of 0.1–0.7 µg/day. Drinking-water generally contains less than 10 µg nickel/litre, but occasionally nickel may be released from the plumbing fittings, resulting in concentrations of up to 500 µg nickel/litre.

Nickel concentrations in food are usually below 0.5 mg/kg fresh weight. Cocoa, soybeans, some dried legumes, various nuts, and oatmeal contain high concentrations of nickel. Daily intake of nickel from food will vary widely, because of different dietary habits, and can range from 100 to 800 µg/day; the mean dietary nickel intake in most countries is 100–300 µg/day. Release of nickel from kitchen utensils may contribute significantly to oral intake. Pulmonary intake of 2–23 µg nickel/day can result from smoking 40 cigarettes a day.

Dermal exposure in the general environment is important for the induction and maintenance of contact hypersensitivity caused by daily skin contact with nickel-plated objects or nickel-containing alloys (e.g., jewellery, coins, clips).

Iatrogenic exposure to nickel results from implants and prostheses made from nickel-containing alloys, from intravenous or dialysis fluids, and from radiographic contrast media. An estimated average intravenous nickel uptake from dialysis fluids is 100 µg per treatment.

In the working environment, airborne nickel concentrations can vary from a few µg/m^3 to, occasionally, a few mg/m^3, depending on the process involved and the nickel content of the material being handled.

Throughout the world, millions of workers are exposed to nickel-containing dusts and fumes during welding, plating and grinding, mining, nickel refining, and in steel plants, foundries, and other metal industries.

Dermal exposure to nickel may occur in a wide range of jobs, either by direct exposure to dissolved nickel, e.g., in refining, electroplating, and electroforming industries or by handling nickel-containing tools. Wet cleaning work may involve exposure to nickel, because of the amounts of nickel that become dissolved in the washing water.

1.5 Kinetics and metabolism in human beings and animals

Nickel can be absorbed in human beings and animals via inhalation or ingestion, or percutaneously. Respiratory absorption with secondary gastrointestinal absorption of nickel (insoluble and soluble) is the major route of entry during occupational exposure. A significant quantity of inhaled material is swallowed following mucociliary clearance from the respiratory tract. Poor personal hygiene and work practices can contribute to gastrointestinal exposure. Percutaneous absorption is negligible, quantitatively, but is important in the pathogenesis of contact hypersensitivity. Absorption is related to the solubility of the compound, following the general relationships nickel carbonyl > soluble nickel compounds > insoluble nickel compounds. Nickel carbonyl is the most rapidly and completely absorbed nickel compound in both animals and human beings. Studies in which nickel was administered via inhalation are limited. Studies on hamsters and rats with insoluble nickel oxide showed poor absorption, with retention of much of the material in the lung after several weeks. In contrast, absorption of soluble nickel chloride or amorphous nickel sulfide was rapid. Nickel is transported in the blood, principally bound to albumin.

Gastrointestinal absorption of nickel is variable and depends on the composition of the diet. In a recent study on human volunteers, absorption of nickel was 27% from water compared with less than 1% from food. All body secretions are potential routes of excretion including urine, bile, sweat, tears, milk, and mucociliary fluid. Non-absorbed nickel is eliminated in the faeces. Transplacental transfer has been demonstrated in rodents. Following parenteral administration of nickel salts, the highest nickel accumulation occurs in the kidney, endocrine glands, lung, and liver: high concentrations are also observed in the brain following administration of nickel carbonyl. Data on nickel excretion suggest a two-compartment model. Nickel concentrations in the serum and urine of healthy non-occupationally exposed adults are 0.2 µg/litre (range: 0.05–1.1 µg/litre) and 1.5 µg/g creatinine (range: 0.5–4.0 mg/g creatinine), respectively. Increased concentrations of nickel are seen in both of these fluids following occupational exposure. The body burden of nickel in a non-exposed, 70-kg adult is 0.5 µg.

1.6 Effects on organisms in the environment

In microorganisms, growth was generally inhibited at nickel concentrations in the medium of 1–5 mg/litre in the case of actinomycetes, yeast, and marine and non-marine eubacteria and at levels of 5–1000 mg/litre in filamentous fungi. In algae, no growth was observed at approximately 0.05–5 mg nickel/litre. Abiotic factors, such as the pH, hardness, temperature, and salinity of the medium and the presence of organic and inorganic particles, influence the toxicity of nickel.

Nickel toxicity in aquatic invertebrates varies considerably according to species and abiotic factors. A 96-h LC_{50} of 0.5 mg nickel/litre has been found for *Daphnia* spp., while, in molluscs, 96-h LC_{50} values were around 0.2 mg/litre in two freshwater snail species and 1100 mg/litre in a bivalve.

In fish, 96-h LC_{50} values generally fall within the range of 4–20 mg nickel/litre, but can be higher in some species. Long-term studies on fish, and fish development, in soft water demonstrated some effects on rainbow trout at levels as low as 0.05 mg nickel/litre. In terrestrial plants, nickel levels above 50 mg/kg dry weight are usually toxic. Copper was found to act toxicologically in a synergistic way, whereas calcium reduced the toxicity of nickel. Data on the effects of nickel on terrestrial animals are limited.

Earthworms seem to be relatively insensitive to nickel, if the medium is rich in microorganisms and organic matter, thus, making the nickel less available to the earthworms. Nickel has not been considered as a broad-scale global contaminant; however, ecological changes, such as decreases in the number and diversity of species, have been observed near nickel-emitting sources. Microecosystem studies have shown that addition of nickel to soil disturbs the nitrogen cycle.

1.7 Effects on experimental animals and *in vitro* test systems

Nickel is essential for the catalytic activity of some plant and bacterial enzymes. Slow weight gain, anaemia, and decreased viability of offspring have been described in some animal species after dietary deprivation of nickel.

Summary and conclusions

The most acutely toxic nickel compound is nickel carbonyl, the lung being the target organ; pulmonary oedema may occur within 4 h following exposure. The acute toxicity of other nickel species is low.

Though contact allergy to nickel is very common in human beings, experimental sensitization in animals is only successful under special conditions. Long-term inhalation exposure to metallic nickel, nickel oxide, or nickel subsulfide caused mucosal damage and inflammatory reaction in the respiratory tract in rats, mice, and guinea-pigs. Epithelial hyperplasia was observed in rats after inhalation exposure to aerosols of nickel chloride or nickel oxide.

High-level, long-term exposure to nickel oxide led to gradually progressive pneumoconiosis in rats. Inflammatory reaction, sometimes accompanied by slight fibrosis, was observed in rabbits after high-level exposure to nickel-graphite dust. Pulmonary fibrosis was seen in rats exposed to nickel subsulfide.

Nickel salts, administered parenterally, induced a rapid transitory hyperglycaemia in rats, rabbits, and chickens. These changes may be associated with effects on alpha and beta cells in the islets of Langerhans. Nickel also decreased the release of prolactin. Nickel chloride, given orally or by inhalation, has been reported to decrease iodine uptake by the thyroid.

Nickel salts, given intravenously, decreased blood flow in the coronary arteries in the dog; high concentrations of nickel decreased the contractility of dog myocardium *in vitro*.

Nickel chloride affects the T-cell system and suppresses the activity of natural killer cells. Parenteral administration of nickel chloride and nickel subsulfide have been reported to cause intrauterine mortality and decreased weight gain in rats and mice. Inhalation exposure to nickel carbonyl caused fetal death and decreased weight gain, and was teratogenic in rats and hamsters. Information on maternal toxicity was not given in any of these studies. Nickel carbonyl has been reported to cause dominant lethal mutations in rats.

Several inorganic nickel compounds were tested for mutagenicity in various test systems. Nickel compounds were generally inactive in bacterial mutagenesis assays, except where fluctuation tests were used. Mutations were observed in several cultured mammalian cell types. Nickel compounds inhibited DNA synthesis in a wide variety of organisms. In addition, nickel compounds induced chromosomal aberrations and sister chromatid exchange (SCE) in both mammalian and human cultured cells. Chromosomal aberrations, but not sister chromatid exchange (except in one study on electrolysis workers), were observed in human beings, occupationally exposed to either insoluble or soluble nickel compounds. Nickel induced cell transformation *in vitro*.

In an inhalation study, nickel subsulfide induced benign and malignant pulmonary tumours in rats. A few pulmonary tumours were seen in rats in a series of inhalation studies with nickel carbonyl. There was no significant increase in lung tumours in rats in an adequate inhalation study with metallic nickel. Inhalation exposure to black nickel oxide did not induce lung tumours in Syrian golden hamsters (a species resistant to lung carcinogenesis). Adequate carcinogenicity studies on inhalation exposure to other nickel compounds were not available. However, nickel subsulfide, metallic nickel powder, and an unspecified nickel oxide induced benign and malignant lung tumours in rats after repeated intratracheal instillations.

Nickel carbonyl, nickelocene, and a large number of slightly soluble or insoluble nickel compounds, including nickel subsulfide, carbonate, chromate, hydroxide, sulfides, selenides, arsenides, telluride, antimonide, various unidentified oxide preparations, two nickel-copper oxides, metallic nickel, and various nickel alloys, induced local mesenchymal tumours in a variety of experimental animals after intramuscular, subcutaneous, intraperitoneal, intrapleural, intraocular, intraosseous, intrarenal, intra-articular, intratesticular or intra-adipose administration. No local carcinogenic response was seen in single-dose studies with some nickel alloys, colloidal nickel hydroxide, or with two specimens of nickel oxide, especially prepared for carcinogenicity testing by calcining at 735 °C or 1045 °C.

Nickel sulfate and nickel acetate, but not nickel chloride, induced tumours of the peritoneal cavity in rats after repeated intraperitoneal administration.

Summary and conclusions

Metallic nickel and a very large number of nickel compounds have been tested for carcinogenicity by parenteral routes of administration; with few exceptions, they caused local tumours.

Only nickel subsulfide has been shown convincingly to cause cancer after inhalation exposure. However, the number of adequate inhalation studies is very small.

In studies using repeated intratracheal instillation, nickel powder, nickel oxide, and nickel subsulfide caused pulmonary tumours.

When nickel sulfate and nickel chloride, which had not induced local tumours in intramuscular studies, were tested using repeated intraperitoneal administration, they elicited a carcinogenic response.

1.8 Effects on human beings

In terms of human health, nickel carbonyl is the most acutely toxic nickel compound. The effects of acute nickel carbonyl poisoning include frontal headache, vertigo, nausea, vomiting, insomnia, and irritability, followed by pulmonary symptoms similar to those of a viral pneumonia. Pathological pulmonary lesions include haemorrhage, oedema, and cellular derangement. The liver, kidneys, adrenal glands, spleen, and brain are also affected. Cases of nickel poisoning have also been reported in patients dialysed with nickel-contaminated dialysate and in electroplaters who accidentally ingested water contaminated with nickel sulfate and nickel chloride.

Chronic effects such as rhinitis, sinusitis, nasal septal perforations, and asthma have been reported in nickel refinery and nickel plating workers. Some authors reported pulmonary changes with fibrosis in workers inhaling nickel dust. In addition, nasal dysplasia has been reported in nickel refinery workers. Nickel contact hypersensitivity has been documented extensively in both the general population and in a number of occupations in which workers were exposed to soluble nickel compounds. In several countries, it has been reported that 10% of the female population and 1% of the male population are sensitive to nickel. Of these, 40–50% have vesicular hand eczema, which, in some cases, can be very severe and lead to loss of working ability. Oral nickel intake may aggravate vesicular hand eczema and, possibly, also eczema arising on other parts of the body where there has not been any skin contact with nickel.

Prostheses, or other surgical implants, made from nickel-containing alloys have been reported to cause nickel sensitization or to aggravate existing dermatitis.

Nephrotoxic effects, such as renal oedema with hyperaemia and parenchymatous degeneration, have been reported in cases of accidental industrial exposure to nickel carbonyl. Transient nephrotoxic effects have been recorded after accidental ingestion of nickel salts.

Very high risks of lung and nasal cancer have been reported in nickel refinery workers employed in the high-temperature roasting of sulfide ores, involving substantial exposure to nickel subsulfide, oxide, and, perhaps, sulfate. Similar risks have been reported in processes involving exposure to soluble nickel (electrolysis, copper sulfate extraction, hydrometallurgy), often combined with some nickel oxide exposure, but with low nickel subsulfide exposure. The risk to miners and other refinery workers has been reported to be much lower. Cancer rates have generally been close to normal in stainless steel welding and nickel-using industries, with the exception of those involving exposure to chromium, particularly electroplating. However, nickel/cadmium battery workers exposed to high levels of both nickel and cadmium may have suffered a slightly increased risk of lung cancer.

Excesses of various cancers other than lung and nasal cancers, such as renal, gastric, or prostatic cancer, have occasionally been reported in nickel workers, but none has been found consistently.

The epidemiological data can be used to address two important questions: (i) whether specific nickel compounds have been shown to be carcinogenic; and (ii) whether low-exposure cohorts provide upper limits of risk at specified exposure levels.

(a) Soluble nickel

There was evidence of a cancer hazard in workers exposed to soluble nickel concentrations of the order of 1–2 mg/m^3, both in electrolysis and in the preparation of soluble salts. These workers were also exposed to other nickel compounds, but often at lower levels than in other high-risk processes. In the absence of historical exposure measurements it is impossible to draw unequivocal conclusions, but the evidence that soluble nickel is carcinogenic is certainly strong. Refinery dust sometimes contains a substantial proportion of nickel sulfate in addition to nickel subsulfide. This raises the possibility that the very high cancer risk observed in workers employed in the high-temperature oxidation of nickel subsulfide may be partly due to soluble nickel.

(b) Nickel subsulfide

In refinery areas where cancer risks were high, exposure to nickel subsulfide almost always occurred together with exposure to the oxide and, perhaps, sulfate (see above). Thus, it is difficult to demonstrate from epidemiological data alone, that nickel subsulfide is carcinogenic, though this seems likely.

(c) Nickel oxide

Nickel oxide was present in almost all circumstances in which cancer risks were elevated, together with one or more other forms of nickel (nickel subsulfide, soluble nickel, metallic nickel). As for nickel subsulfide, it is difficult to either demonstrate or disprove its suspected carcinogenicity on the basis of epidemiological data alone.

(d) Metallic nickel

No increased cancer risk has been demonstrated in workers exposed exclusively to metallic nickel. The combined data on nickel alloy workers and gaseous diffusion workers, all of whom were exposed to average concentrations of the order of 0.5 mg nickel/m^3, show no excess risk, though the total number of lung cancers in these cohorts was too small to exclude a small increase in risk at this level.

(e) Conclusion

Although some, and perhaps all, forms of nickel may be carcinogenic, there is little or no detectable risk in most sectors of the nickel industry at current exposure levels; this includes some processes that were associated, in the past, with very high lung and nasal cancer risks. Long-term exposure to soluble nickel at concentrations of the order of 1 mg/m^3 may cause a marked increase in the relative risk of lung cancer, but the relative risk among workers exposed to average metallic nickel levels of about 0.5 mg/m^3 is approximately 1. The cancer risk at a given exposure level may be higher for soluble nickel compounds than for metallic nickel and, possibly, than for other forms as well. The absence of any marked lung cancer risk among nickel platers is not surprising, as the average exposures to soluble nickel are very much lower than those in electrolytic refining or nickel salt processing.

2. IDENTITY, PHYSICAL AND CHEMICAL PROPERTIES, ANALYTICAL METHODS

2.1 Identity, and physical and chemical properties of nickel and nickel compounds

Nickel is a silvery white metal belonging to Group VIIIb of the periodic table. Nickel is slightly more resistant to oxidation than iron and cobalt, with a standard potential of -0.236 V relative to the hydrogen electrode (Stoeppler, 1980). Several hundreds of nickel compounds have been identified and characterized. Nickel has a specific density of 8.90 g/cm^3, a melting point of 1555 °C, and a boiling point of 2837 °C (Table 1). It is insoluble in water, soluble in dilute nitric acid and aqua regia, and slightly soluble in hydrochloric and sulfuric acid. Nickel usually has an oxidation state of two, but also occurs as relatively stable tri- and tetravalent ions (Stoeppler, 1980). Several binary nickel compounds are commercially and environmentally significant. A brief description of the chemistry of some of these compounds is given below. Physical and chemical properties of nickel and its compounds are summarized in Table 1.

Nickel forms complexes (chelates) that are insoluble in water, but soluble in organic solvents. These compounds are often very stable and play an important role in trace analysis. For example, nickel dimethylglyoxime is the compound that makes possible the separation of nickel from cobalt, which is similar in its chemical and analytical behaviour (Stoeppler, 1980). Divided nickel (Raney nickel) absorbs up to seventeen times its volume of hydrogen and can act as an catalyst (Lewis & Ott, 1970).

2.1.1 Nickel carbonate hydroxide

Nickel carbonate hydroxide ($2NiCO_3.3Ni(OH)_2.4H_2O$) is insoluble in water, but soluble in ammonia and in dilute acids. The composition of basic nickel carbonate can vary. The most common forms range from $2NiCO_3.3Ni(OH)_2.XH_2O$ to $NiCO_3.Ni(OH)_3.XH_2O$. The tetrahydrate occurs in nature as zaratite. It is used in nickel

Table 1. Physical properties of nickel and nickel compounds[a]

Name	Chemical formula	Relative molecular mass	Appearance	Density (g/cm^3)	Melting point (°C)	Boiling point (°C)	Solubility (water; other solvents)
Nickel	Ni	58.70	lustrous, white, face-centered cubic crystals	8.90	1555 1455[b]	2837	insoluble
Nickel acetate	Ni(CH$_3$CO$_2$)$_2$	176.80	green crystalline mass or powder	1.744	[c]	-	soluble; soluble in alcohol
Nickel arsenate	Ni$_3$(AsO$_4$)$_2$	453.97	yellow-green powder	4.982	-	-	insoluble; soluble in acids
Nickel bromide	NiBr$_2$	218.53	yellow-green, deliquescent crystals	-	loses H$_2$O at 200	-	soluble; soluble in alcohol
Nickel carbonate	2NiCO$_3$	118.70	light-green crystals	-	decomposes	-	insoluble; soluble in acids
Nickel carbonyl	Ni(CO)$_4$	170.73	colourless, volatile liquid	1.318 (17 °C)	-19.3	43	insoluble; soluble in organic solvents

Table 1 (continued)

Nickel chloride	$NiCl_2$	129.61	yellow, deliquescent crystals	3.55^d		987^d	soluble
Nickel chloride hexahydrate	$NiCl_2 \cdot 6H_2O$	237.70^d	green, monoclinic, deliquescent crystals	-	-	-	soluble; soluble in alcohol
Nickel fluoride	NiF_2	96.69	yellow-green, tetragonal crystals	4.72	-	-	slightly soluble
Nickel hydroxide	$Ni(OH)_2$	92.72	green powder	-	decomposes above 200	-	insoluble; soluble in acids and ammonia
Nickel hydroxy-carbonate tetrahydrate	$2NiCO_3 \cdot 3Ni(OH)_2 \cdot 4H_2O$	587.67^b	green powder	-	-	-	insoluble; soluble in acids
Nickel nitrate	$Ni(NO_3)_2$	182.72	green, deliquescent crystals	2.05	56.7	137	soluble; soluble in alcohol
Nickel oxide	NiO	74.69	green or black powder	6.67^d	1990^d	-	insoluble; soluble in acid
Nickel phosphate	$Ni_3(PO_4)_3$	366.07	light-green powder	-	-	-	insoluble; soluble in acid

Table 1 (continued)

Name	Chemical formula	Relative molecular mass	Appearance	Density (g/cm^3)	Melting point (°C)	Boiling point (°C)	Solubility (water; other solvents)
Nickel sulfate	NiSO$_4$	154.77	α blue-green, tetragonal crystals β green, monoclinic crystals	-	53.3 (a-b) loses water at 280	-	soluble soluble
β-Nickel sulfide	NiS	90.77 [d]	trigonal crystals [e]	5.3 [d]	797 [d]	-	insoluble
Nickel subsulfide	Ni$_3$S$_2$	240.26 [b]	pale, yellowish bronze metallic [b]	5.82 [b]	790 [b]	-	insoluble; soluble in nitric acid [b]

[a] From: Windholz et al. (1983).
[b] From: Weast (1981).
[c] Data not available.
[d] From: Blankenstein & Starck (1979).
[e] From: Neumuller (1985).

plating, as a catalyst for the hardening of fats, and in colours and glazes for ceramics (Windholz et al., 1983). High purity nickel carbonate is used in electronic components (IARC, 1976).

2.1.2 Nickel carbonyl

Nickel carbonyl (Ni(CO)$_4$) is a colourless volatile liquid and is formed when nickel powder is treated with carbon monoxide at about 50 °C. It is used for the production of pure nickel by thermal deposition at atmospheric pressure and at 200–250 °C (Stoeppler, 1980). The carbonyl is insoluble in water, but soluble in most organic solvents (Windholz et al., 1983).

2.1.3 Nickel chloride and nickel chloride hexahydrate

Nickel chloride (NiCl$_2$) and nickel chloride hexahydrate (NiCl$_2$.6H$_2$O) are both soluble in water. The anhydrate salt is used as an absorbent for ammonia in gas masks and in nickel plating (Windholz et al., 1983).

2.1.4 Nickel hydroxide

Nickel hydroxide (Ni(OH)$_2$) is insoluble in water but soluble in acids (Windholz et al., 1983). When dissolved in ammonia it forms complexes. It is used as electrode material for secondary cells (Blankenstein, 1979).

2.1.5 Nickel nitrate

Nickel nitrate (Ni(NO$_3$)$_2$) dissolves easily in water and alcohol. It is used in nickel plating and nickel-cadmium batteries (Neumüller, 1985).

2.1.6 Nickel oxide

Nickel oxide (NiO) includes several nickel-oxygen compounds, which differ in stoichiometry, and chemical and physical properties (see Table 32 in section 8.4.3). The different nickel oxides, and also the nickel-copper oxides present in the nickel refining industry, have different biological properties (Sunderman et.al., 1987a).

Nickel oxide is insoluble in water. The solubility in acids and other properties depend on the method of preparation. Nickel oxide is an important raw material for smelting and alloy-producing processes. It is also used as a catalyst and in glass colours (Blankenstein & Starck, 1979). Nickel oxide exists in two forms. Black nickel oxide is chemically reactive and forms simple salts in the presence of acids. Green nickel oxide is an inert and refractory material. It is used primarily in metallurgical operations.

2.1.7 Nickel sulfate

Nickel sulfate ($NiSO_4$), which exists as a hexahydrate in the α-form, changes into the β-form at 53.3 °C (Windholz et al., 1983). It is produced by dissolving nickel oxide or hydroxide in sulfuric acid (Neumüller, 1985). It is the main component of the electrolyte solution in electrolytic refining and is a raw material for the production of catalysts. It is also used in fabrication of jewellery.

2.1.8 Nickel sulfide

Nickel sulfide (NiS) occurs naturally as millerite. It is insoluble in water and is of importance in catalyst production and in the hydrogenation of sulfur compounds in the oil industry (Blankenstein, 1979).

2.1.9 Nickel subsulfide

Nickel subsulfide (Ni_3S_2) exists at high-temperatures in a bronze-yellow form (β-Ni_3S_2). At lower temperatures, it transforms to the green β-form, which is stable at normal temperature, and may be formed electrolytically. The grey mineral heazlewoodite is the same modification, but has been named α-nickel subsulfide. Nickel subsulfide may be formed during the production of nickel from sulfide ores.

2.2 Analytical methods

A variety of methods has been used to determine nickel concentrations in different media. Methods are summarized in Table 2.

Table 2. Analytical methods for nickel determination [a]

Medium	Sample treatment	Analytical method	Detection limit	Comment	Reference
Biological materials					
Serum, urine	urine sampling in PE-bottles; blood collection with PE-catheter and PE-syringe; wet digestion; extraction as furildioxime into MIBK	analysis of extract by EAAS, with graphite atomizer, using a deuterium background corrector	0.4 µg/litre	suitable for monitoring occupational exposure; interference from high iron contents possible	Mikac-Devic et al. (1977)
Liver, kidney (animal)	wet digestion; evaporation to dryness; dissolution; extraction as hexamethylenedithio-carbamate–chelate into diisopropylketone and xylene	analysis of extract by EAAS, with graphite atomizer	ns	removal of iron and copper as N-nitroso-phenylhydroxylamine–chelates	Dornemann & Kleist (1980)
Food	filtration; wet digestion extraction as DMG–chelate	analysis of extract by DPV, with HMDE	1 ng/litre	rapid and inexpensive method; higher sensitivity than AAS	Pilhar et al. (1981)
Tissues, body fluids	wet digestion; evaporation to dryness; extraction as DDTC–chelate into MIBK	analysis of extract and of digested sample by EAAS, with graphite atomizer; analysis of extract by GC with FID	ns	good agreement between results from EAAS determination and results from GC-determination	Szathmary & Daldrup (1982)

Table 2 (continued)

Medium	Sample treatment	Analytical method	Detection limit	Comment	Reference
Whole blood, urine, saliva, liver, nails	wet digestion; evaporation to dryness and dissolution; extraction as DMG-chelate	analysis of extract by DPV, with HMDE	0.1 ng/litre	suitable for routine determination in a variety of biological materials	Ostapczuk et al. (1983)
Hair	washing in redistilled acetone, then in deionized water and again in redistilled acetone; repeat twice	analysis by AAS	ns	more convenient for sampling and storage than other biological materials	Bencko et al. (1986)
Serum, whole blood	blood collection with PE-cannula and PP-syringe; wet digestion	analysis by EAAS, with Zeeman background corrector	0.05 µg/litre	suitable for routine determination	Sunderman et al. (1984a)
Body fluids, tissues	wet digestion; extraction as DMG-chelate	analysis by SWV at HMDE	ns	more sensitive method compared with DPV	Ostapczuk et al. (1985a)
Tissue	sampling with plastic forceps and obsidian scalpel; wet digestion	analysis of sample by EAAS, with Zeeman background corrector	10 ng/g dry weight	minimal nickel contamination	Sunderman et al. (1985)

Table 2 (continued)

Urine	sampling in PE-bottles; wet digestion using nitric acid, perchloric acid, and ascorbic acid; extraction as APDC-chelate into MIBK	analysis of extract by EAAS, with graphite atomizer	ns	suitable for monitoring occupational exposure	Long-zhu & Zhe-ming (1985)
Urine	filtration; acidification; complexation with HPDC within automated analytical system	automated determination by electrochemical determination, following separation by HPLC (reversed phase)	0.1 ng/ 10 µlitre sample	convenient for determination of multimetals (normal to occupational exposure levels) after direct injection of sample	Bond et al. (1986)
Urine	sampling in PE-bottles; acidification; centrifugation	analysis by EAAS, with Zeeman background corrector	0.45 µg/ litre	direct analysis of sample	Sunderman et al. (1986a)
Plasma	dilution of sample with nitric acid and Triton X-100	analysis by EAAS, with Zeeman background corrector	0.09 µg/ litre	no sample pretreatment necessary	Andersen et al. (1986)
Lung tissue	freeze drying following collection; wet digestion; evaporation to dryness; dilution	analysis by EAAS, with graphite atomizer	lower ng/g amounts	suitable method for the routine determination of trace elements	Baumgardt et al. (1986)

Table 2 (continued)

Medium	Sample treatment	Analytical method	Detection limit	Comment	Reference
Blood, serum, sweats, urine	enzymatic digestion of blood and serum; ultrasonic treatment of urine and sweat	analysis by EAAS, with Zeeman background correcter	0.1 µg/litre	lower risk of contamination with pretreatment method used	Christensen & Pedersen (1986)
Food					
Foodstuffs	wet digestion; dilution; extraction as APDC-DDDC-chelate into n-methylpentan-2-one	analysis of extract by FAAS	0.048-0.061 mg/kg	interference by high copper content	Evans et al. (1978)
Fish, shellfish	freeze-drying; low temperature ashing	analysis of dissolved sample by FAAS	ns		Ikebe & Tanaka (1979)
Foodstuffs	wet digestion; exporation to dryness; dissolution; extraction as DMG-chelate	analysis of extract by DPASV, with HMDE	ns	acurate and inexpensive method	Valenta et al. (1981)
Dried milk powder	dry ashing; extraction as DMG-chelate; dilution	analysis of diluted extract by DPV, with HMDE	5 ng/sample	sensitive and accurate method, less interference compared with FID-GC	Meyer & Neeb (1985)

Table 2 (continued)

Dried milk powder	dry ashing; extraction as Na-FDEDTC-chelate in chloroform; evaporation and dissolution in chloroform	analysis of solution by GC, with FID	100 ng/ µlitre sample	interference by higher iron content in presence of low copper content is possible	Meyer & Neeb (1985)
Citrus leaves, rice flour (both standard reference materials)	wet digestion; extraction as HMDE chelates into chloroform	analysis of extract by HPLC	µg/g	simultaneous determination of nickel, molybdenum, zinc, and copper	Ichinoki & Yamazaki (1985)
Water					
Sea water	extraction with APDC and DDTC into Freon TF and back-extraction into nitric acid	analysis of extract by EAAS with graphite atomizer using a deuterium background corrector	ns	probably only ionic form of total nickel is measured; stability of extract is good	Danielsson et al. (1978)
Sea water	PE bottles or Teflon-coated PVC ball-valve samplers; a) double extraction with APDC and DDTC into chloroform; back-extraction into nitric acid; evaporation of back-extract and redissolution into nitric acid	analysis of extract by EAAS with graphite furnace	10 ng/litre (instrumental detection limit)	concentration factor 200	Bruland & Franks (1979)

Table 2 (continued)

Medium	Sample treatment	Analytical method	Detection limit	Comment	Reference
Sea water (continued)	b) concentration on Chelex-100 resin		15 ng/litre (instrumental detection limit)	inefficient concentration factor by Chelex-100	Bruland & Franks (1979)
Sea water	PVC samplers; storage in PE containers; filtration; extraction with APDC into MIBK	analysis of extract by FAAS	ns	probably only ionic form of total nickel is measured	Frache et al. (1980)
Sea water	adsorption on PAN-resin	analysis of resin phase by ion-exchange calorimetry	0.077 µg/litre	rapid and inexpensive method	Yoshimura et al. (1980)
Fresh water			0.34 µg/litre		
Fresh water, drinking-water	buffered extraction as DMG-chelate	analysis of extract by DPP, with HMDE	2 µg/litre	suitable for routine water analysis	Flora & Nieboer (1980)
Sea water, fresh water, waste water	UV-irradiation; extraction as DMG-chelate	analysis of extract by DPV, with HMDE	1 µg/litre	rapid and inexpensive method; high sensitivity	Pilhar et al. (1981)

Table 2 (continued)

Fresh water (river)	UV-irradiation; enrichment by Donnan dialysis	analysis of electrolyte by FAAS	ns	enrichment factor decreases at higher calcium concentrations	Wilson & DiNunzio (1981)
Aqueous solution	extraction as heptoxime-chelate; evaporation and redissolution in toluene/methanol/LiCl	analysis of extract by DPP with HMDE	1-2 µg/litre	in case of high copper and iron concentrations, extraction with NH_4OH is necessary to prevent interference	Gemmer-Colos et al. (1981)
Sea water	preconcentration by complexation with 8-hydroxyquinoline; adsorption on C_{18}-bonded silica gel; evaporation of eluate to dryness; dissolution in nitric acid	analysis of diluted extract by ICP-AES	0.05 mg/litre	concentration factor 200	Watanabe et al. (1981)
Sea water	preconcentration by complexation with 8-hydroxyquinoline; adsorption on C_{18}-bonded silica gel; evaporation of eluate to dryness, dissolution in nitric acid	analysis by EAAS with graphite furnace	ns	concentration factor 50	Sturgeon et al. (1982)

Table 2 (continued)

Medium	Sample treatment	Analytical method	Detection limit	Comment	Reference
Industrial plant solutions	complexation with APDC and DDTC with automated analytical system	separation of chelate by HPCC (reversed phase) followed by electrochemical and spectrophotometric detection within automated system	0.5 ng (electrochemical) 0.1 ng (spectrophotometric)	suitable for automated monitoring of nickel and copper	Bond & Wallace (1983)
Sea water	preconcentration by adsorption on immobilized 8-hydroxyquinoline	analysis of eluate by ICP-MS	1 µg/litre	50-fold preconcentration	McLaren et al. (1985)
Sea water	coprecipitation with gallium	analysis of dissolved precipitate by ICP-AES	60 ng/litre	200-fold preconcentration, appropriate for multi-element analysis	Akagi et al. (1985b)
Fresh water (river)	filtration; oxidative UV-photolysis	analysis by DPCSV, with HMDE	0.4 µg/litre	elimination of interference caused by dissolved organic matter by UV-photolysis	Weidenauer & Lieser (1985)

Table 2 (continued)

Drinking-water	evaporation onto cellulose matrix, grinding and pelletizing of residue	analysis by PIXE	1.2 µg/litre	suitable for multi-element analysis	Ali et al. (1985)
Waste water, plating solution	dilution; separation of metal ions as EDTA-complexes	ion-chromatographic analysis, with anion separator	ns		Tanaka (1985)
Sea water (synth.)	extraction with DDTC into chloroform	analysis of diluted extracts by GC, with FID by GC with ECD	17 µg/litre to 0.2 µg/litre (depending on type of column)	inexpensive method; appropriate for multimetal analyses	Carvajal & Zienius (1986)
Rain water	filtration; acidification; extraction as DMG-chelate	analysis by DPSV, with HMDE	0.24 mg/litre	rapid, inexpensive and sensitive method for multi-element analysis	Vos et al. (1986)

Soil

Rock material (standard reference material)	wet digestion with HF and HNO_3; extraction as DDTC-chelate into MIBK	analysis of extract by FAAS	5–200 mg/kg	appropriate for iron, molybdenum, and calcium-rich geological materials	Sanzolone et al. (1979)

Table 2 (continued)

Medium	Sample treatment	Analytical method	Detection limit	Comment	Reference
River sediments, rock material, plants	wet digestion; filtration and dilution	analysis of diluted sample by FAAS using a deuterium background corrector	0.1 mg/kg	elimination of interference of matrix effects by use of deuterium background detector	Abo-Rady (1979a)
Soil	wet digestion; extraction as APDC-chelate into MIBK; reextraction with nitric acid	analysis of re-extracts by EAAS with zirconium coated graphite atomizer	4 mg/kg re-extract	concentration factor 5; reduction of interference by re-extraction	Schmidt & Dietl (1981)
Soil	acid digestion	analysis by ICP-AES	0.010-0.015 mg/kg (depending on spectral path)	suitable for rapid multielement analysis	Church (1981)
Soil	wet digestion; extraction as DMG-chelate	analysis by ASWV	0.08 μg/ml analyte solution	more sensitive and rapid method for determination of heavy metals than DPV	Ostapczuk et al. (1985b)

Table 2 (continued)

Air				
Air	adsorption on cellulose ester membrane filter; wet digestion	analysis by FAAS	1 µg/sample	suitable for determining occupational exposure	US NIOSH (1977b)
Air	adsorption on cellulose ester membrane filter; wet digestion; evaporation to dryness; dilution	analysis by ICP-AES	1 µg/sample	suitable for simultaneous multi-element analysis	Mackenzie Peers (1986)
Air	adsorption in alcoholic iodine solution; extraction as furildioxime chelate into chloroform	analysis by colorimetry	1 µg/m^3	nickel carbonyl is measured, interference by gaseous nickel compounds	Stedman (1986a)
Air	direct sampling into chemiluminescence detector; mixing of sample with carbon monoxide	analysis by chemiluminescence	0.2 µg/m^3	allows continuous measuring of nickel carbonyl	Stedman (1986b)
Various materials				
Steel	extraction as DMG-chelate; complexation of Fe^{3+} and Mn^{2+} with triethanolamine solution	analysis by DPP, with HMDE	1 µg/kg	Copper can be determined simultaneously	Weinzierl & Umland (1982)

Table 2 (continued)

Medium	Sample treatment	Analytical method	Detection limit	Comment	Reference
City waste incinerator ash (standard reference material)	wet digestion; evaporation to near dryness; dilution filtration	analysis by ICP-AES	25 µg/litre analyte solution	multistep digestion procedure necessary because of difficult matrix	Taylor, et al. (1985)

[a] *Abbreviations:*

APDC	ammonium pyrolidinedithiocarbamate	
ASWV	adsorption square wave voltammetry	
DDDC	diethylammonium diethyldithiocarbamate	
DDTC	diethyldithiocarbamate	
DMG	dimethylgyoxime	
DPASV	differential pulse anodic stripping voltammetry	
DPCSV	differential pulse cathodic stripping voltammetry	
DPP	differential pulse polarography	
DPV	differential pulse voltammetry	
EAAS	electrothermal atomic absorption spectroscopy	
ECD	electron-capture detector	
FAAS	flame atomic absorption spectroscopy	
PP	polypropylene	
FID	flame ionization detector	
GC	gas-chromatography	
HMDE	hanging mercury drop electrode	
HPLC	high-performance liquid chromatography	
ICP-AES	inductively coupled plasma atomic emission spectroscopy	
ICP-MS	inductively coupled plasma mass spectroscopy	
MIBK	methyl isobutyl ketone	
Na FDEDTC	natrium (ditrifluorethylene)dithiocarbamate	
ns	not specified	
PAN	[1-(2-pyridylazo)-2-naphthol]	
PE	polyethylene	
PIXE	particle-induced X-ray emission	
PVC	polyvinyl chloride	
TPP	tetraphenylporphine	

2.2.1 Determination of trace amounts

The use of very sensitive instrumental methods has shown that detection limits are not so much set by the capabilities of the instrument as by contamination from different sources. Sources of contamination include laboratory air, laboratory equipment and construction material, reagents and the analyst. In order to obtain reliable results, especially when determining trace (mg/kg) and ultratrace (μg/kg) amounts, specific procedures concerning contamination control during sample collection, storage, processing, and analysis must be adhered to.

Besides contamination control during sample processing, the establishment of the level of accuracy of the analytical procedure is of great importance. Thus, analysis of certified reference materials is recommended. Recovery experiments to check the analytical procedure include the spiking of samples with known amounts of nickel.

2.2.2 Sample collection

Great care must be taken to minimize the risk of contamination during sample collection by the use of suitable procedures (Nieboer & Jusys, 1983; Boyer & Howitz, 1986). Persons who handle samples should wear talc-free gloves to avoid nickel contamination from sweat. When collecting liquid samples, e.g., sea water, fresh water, or urine, acid-washed polyethylene containers should be used. As stainless steel is a source of nickel contamination, Teflon®, intravenous catheters are recommended for blood collection. Tissues should be dissected with plastic forceps and obsidian scalpels (Sunderman et al., 1985).

Collection of airborne particulate nickel involves pumping a known volume of air through a membrane filter, which usually consists of cellulose, PVC, or glass fibre (Mackenzie Peers, 1986). Equipment of the air sampling system with a cyclone, cascade, or cascade impactor allows sampling of respirable particulate nickel (Roy, 1985).

Volatile nickel compounds, such as nickel carbonyl, can be absorbed in an alcoholic iodine solution through which the air being sampled is passed (NIOSH, 1977a; Stedman, 1986a).

2.2.3 Sample pretreatment

Prior to the determination of nickel in biological and environmental materials, the organic constituents must be oxidized or removed to avoid interference during analysis. The most common methods include wet digestion, i.e., oxidation of organic matter by reagents, such as nitric acid, sulfuric acid, perchloric acid, or hydrogen peroxide, or combinations of these compounds, and dry ashing, which ensures oxidation of organic matter by the action of oxygen and high temperatures. Puchyr & Shapiro (1986) developed an extraction method for food samples that involved low-temperature HCl/HNO_3-leaching followed by filtration. This method proved to be very efficient and less hazardous and less time-consuming than common wet or dry digestion techniques. Organic substances, dissolved in natural waters, and certain liquid foods are successfully decomposed by oxidative ultraviolet (UV) photolysis (Pilhar et al., 1981; Weidenauer & Lieser, 1985).

As nickel concentrations are often low in relation to analytical detection limits, preconcentration steps are introduced, which may also separate nickel from substances interfering with analysis. Techniques very frequently employed include chelate extraction with dithiocarbamates, dimethylglyoxime, furildioxime, or 8-hydroxyquinoline into organic non-polar solvents. Tanaka (1985) used EDTA as a complexing agent prior to determination of nickel in waste water and plating solution: Gemmer-Colos et al. (1981) reported complete extraction of nickel-heptoxime from an aqueous nickel solution at low pH values. Interfering cobalt and iron ions were eliminated by treatment of the extract with ammonia. Another preconcentration technique, prior to analysis of nickel in fresh and sea water, is the use of chelating ion-exchanged resins, e.g., Chelex 100®, (Bruland et al., 1979) or 1-(2-pyridylazo)-2-naphthol (PAN) (Yoshimura et al., 1980). Brajter & Slonawska (1986) considered Chelex-P®, a dibasic phosphate ester of cellulose, as very efficient for the preconcentration of nickel in water samples. A less time-consuming method for the preconcentration of nickel in sea water was developed by Watanabe et al. (1981), Sturgeon et al. (1982), and McLaren et al. (1985). It involved complexation of the trace metals by 8-hydroxyquinoline followed by adsorption on C_{18} chemically bonded silica gel. Wan et al. (1985) achieved a greater enrichment factor, smaller sample volume, and removal of interfering humic substances when preconcentrating nickel and other trace metals in natural waters on XAD-7 regions (cross-linked polymer of methylmethacrylate) in a two-step procedure at two different pH values. A very efficient preconcentration method was developed by Burba & Willmer (1985) in which trace metals in natural waters were enriched on metal hydroxide coated cellulose, using iron hydroxide and indium hydroxide. The use of gallium hydroxide as a coprecipitation agent for multi-element determination in sea water, and zirconium hydroxide as a coprecipitation agent for multi-element determination in sea and fresh water has been described (Akagi et al., 1985a,b). Zirconium caused spectral interferences in the inductively coupled plasma atomic emission spectrometry, whereas coprecipitation with gallium proved to be more efficient with lower limits of detection in subsequent analysis.

2.2.4 Analytical methods

The two most commonly used analytical methods for nickel are atomic absorption spectroscopy and voltammetry.

In biological samples, such as tissues and body fluids, nickel concentrations are routinely determined by electrothermal atomic absorption spectroscopy (EAAS). Acid digestion is required before analysis of biological samples, which is commonly followed by an enrichment step. The IUPAC Subcommittee on the Environmental and Occupational Toxicology of Nickel (Sunderman, 1980) developed a reference method for the determination of nickel in serum or urine by EAAS, after acid digestion and the subsequent extraction of nickel with ammonium pyrrolidine dithiocarbamate (APDC) into methyl isobutyl ketone (MIBK).

The introduction of a Zeeman-compensated system improved background compensation and permitted a more rapid and direct determination of nickel levels with considerably lower detection limits, which was suitable for routine use. Sunderman et al. (1984a,1985) applied EAAS with Zeeman background correction for the direct determination of nickel in acid-digested serum (detection limit, 0.05 µg/litre), in whole blood, and in acid-digested tissue homogenates (detection limit, 10 ng/g dry weight). The suitability of this method for the direct determination of nickel in acidified urine with a detection limit of 0.5 µg/litre has been demonstrated (Sunderman et al., 1986a). Andersen et al. (1986a) presented an even more direct method, which only required dilution of the human plasma prior to quantification by Zeeman-corrected EAAS. The limit of detection was 0.09 µg/litre. Recent progress in voltammetry has made this method the most sensitive. Ostapczuk et al. (1983) used a new voltammetric method for the determination of nickel in a variety of biological materials following acid digestion of the sample. The method was based on the application of differential pulse voltammetry (DPV) after prior interfacial accumulation by an adsorption layer of nickel-dimethylglyoxime chelate at the hanging mercury drop electrode (HMDE). The measurement of nickel concentrations as low as 1 ng/litre was possible using this method, which was also suitable for analysing food samples (Meyer & Neeb, 1985). Though it requires time-consuming sample digestion procedures, voltammetry is more sensitive, more rapid, and less costly than EAAS (Ostapczuk et al., 1983). An isotope dilution gas chromatography-mass spectrometric method for the detection of nickel in biological materials at the ng/litre level was recently introduced by Aggarwal et al. (1988). The method depends on the preparation of a thermally stable and volatile chelate (chelating agents: sodium diethyldithiocarbonate or lithium bis(trifluoroethyl) dithiocarbamate) followed by on-column injection into a gas chromatographic column and electron ionization of the eluted chelate in the mass spectrometer.

Analysis for nickel in natural water is frequently performed by EAAS following preconcentration. Large concentration factors (200:1) provide detection limits as low as 10 ng/litre in sea-water analysis (Bruland et al., 1979). Inductively-coupled plasma atomic emission spectroscopy (ICP-AES) is gaining importance in simultaneous multi-element determination. Provided that there is sufficient enrichment, nickel concentrations as low as 60 ng/litre can be determined in natural waters (Akagi et al., 1985a).

Pilhar et al. (1981) presented DPV-HMDE with prior chelate adsorption at the electrode as a simple, rapid, and inexpensive procedure for determining nickel levels in natural waters and waste water, with a detection limit of 1 ng/litre. This method is also suitable for determining the nickel contents of various kinds of food

(Valenta et al., 1981; Meyer & Neeb, 1985). Particle-induced X-ray emission makes possible the detection of various trace metals in water at the ng/litre level (Ali et al., 1985).

Atomic absorption spectroscopy is the most widely used method of analysis for nickel in soil. The sample must undergo acid digestion and may be submitted to enrichment procedures. Detection limits are in the mg/kg range (Abo-Rady, 1979a; Sanzolone et al., 1979; Schmidt & Dietl, 1981). Voltammetry, which has been successfully used for the determination of nickel in a variety of biological samples, has also been applied in the analysis of acid-digested soil samples, using square wave voltammetry as the more efficient method (Ostapczuk et al., 1985b).

Determination of nickel in air samples has been performed using different methods (NIOSH, 1977b). However, flame atomic absorption spectroscopy (FAAS) is the most commonly used analytical technique for measuring the nickel concentration in air samples. Following an acid digestion procedure, 1 μg of nickel in 1 ml sample can be detected by this method. Interference by a 100-fold excess of iron, manganese, chromium, copper, cobalt or zinc can be minimized by proper burner elevation and the use of an oxidizing flame.

A technique suitable for the simultaneous determination of several metals in air has been reported (Mackenzie Peers, 1986). Following acid digestion of the absorbing cellulose ester membrane filter the extracted sample was analysed by ICP-AES with a detection limit of 1 μg/sample.

Volatile nickel carbonyl in air can be determined by colorimetry, as a coloured furildioxime-chelate (Stedman, 1986a), or directly, by photometric detection of chemiluminescence (Stedman, 1986b). Detection limits are 1 μg/m^3 and 0.2 μg/m^3, respectively.

Electron microscopy and X-ray microanalysis can be used for the determination of nickel in single dust particles, such as welding fumes and grinding dusts.

3. SOURCES OF HUMAN AND ENVIRONMENTAL EXPOSURE

3.1 Natural occurrence

Nickel is a ubiquitous element and has been detected in different media in all parts of the biosphere. It is the fifth most abundant element by weight after iron, oxygen, magnesium, and silicon, and the 24th most abundant element in the earth's crust. However, the average concentration of nickel in the earth's crust is only about 0.008% (Mason, 1952). Meteorites have been found to contain 5-50% nickel. Nickel-enriched nodules have been discovered on the ocean floor (NAS, 1975).

3.1.1 Rocks

Most nickel occurs in the ferromagnesium minerals of igneous and metamorphic rocks, e.g., olivine [$(MgFe)_2.SiO_4$]. Normal nickel concentrations in igneous rocks range from 2 to 60 mg/kg (acidic rocks), 50-200 mg/kg (basic rocks), and 10-2000 mg/kg (ultramafic rocks) (Boyle, 1981). Among the major sedimentary rocks, shale and carbonate rocks contain an average of 50 mg nickel/kg; sandstone contains only 1 mg nickel/kg (NAS, 1975).

The most commercially important nickel ore deposits are accumulations of nickel sulfide minerals in ultramafic igneous rocks. Such deposits are found in Australia, Canada, and the USSR. The ores are composed almost entirely of pentlandite [$(Fe,Ni)_9S_8$], chalcopyrite ($CuFeS_2$), and pyrrhotite (Fe_xS_{x+1}), and usually contain 1-4% nickel (Duke, 1980). Other nickel ore deposits are formed by the weathering of ultramafic ferromagnesium silicate rocks in humid tropical areas. The residual soil (laterite) developing during the weathering process may contain up to 10 times the amount of nickel in the original rock (Duke, 1980). The nickeliferous lateritic weathering profile is characterized by two deposits, an upper oxide zone and a silicate zone, both varying in proportion. The oxide zone is composed of iron oxides containing nickel in solid solution. In the silicate zone, also called the garnierite zone, nickel is found in

Sources of exposure

the mineral serpentine $Mg_3Si_2O_5(OH)_5$ substituting for magnesium. The nickel content of lateritic ores is approximately 1–3% (Duke, 1980). Important deposits are located in Brazil, Cuba, Dominican Republic, Guatemala, Indonesia, New Caledonia, and the Philippines.

3.1.2 Soils

In glacial areas, nickel-containing components may have been dispersed over wide areas; thus, the nickel contents of the soil can differ considerably from the nickel content of the underlying bedrock.

In unweathered glacial sediments, nickel occurs in the same mineral phases as those in which it is found in the rocks, i.e., sulfides and silicates. Weathering of rocks and soils leads to nickel release from nickeliferous minerals. The nickel released is largely retained in the weathered material in association with clay particles and therefore not considered to be very mobile in the superficial environment (Duke, 1980).

Nickel can exist in soils in several forms (Hutchinson et al., 1981) including:

(*a*) inorganic crystalline minerals or precipitates (e.g., in the lattice of aluminium silicates);

(*b*) complexed or adsorbed on organic cation surfaces (e.g., organic matter) or on inorganic cation exchange surfaces (e.g., clay minerals); and

(*c*) water-soluble, free-ion, or chelated metal complexes in soil solution.

In a soil-water system, nickel may form complexes with inorganic ligands (Cl^-, OH^-, SO_4^{2-}, or NH_3,) (Richter & Theiss, 1980) and organic ligands (e.g., humic or fulvic acids) (Nriagu, 1980). Agricultural soils of the world contain between 3 and 1000 mg nickel/kg (NAS, 1975). In forest floor samples collected from 78 sites in 9 states in the northeastern USA, nickel was present at concentrations in the range of 8.5–15 mg/kg (Friedland et al., 1986).

3.1.3 Water

Nickel occurs in aquatic systems as soluble salts adsorbed on clay particles or organic matter (detritus, algae, bacteria), or associated with organic particles, such as humic and fulvic acids and proteins.

Nickel may enter surface waters from three natural sources (Boyle, 1981), i.e., as particulate matter in rainwater, through the dissolution of primary bedrock minerals, and from secondary soil phases.

The fate of nickel in freshwater and sea water is affected by several factors including pH, pE, ionic strength, type and concentration of organic and inorganic ligands, and the presence of solid surfaces for adsorption (Snodgras, 1980).

In natural waters, at a pH range of 5–9, the divalent ion Ni^{2+} ($Ni(H_2O)_6^{2+}$) is the dominant form. In this pH range, nickel may also be adsorbed on iron and manganese oxides, or form complexes with inorganic ligands (OH^-, SO_4^{2-}, Cl^- or NH_3) (Richter & Theiss, 1980).

If sulfate concentrations are sufficiently high, nickel sulfate may be the predominant soluble form; under anaerobic conditions, sulfide is the major factor controlling the solubility of nickel (Richter & Theiss, 1980). Nickel concentrations of 0.228–0.693 µg/litre, determined for a vertical open-ocean water profile, were considered to reflect the actual nickel concentration in this medium (Bruland et al., 1979). Concentrations of nickel in freshwater systems are generally less than 2–10 µg/litre (Stokes, 1981). For nickel levels in drinking-water see section 5.1.2.

3.1.4 Fossil fuels

Nickel occurs in both coal and crude oil in minor quantities, originating from vegetation and from percolating waters containing nickel leached from rocks. The average value in some Canadian coals was found to be 15 mg nickel/kg (Hawley, 1955). The nickel contents in some Western Canadian crude oils, analysed by Hodgson (1954), were in the range of 0.09–76.6 mg/kg.

Sources of exposure

3.1.5 Air

Atmospheric nickel is considered to exist mainly in the form of aerosols with different nickel concentrations in particles depending on the type of source (Schmidt & Andren, 1980). Major natural sources include the aerosols constantly produced by the oceanic surface, windblown soil dusts, and volcanic ash. Nickel is released from plants during growth, at different levels, depending on soil composition. Forest fires produce nickel-containing smoke particles. A part of atmospheric nickel originates from meteoric dusts.

Atmospheric nickel concentrations for remote areas that are considered to be relatively free from man-made nickel emissions are in the range of $<0.1–1$ ng/m^3 (marine) and $1–3$ ng/m^3 (continental) (Schmidt & Andren, 1980). The wide variation in ambient nickel concentrations reflects the influence of nickel emissions from distant sources being transported by means of meteorological processes.

Nickel from natural sources, excluding volcanic dust and forest fires, is probably in the form of the oxide (Barrie, 1981).

3.2 Man-made sources

3.2.1 Production, use, and disposal

3.2.1.1 Primary production

The methods for the extraction and refining of nickel minerals depend on the mineralogical and geological characteristics of the ore. To date, nickel has mainly been extracted from sulfide and laterite ores.

Nickel sulfide ores are mostly mined underground using drilling, blasting, and other techniques. Milling procedures include liberation, flotation, and magnetic separation. Liberation of the sulfides from the gangue includes grinding of the rock material. Then the sulfides are concentrated by flotation processes. Flotation involves streaming air bubbles through an aqueous slurry of the ore particles in a flotation cell. The particles that are not wetted by the liquid adhere to the air bubbles, rise to the surface of the slurry, and can be removed. The addition of different chemicals to the flotation medium allows the selective flotation of nickel- and copper-rich fractions.

Most of the pyrrhotite (both lump ore and ground ore) can be separated magnetically, because of its magnetic properties.

Laterite nickel deposits are mined from surface pits using earth-moving equipment.

Both sulfide ore concentrates and laterite ores are subjected to pyro- and hydrometallurgical processes. The pyrometallurgical processing basically involves three operations, i.e., roasting, smelting, and converting.

During roasting, the concentrate is oxidized by hot air. Most of the iron is oxidized, while nickel, copper, and cobalt remain combined with sulfur. Part of the sulfur is removed as gaseous sulfur dioxide.

The roasted product is smelted in a furnace together with a siliceous flux to obtain two immiscible phases, an iron-rich silicate slag and a nickel-rich sulfide matte, which also contains iron, copper, and cobalt.

The matte is treated in a "converter" where more sulfur is driven off and the remaining iron is oxidized and removed as slag. The matte is allowed to cool and treated in different ways. It may be, for example, cast into anodes for electrolytic refining or cooled slowly to facilitate crystallization to nickel sulfide, copper sulfide and a nickel-copper alloy containing the desired metals. These three phases can then be separated by flotation and magnetic separation. The species of nickel likely to be present during roasting, smelting, and converting include the ore, nickel subsulfides, nickel copper sulfides, nickel oxides, nickel-copper oxides, arsenides, and anhydrous nickel sulfate. The extraction of nickel from laterite ores is similar to the extraction of nickel from sulfide ores with the exception that sulfur (commonly gypsum) has to be added. The molten matte is charged into a converter where the iron is oxidized and the sulfur combines with nickel to form Ni_3S_2.

Smelting to ferronickel is essentially the same as matte smelting, except that no sulfur is added. It is often applied to laterite ores.

The resulting iron-nickel alloy contains 20–50% nickel (Duke, 1980).

Most of the nickel matte obtained from sulfide or laterite ore smelting undergoes further refining techniques, such as electro-, vapo-, or hydro-metallurgical refining, but a part of the matte is roasted to marketable nickel oxide sinter.

Hydrometallurgical refining can be applied both to laterite ore and sulfide ore or sulfide ore concentrates. Soluble nickel amines are formed during pressure leaching of the sulfide ore concentrate with strong ammoniacal solution at a moderately elevated temperature. The saturated solution is boiled to drive off ammonia and precipitate copper as sulfide. Sulfur is oxidized. Nickel and cobalt are recovered as pure metal powders by reduction with hydrogen under pressure.

Laterite ores must first be reduced. The reduced ore is leached with an ammonia-ammonium carbonate solution. Nickel dissolves as nickel amine. The saturated solution is heated by steam, ammonia is driven off, and nickel is precipitated as a basic carbonate.

Pure nickel (99.9%) can be produced by electrolytic refining. Generally, an impure metal anode (produced by reducing nickel oxide) and a cathode starting sheet are placed in an acidic electrolytic solution. When a current flows, nickel and other metals are dissolved from the anode. The electrolyte is then removed, purified and returned to the cathode compartment, where nickel is deposited on the cathode.

During vapometallurgical refining, impure metal obtained by the reduction of nickel oxide is subjected to the action of carbon monoxide forming volatile nickel carbonyl [$Ni(CO)_4$] (carbonyl or Mond process). This reaction is reversed by heat and the nickel carbonyl decomposes to pure nickel metal and carbon monoxide. The carbonyl process produces the purest nickel (99.97% or more).

The smelting and refining processes yield various marketable forms of nickel of different purities (Table 3).

Table 3. Commercial forms of primary nickel [a]

Type	Composition (%)								
	Nickel	Carbon	Copper	Iron	Sulfur	Cobalt	Oxygen	Silicon	Chromium
Pure unwrought nickel									
Cathode	>99.9	0.01	0.005	0.002	0.001	-	-	-	-
Pellets	>99.97	<0.1	0.001	0.0015	0.0003	5×10^{-5}	-	-	-
Powder	99.74	<0.1	-	<0.1	<0.01	-	<0.15	-	-
Briquettes	99.9	0.01	0.001	0.002	0.0035	0.03	-	-	-
Rondelles	99.25	0.022	0.046	0.087	0.004	0.37	0.042	-	-
Ferronickel [b]	20-50[c]	1.5-1.8	-	Rest	<0.3	_[c]	-	1.8-4	1.2-1.8
Nickel oxide	76.0	-	0.75	0.3	0.006	1.0	Rest	-	-

[a] Modified from: Corrick (1977).
[b] Ranges used to denote variable grades produced.
[c] Cobalt included with nickel (1-2%).

3.2.1.2 Intermediate products and end-use

Most of the nickel produced is used in the production of alloys (Table 4). In the production of nickel steel alloys, steel scrap, limestone, iron oxide ore, and nickel are charged into a furnace (open-hearth furnace, electric arc furnace and cupola) where the steel and iron alloys are melted. After final adjustment of the carbon and alloy contents, the steel is cast into moulds. Non-ferrous melting is commonly performed in a reverberatory furnace.

Table 4. Consumption of nickel by intermediate product and end-use industry in 1985 in the USA

Index	Consumption [a] (% of total)
Intermediate product	
Stainless and alloy steels	42
Nonferrous alloys	36
Electroplating	18
End-use industry	
Transportation	23
Chemical industry	15
Electrical equipment	12
Construction	10
Fabricated metal products	9
Petroleum	8
Household appliances	8
Machinery	8
Other	7

[a] Data from: US Bureau of Mines (1986).

The forming and shaping of ingots, after the casting of the alloy, is performed by hot-working, grinding, and welding. Hot-working includes the reduction of the cross-section, e.g., by forging or rolling. The resulting product may be cut and then extruded to the desired form. Grinding is necessary to condition the metal surface for further processing, e.g., welding. Welding techniques, such as electric-arc, electric-spot oxyacetylene-torch, or furnace-brazing, are used to fabricate assembled shapes. In special cases, forming of parts may also be performed by sintering, e.g., by sintering nickel powder from the Mond process.

The addition of nickel to steel and cast iron yields an alloy with increased strength and toughness and resistance to corrosion. Stainless steel is used in the chemical and food-processing industries. Because of their ferromagnetic properties, iron-nickel alloys are important materials for the electrical industry.

Various medical devices, such as prostheses or orthopaedic implants, are made from stainless steel.

Nickel-copper alloys exhibit the highest mechanical strength and resistance to corrosion and are used in the chemical and machine industries (pipes, nozzles, machine parts for the food and textile industries). Their high resistance to corrosion makes these alloys a valuable material for the shipbuilding industry. The nickel-copper alloy containing 77–63% nickel is known as Monel metal.

Nickel alloys containing chromium, molybdenum, aluminium, cobalt, titanium, or combinations of these elements are of special industrial importance, because of their high-temperature resistance.

Nickel-chromium alloys are used for jet engine components, in nuclear reactors, and for turbine blades. Superalloy is an extremely high-temperature resistant alloy containing 10–20% cobalt. It is used in turbine blades and other engine components of jets, ships, and racing vehicles, where extreme mechanical and high-temperature resistance is required.

In plating, nickel gives a hard, tarnish resistant surface that can be polished, which makes the finished product suitable for consumer items, such as automobile components, household furniture, and plumbing fixtures. Normally nickel-plated consumer items are covered with a thin layer of chromium plating.

Other important uses of nickel are in nickel-cadmium batteries, electronic equipment, and computers. Nickel compounds are used as catalysts in the manufacture of organic chemicals, petroleum refining, and edible oil hardening. They are also constituents of pigments and colours for ceramics and glassware, and of marine anti-fouling paint. In the glass industry, nickel is used in moulds for bottles. Nickel compounds are also used as a coating for pressure sensitive papers. In the United Kingdom, "silver" coinage (5 p, 10 p, and 20 p) is based on cupro-nickel alloys containing approximately 20% nickel. Coinage from other countries contains higher levels, e.g., Canadian 10 cents (99.8% nickel), and French 1 and 2 francs (99.8–99.9% nickel).

The production of secondary nickel in the form of scrap recovery is a major source of nickel. Recycled scrap is generally melted and refined and subsequently used for the production of steels and alloys, similar in composition to those in which it entered the recycling process. Thus, scrap recycling processes are analogous with those used in primary production.

3.2.1.3 World production levels and trends

The development of global mine production during this decade is shown in Table 5.

The nickel market weakened considerably from 1981 to 1983, because of a reduction in demand arising from a recession in the economy. In 1984, production and demand increased again. From a 1983 base, the US Bureau of Mines (1986) estimated that there would be an increase in the average annual demand of about 2.5%, up to 1990.

The identified world deposits with an average nickel content of approximately 1% or more, contain 143 million tonnes of nickel (US Bureau of Mines, 1986). In addition, there are extensive deep-sea resources of nickel in manganese nodules, particularly in the Pacific Ocean (US Bureau of Mines, 1986).

At present, there are only a few actual and potential substitutes for nickel, e.g., aluminium, coated steel, titanium, and plastic for industrial purposes, and platinum, cobalt, and copper for catalytic uses. However, the use of these substitutes results in increased costs and a lower quality end-product (US Bureau of Mines, 1986).

Table 5. Global mine production of nickel, by country [a,b] (short tonnes of nickel)

Country or territory	1980	1981	1982	1983 [c]	1984 [d]	1985 [d]
Albania (content of ore)[d]	6 100	6 200	6 400	6 400	6 600	6 600
Australia (content of concentrate)	81 927	81 963 [e]	96 510	84 465	82 900	81 000
Botswana (content of matte)	17 022	18 200	19 573	20 079	19 300	19 000
Brazil (content of ore)	2 504 [e]	2 573 [e]	5 306	11 840	12 100	12 000
Burma (content of speiss)	15	22	22 [d]	22 [d]	22	-
Canada[f]	203 709	176 642	97 824	134 300	192 000	195 000
China[d]	12 000	12 000	13 200 [e]	14 300 [e]	15 400	16 000
Colombia (content of ferroalloys)	-	-	1 100	15 000	15 400	10 000
Cuba (content of oxide, sinter, sulfide)	40 338	42 489	39 790	41 500 [d,e]	35 050	40 000
Dominican Republic	18 019	20 601	5 838	23 369	26 698 [g]	27 000
Finland (content of concentrate)	7 199	7 566	6 852	5 418	5 500	6 000
German Democratic Republic[d]	3 000	3 000	2 800	2 400	2 300	-
Greece (recoverable content of ore)[h]	16 796	17 200	5 500 [d,e]	18 500 [d,e]	18 400	16 000
Guatemala	7 650	-	-	-	-	-
Indonesia (content of ore)[h]	58 738	53 848	50 578	54 430	68 900	70 000
Morocco (content of nickel ore and cobalt ore)	148	144 [e]	140	-	-	-
New Caledonia (recoverable content of ore)	95 451	86 079	66 250	43 542	45 200	44 000
Norway (content of concentrate)[d]	2 200 [e]	7 700 [e]	3 900 [e]	4 000 [e]	3 900 [e]	-
Philippines	51 934	32 239	22 183	17 522	18 300	25 000
Poland (content of ore)[d]	2 300	2 300	2 300	2 300	2 300	-

Table 5 (continued)

South Africa, Republic of	28 329	29 100	24 250 [d]	22 600 [d]	27 600	27 000
USSR (content of ore)[d]	170 000	174 000	182 000	187 000	192 000	197 000
USA (content of ore shipped)	14 653	12 099	3 203	-	14 540 [g]	6 900
Yugoslavia (content of ore)[d]	2 200	4 400	4 400 [e]	3 300 [e]	4 400	3 000
Zimbabwe (content of concentrate)	16 617	14 350	14 671	11 186	11 080	11 000
Total	858 850 [e]	804 715 [e]	674 590	723 473	819 890	821 000

[a] From: US Bureau of Mines (1985; 1986).
[b] As far as possible, this table represents recoverable mine production of nickel. Where data relate to some more highly processed form of nickel, the figure given has been used in place of an unreported actual mine output, to provide some indication of the magnitude of mine output. See notes in parentheses and footnotes.
[c] Preliminary.
[d] Estimated.
[e] Revised.
[f] Refined nickel and nickel content.
[g] Reported figure.
[h] Includes a small amount of cobalt not reported and not recovered separately.

3.2.1.4 Emissions from the primary nickel industry

Data on the loss of nickel into the environment during production are limited. The smelting and roasting stages of ore refining and alloy production may be considered as the more important sources of nickel emission, because these processes generate flue dust, i.e., fine particulate matter that is swept from roasters and reverberatory furnaces by air and combustion gases that pass through these units.

During an environmental study initiated by the Ontario Ministry of Environment, trace metal emission rates from two nickel smelters were calculated on the basis of the results of chimney stack emission tests (Chan & Lusis, 1986).

The annual emissions of nickel during the study period are given in Table 6. Annual emissions from a 381-m stack of one smelter that emits particulates and gases from pyrometallurgical smelting processes are listed in Table 7. This was considered the most significant emission source.

Table 6. Yearly emissions (in tonnes) of nickel (Sudbury Basin, Canada) for the period 1973-81 [a,b]

Source	Variation	Nickel
INCO 381-m stack	Maximum	342
	Average	228
	Minimum	53
INCO 194-m stack	Maximum	
	Average	226
	Minimum	
INCO Smelter	Maximum	40
(low level)	Average	31
	Minimum	15
Falconbridge	Maximum	
93-m stack	Average	9.6
	Minimum	

[a] From: Chan & Lusis (1986).
[b] Basis: 365 × 24 h/day production.

Table 7. Average measured emissions of nickel from a 381-m stack (Canada) (in kg/h) [a]

Year	Emission
1973	48
1974	55
1975	15
1976	22
1977	33
1978	20
1979	12
1980	44
Average	31

[a] From: Chan & Lusis (1986).

Data on the chemical forms of nickel released into the atmosphere from production processes are practically non-existent. In most cases, statements are based on assumptions.

Species of nickel emitted into the air from mining garnierite and processing it to produce ferronickel at a facility in the USA, were assumed to be in the form of silicates, as in the ore, but were expected to be minimal (Radian Corporation, 1984). Depending on the temperature reached during drying and calcining, some nickel on the surface of ore fragments may become oxidized and emitted as iron-nickel oxide (Radian Corporation, 1984). Emissions during roasting and smelting would probably be in the form of nickel oxide combined with iron oxide as a ferrite (Radian Corporation, 1984; Warner, 1984).

When producing nickel from the sulfide ore, the process of roasting the concentrated ore may lead to the formation of small amounts of nickel sulfate and the emission of fine particles that are sulfated as they are carried through the flues (Warner, 1984).

According to the investigations of Radian Corporation (1984), emissions from matte refining processes at a US nickel refinery are expected to be predominantly in the form of subsulfide, as the processed matte is sulfide, and metallic nickel. A refinery dust sample from a Canadian nickel refinery was calculated to contain 20% nickel sulfate, 57% nickel sulfide, and 6.3% nickel oxide (Gilman & Ruckerbauer, 1962). Warner (1984) reported a nickel content of 5–10% (10% of which was water-soluble) in flue dusts from a Canadian smelter; most of these dusts are captured and recycled.

3.2.1.5 Emissions from the intermediate nickel industry

Fumes from stainlesss steel melting processes were found to contain 5% of total nickel in a water-soluble form. Chemically, it occurs in fumes from stainless steel manufacturing mainly as the metallic alloyed element in the iron matrix or in small amounts as nickel oxide (Koponen et al., 1981). Nickel emissions into the atmosphere can occur potentially from electroplating, and from grinding, polishing, and cutting operations performed on the finished product and scrap metal. However, in the case of electroplating, they are considered to be very low or non-existent, or are retained in the workplace area (Radian Corporation, 1984).

Grinding, polishing, and cutting operations could release metallic nickel into the working environment with possible emission to the outside atmosphere as a result of work-area ventilation (Radian Corporation, 1984).

3.2.1.6 Emissions from the combustion of fossil fuels

The major source of airborne nickel is the combustion of fossil fuels containing trace amounts of nickel (section 3.1.4). Combustion sources include facilities burning coal and oil for power generation or space heating.

Krishnan & Hellwig (1982) estimated emissions of trace metals in the USA from various coal and oil combustion sources (Table 8) and showed that nickel was a substantial trace pollutant. Nickel was the only trace metal emitted at a significant rate from domestic oil-fired boilers. The combustion of oil is a much more significant source of nickel emissions than the combustion of coal and is estimated to contribute 76–98% of the total nickel emissions from coal and oil combustion in the USA (Krishnan & Hellwig, 1982). A quantitative assessment of source contributions to inhalable

Table 8. Emissions of trace metals in the USA from coal and oil combustion (metric tonnes per year)[a]

Trace	Utility boilers[b] (>264 GJ/h input)		Industrial boilers[c] (>26 GJ/h input)		Commercial boilers[c] (>26 GJ/h input)		Residential boilers[b] (>422 MJ/h input)	
	Coal	Oil	Coal	Oil	Coal	Oil	Coal	Oil
Arsenic	149.1	144.7	214.8	54.7	99.3	84.6	60.3	3.2
Beryllium	19.2	7.0	6.2	2.2	3.7	0.1	0.8	4.1
Cadmium	7.7	219.7	4.9	83.9	2.7	128.4	1.6	23.1
Chromium	561.3	87.5	33.3	33.1	7.9	76.8	7.8	2.3
Lead	360.0	61.1	113.3	23.6	58.2	36.5	36.8	19.9
Manganese	407.2	33.7	31.5	7.5	22.8	11.6	163.9	1.2
Mercury	86.9	3.1	4.6	1.0	1.3	1.5	1.0	2.5
Nickel	281.1	877.3	34.0	363.1	14.1	818.0	7.8	216.4
Selenium	120.9	30.9	44.4	11.7	17.4	18.1	14.5	21.2
Vanadium	390.0	4637.0	29.4	1505.0	12.0	3293.0	7.8	6.1

[a] From: Krishnan & Hellwig (1982).
[b] 1978.
[c] 1977.

particulate matter in metropolitan Boston revealed a high correlation between inhalable nickel particles (aerodynamic diameter, 2.5–15 μm) and residual oil combustion (Thurston & Spengler, 1985).

Cass & McRae (1983) evaluated routine air monitoring data from sites in the South Coast Air Basin of California, in order to relate sources to particular trace elements determined in the samples. Eighty-one percent of fine nickel emissions (aerodynamic diameter < 10 μm) were calculated to arise from fuel oil fly ash. However, a similar study by Kowalczyk et al. (1982) failed to assign nickel particulate to any specific type of source.

Fly ash emitted from combustion sources has been analysed, in order to gain information on the chemical species of nickel present in air. Henry & Knapp (1980) analysed fly ash samples from the stacks of oil-fired and coal-fired power plants. In fly ash samples from oil-fired plants, 60–100% of the nickel components were water soluble, whereas, with one exception, samples from coal-fired plants contained 20–80% water-soluble material. As the sulfate ion was the only major ion detected in the water-soluble phase, it was concluded that nickel sulfate is the predominant form of nickel in emissions from oil-fired and coal-fired power plants. This conclusion was confirmed by Fourier transform infrared analysis (Gendreau et al., 1980).

Analysis of filter-collected fly ash from five oil-fired units revealed the presence of metals, including nickel, as sulfates in the soluble phase (Dietz & Wieser, 1983). As the sulfate amount measured by ion chromatography was, on average, 17% less than the sulfate amount expected from stoichiometric considerations, Dietz & Wieser (1983) suggested that some of the soluble nickel might have been present as partially soluble oxide or very finely dispersed particles of metal oxide.

Major components in the insoluble phase of fly ash samples from oil-fired utility boilers were determined by X-ray diffraction to be oxides of iron, aluminium, calcium, and silicon, and possibly nickel oxide (Henry & Knapp, 1980).

Hulett et al. (1980) studied the 100–200 μm fractions of fly ash specimens from 4 coal-fired power plants. They separated the ash magnetically into 3 insoluble fractions, i.e., glass, mullite-quartz, and magnetic spinel. Chemical determination showed that 90% of the nickel was present in the magnetic spinel phase. The nickel was assumed to be in the form of a substituted spinel, $Fe_{3-x}Ni_xO_4$.

The results of studies by Hansen & Fisher (1980) and Hansen et al. (1984) indicated that most of the nickel, present in coal combustion fly ash particles, was soluble and associated primarily with sulfate.

Thus, nickel emissions into the atmosphere from coal and oil combustion are considered to be composed predominantly of nickel sulfate, with smaller amounts of nickel oxide and nickel combined with other metals in complex oxides.

Another potentially important source of nickel in the environment is the combustion of diesel oil, which can contain 2 mg nickel/litre (2 ppm) (Fishbein, 1981). The vapour phase of diesel engine exhaust may also contain nickel carbonyl. In urban air near a busy intersection, Filkova & Jäger (1986), using EAAS, measured nickel carbonyl concentrations in the range of 0–14.1 ng/m^3.

3.2.1.7 Emissions from sewage sludge and waste incineration

Estimates made by Schmidt & Andren (1980) (section 4.1.1) indicated that, after fuel combustion, and nickel mining and refining, waste and sewage sludge incineration is the next major source of nickel emissions.

Evaluation of emission data from sewage sludge incinerators indicated that less than 1% of the nickel contained in the sludge feed was emitted as a fume, while the major part was emitted as fly ash (Gerstle & Albrinck, 1982). Dewling et al. (1980) noted that 80% of the nickel in the feed sludge of a fluidized bed, waste-water sludge incinerator in north-west Bergen was retained in the ash. Emission rates may vary widely, depending on combustion temperature, sludge composition, pollution control devices, and type of incinerator (Gerstle & Albrinck, 1982; Samela et al., 1986). Nickel species present in emissions from sewage sludge and waste incineration were analysed by Henry and co-workers (1982). The water-soluble phase of sewage sludge incinerator emissions contained mainly sulfate ions, indicating that the water-soluble nickel existed in the sulfate form. The soluble phase of refuse incinerator emissions also contained chloride ions, suggesting that nickel can be present in this phase as the chloride or sulfate. The insoluble phases of emissions from the two sources were similar and it is highly probable that the nickel may exist as complex oxides and iron spinels.

3.2.1.8 Miscellaneous emission sources

Nickel can be emitted during cement manufacturing and asbestos mining and milling, because nickel is a natural component of the minerals used in these operations.

During cement manufacturing, nickel is emitted, either as a component of the clays, limestones, and shales, used as raw materials, or as an oxide formed in the high temperature process kilns. Swedish cement was found to contain 5–59 mg nickel/kg (Wahlberg et al., 1977).

Crude chrysotile asbestos fibres from different mines in Canada (Quebec) contained 63–389 mg nickel/kg. In the host rock, the nickel content was 265–3075 mg/kg. Milled fibres are enriched by a factor of 4 (Barbeau et al., 1985). Nickel emitted into the air is expected to be in the form of silicates.

3.2.1.9 Waste disposal

Nickel from various industrial processes and other sources reaches waste water.

Klein et al. (1974) examined the major sources of nickel flowing into the New York City municipal waste-water collection system. The electroplating industry was found to be the dominant source of nickel (62%) in waste-water treatment plants. The total daily amount of nickel discharged into the sewers by electroplating firms was estimated to be 508 kg. The effluent of an electroplating factory in India contained 578.12 mg nickel/litre (Ajmal & Khan, 1985). Residential sources contribute 25% of the nickel in waste water, 3% comes from other industrial sources, and 10% from run-off. The total amount of nickel reaching the harbour of New York City, estimated to be 978 kg/day, originated as 43% from the treatment plant effluents, 30% from run-off, 20% from untreated waste water and 7% from sludge. Nickel concentrations in the influents of 12 waste-water treatment plants ranged from 0.05 to 0.31 mg/litre.

In the raw sewage of 25 full-scale municipal sewage-treatment plants, the nickel concentration varied between undetectable and 0.69 mg/litre (Sung et al., 1986).

Conventional treatment of mixed waste water consists of hydroxide precipitation of the metals at an alkaline pH, followed by removal of the resulting solids by sedimentation and, sometimes, filtration. Chen et al. (1974) reported a removal efficiency of the secondary treatment process (sludge activation and sedimentation) of 25–57%. The final effluent contained 0.14–0.177 mg nickel/litre. Sung et al. (1986) measured nickel levels in the influents and effluents of 25 sewage-treatment plants in the USA. In treatment plants with 50% minimum removal efficiency, 50% or more of the nickel was removed in the primary effluent at 5% of the plants; 50% or more was removed in the secondary effluent in 11% of the plants, and 50% or more was removed in the discharge of 10% of the plants.

Finally, residues from waste-water treatment are disposed of by deep-well injection, ocean dumping, land treatment, landfill, or incineration. Deep-well disposal is limited to residual liquids containing low levels of suspended solids and is most applicable to scrubber water blow-down. Ocean dumping is a source of contamination for coastal areas. Land treatment, such as sewage sludge treatment of agricultural soils, is a potential source of soil, and subsequent food plant, contamination.

Leachates from landfills may contaminate ground water and may contain 1.85–8.2 mg nickel/litre (Hrudey, 1985). Incineration of sewage sludge gives rise to considerable air emissions.

4. ENVIRONMENTAL TRANSPORT, DISTRIBUTION, AND TRANSFORMATION

4.1 Transport and distribution between media

Nickel is introduced into the environment from both natural and man-made sources (section 3). It is circulated throughout all environmental compartments (atmosphere, pedosphere, hydrosphere, and biosphere) by means of chemical and physical processes, such as wet and dry deposition, and, to a much lesser extent, by means of the biological transport mechanisms of living organisms.

4.1.1 Air

Nickel is emitted into the atmosphere from various natural sources, as indicated in Table 9. As only limited data are available concerning the relative quantities emitted, estimates have been made. Estimated emission values vary depending on the impact that is attributed to individual sources. Barrie (1981) considered sea spray to be a major contributor of atmospheric nickel. Generally, soil and volcanoes appear to be major sources and may contribute 40–50% of airborne nickel from natural sources.

Estimated man-made inputs into the atmosphere exceed the natural inputs (Tables 9 and 10). It has been estimated that the total amount of nickel that has been dispersed into the world ecosystems through the atmosphere is 1.0×10^9 kg (Nriagu, 1979). As indicated in Table 10, combustion of oil and incineration of waste contribute more than 70% of the nickel from man-made sources, followed by nickel mining and refining with 17%.

The transport and distribution of nickel particulates to, or between, different environmental compartments is strongly influenced by particle size and meteorological conditions. Particle size is primarily a function of the emitting source. Generally, particles from man-made sources are finer than dust particles of natural origin, e.g., soil (Beijer & Jernelöv, 1986). As the highest nickel concentrations

Table 9. Global emission of nickel from natural sources to the atmosphere. Emission rate (10^6 kg/year)

Source	Nriagu (1980)	Schmidt & Andren (1980)	Barrie (1981)
Soil dust	20	4.8	7.5-37.5
Volcanoes	3.8	2.5	10-60
Vegetation	1.6	0.82	1.5-20
Forest fires	-[a]	0.19	0.3-15
Meteoric dust	-	0.18	-
Sea salt	-	-	27
Sea aerosol	-	0.009	-
Total	26[b]	8.5	46-160

[a] No data available.
[b] Total includes 0.6 for "others".

Table 10. Global emission of nickel from man-made sources to the atmosphere [a]

Source	Emission rate (10^6 kg/year)
Residual oil combustion	17
Fuel oil combustion	9.7
Nickel mining and refining	7.2
Municipal incinerators	5.1
Steel production	1.2
Gasoline and diesel fuel combustion	0.9
Nickel alloy production	0.7
Coal burning	0.66
Cast iron production	0.3
Sewage sludge incineration	0.048
Copper-Nickel alloy production	0.04
Total	42.85

[a] Adapted from: Schmidt & Andren (1980).

are found in the smallest particles collected from ambient air (Lee & von Lehmden, 1973; Natusch et al., 1974), these particles are of special environmental and toxicological significance. There is evidence that fine particulate matter, which has a longer residence time in the atmosphere, is carried a long distance, whereas larger particles are deposited near the emission source (Beijer & Jernelöv, 1986). Schmidt & Andren (1980) estimated an atmospheric residence time for nickel particulates of 5.4–7.9 days.

There are no data on the chemical forms of nickel from natural sources in the atmosphere. When considering the composition of the source, a part of airborne nickel may exist as pentlandite ($(FeNi)_9S_8$) and garnierite (a silicate mixture) (Schmidt & Andren, 1980). The chemical composition of nickel compounds released from man-made sources differs from that of compounds from natural sources because of the different processes involved. Flue dust analysis revealed a predominance of oxides and sulfates (section 3). Alterations in the chemical forms, following distribution in the atmosphere, have not been investigated.

The fate of nickel carbonyl, the only gaseous nickel compound of environmental importance, may be deduced from its chemical properties. Nickel carbonyl is unstable in air and decomposes to form nickel carbonate. At 25 °C, the lifetime of ng/m^3 concentrations of nickel carbonyl is about one minute, increasing by one minute for each mg/m^3 of carbon monoxide present (Stedman & Hikade, 1980).

4.1.2 Water

Nickel is introduced into the hydrosphere by removal from the atmosphere (wet and dry deposition of naturally and anthropogenically released nickel), by surface run-off, by discharge of industrial and municipal waste, and following the natural erosion of soils and rocks.

In surface or ground waters not polluted by human beings, the nickel content often reflects the weathering process of the parent soil or rock. However, data are insufficient to separate natural geochemical effects from man-made influences. Keller & Pitblado

(1986) demonstrated a relationship between elevated nickel levels in Sudbury area lakes and nickel-emitting sources.

In rivers, nickel is transported mainly as a precipitated coating on particles and in association with organic matter; in lakes, the ionic form and the association with organic matter are predominant (Snodgras, 1980).

Nickel may be deposited in the sediment by such processes as precipitation, complexation, adsorption on clay particles, and via uptake by biota. Because of microbial activity or changes in physical and chemical parameters, including pH, ionic strength, and particle concentration, sorption processes may be reversed (Di Toro et al., 1986) leading to release of nickel from the sediment.

Part of the nickel is transported via rivers and streams to the ocean. Riverine suspended particulate input is estimated to be 135×10^7 kg/year (Nriagu, 1980). Industrial and municipal waste and atmospheric fallout contribute 0.38×10^7 kg/year and 2.5×10^7 kg/year, respectively (Nriagu, 1980). Possible routes of removal of nickel particulate are scavenging by ferromanganese phases on the ocean floor or algae. The average residence time of nickel in the deep ocean is calculated to be 2.3×10^4 years (Nriagu, 1980).

4.1.3 Rocks and soil

Nickel occurs naturally in several types of rocks (section 3.1) and it may enter the surface environment by the chemical and mechanical degradation of the rock to soil. Nickel is fractionated within the different components of the soil profile, depending on the type of soil and soil chemical conditions. In residual soils nickel is preferentially adsorbed on alkali and alkaline earth cations in the clay minerals (Boyle, 1981).

Depending on soil type, nickel may exhibit a high mobility within the soil profile as demonstrated by Heinrichs & Mayer (1980) for a brown forest soil in the Solling ecosystem, a relatively unpolluted area. The soil profile in the study area revealed an accumulation of nickel in the top organic layer, and a concentration increasing with depth in the subsequent mineral layer, indicating a high mobility of nickel. Similar results were obtained for peat muck soil profiles (Sapek & Sapek, 1980). In the Solling forest ecosystem study, comparison of flux data and ecosystem output (seepage) and input (precipitation) showed that nickel was balanced within the ecosystem and that the forest ecosystem was neither a source nor a sink in the geochemical cycle of nickel. However, this balance may be disturbed in case of heavy pollution.

Water solubility, and thus bioavailability to plants, is affected by soil pH; decreases in pH generally mobilize nickel. Most nickel compounds are relatively soluble at pH values < 6.5, whereas nickel exists predominantly as insoluble nickel hydroxides at pH values > 6.7. Therefore, acid rain has a pronounced tendency to mobilize nickel from soil and increase nickel concentrations in ground water, leading eventually to increased uptake and potential toxicity for microorganisms, plants, and animals (Sunderman & Oskarsson, 1988). Other factors are the numbers of organic and inorganic cation exchange sites, the adsorption strength, and the relative numbers of other cations competing with nickel for cation exchange sites (Hutchinson et al., 1981). Depending on the geochemical characteristics of soils, nickel may become distributed among the different soil compartments finally reaching ground water and, thus, rivers and lakes. Transport of nickel to river systems and oceans is also mediated by erosion and run-off processes.

The main man-made sources of nickel contamination in soils are emissions from nickel smelting and refining and disposal of contaminated sewage sludge. Atmospheric input of nickel into the soil and inputs by waste disposal and through the application of fertilizers are estimated to be 5.5×10^4 kg/year and 1.4×10^4 kg/year, respectively (Nriagu, 1980). On the basis of the nickel pool in soils and the denudation of nickel from continents, the residence time of nickel in soils is calculated to be approximately 3500 years (Nriagu, 1980).

4.1.4 Vegetation and wildlife

Terrestrial plants take up nickel from soil primarily via the roots. The nickel concentrations in most natural vegetation range from 0.05 to 5 mg/kg dry weight (NAS, 1975). The amount of nickel uptake from the soil depends on various geochemical and physical parameters including the type of soil, soil pH, humidity, the organic matter content of the soil, and the concentration of extractable nickel.

Although nickel concentrations exceeding 50 mg/kg (on a dry weight basis) are usually toxic for plants (NAS, 1975), nickel-tolerant species growing on serpentine soils can accumulate nickel levels that are orders of magnitude higher.

Aquatic plants are also known to take up and accumulate nickel (Jenkins, 1980a). As algae are at the lower end of many foodchains, this fact needs special consideration. The highest nickel levels found in aquatic algae and spermatophytes in contaminated areas were 150.9 mg/kg dry weight or 690 mg/kg wet weight, respectively, exceeding normal levels by more than 10 times (Jenkins, 1980a).

Investigations of nickel distribution and cycling in the Solling ecosystem indicated a steady state, i.e., a balance between atmospheric input and binding in the soil and phytomass (Heinrichs & Mayer, 1980). This steady state may be disturbed by human impact, e.g., increase of atmospheric nickel input or acidity of rainfall. Nickel is released into the atmosphere from plant exudates and forest fires. It migrates into soil and surface waters following the decay and mineralization of plants.

Since nickel occurs in virtually all plants at different levels, it may be taken up by herbivorous terrestrial and aquatic animals and their predators. Increased nickel levels in vegetation can give rise to increased nickel contents in grazing animals and their predators. For example, nickel levels were found to be higher in the organs of wild ruminants than in those of domestic animals, because of the higher nickel content in their grazing areas (Groppel et al., 1980).

4.2 Uptake and bioaccumulation

4.2.1 Terrestrial organisms

The best known phenomenon of nickel accumulation in plants is the considerably increased nickel levels found in certain species growing on infertile serpentine soils. Approximately 70 nickel hyperaccumulators, i.e., species with nickel concentrations exceeding 1000 mg/kg, are known, most of them belonging to the genus *Allyssum*. Investigations of nickel-accumulating Flacourtiaceae in

New Caledonia revealed nickel levels in the range of 1000–50 000 mg/kg dry weight (Jaffre et al., 1979). Yang et al. (1985) found a significant inverse correlation between the contents of nickel and the nutrients manganese, boron, and sodium in plant material, thus indicating the role of high nickel concentrations in serpentine substrates as controlling factors in nutrient uptake. Nickel hyperaccumulators are of scientific interest, because of the possible use of vegetation, regrowing on mine dumps, for nickel exploitation (Brooks, 1980).

The major source of nickel accumulation in terrestrial plants is the increased occurrence of nickel in soils. High levels of nickel in soils result from nickel-emitting industrial sources (Rutherford & Bray, 1979; Polemio et al., 1982; Alloway & Morgan, 1986; Gignac & Beckett, 1986) and sewage sludge treatment (Berrow & Burridge, 1981; Chang et al., 1984). Nickel was found to be more available to plants from soils treated with sewage sludge than from inorganically polluted soils (Alloway & Morgan, 1986). The total nickel content in sludge samples collected from different municipal sewage treatment plants in the USA was in the range of 21–1990 mg/kg dry weight (Sung et al., 1986). Coker & Matthews (1983) reported a nickel content of 10–2000 mg/kg dry weight in sewage sludge applied to land in the United Kingdom in 1977. The persistence of nickel in acetic acid-extractable form in soil was found to be 10 years (Berrow & Burridge, 1981). Nickel in polluted soils was found to be highly concentrated in the upper organic soil horizon (Chang et al., 1982; Brown et al., 1983a; Chang et al., 1984), probably because of the high cation exchange capacity of the surface organic layer (Hutchinson et al., 1981).

In a research project initiated by the Federal Environmental Protection Agency of Germany, vegetables and plant crops were grown on soils that were polluted with nickel (average 558 mg/kg soil) through sewage sludge application (Grössman, 1988). Levels in green vegetables, different types of cabbage, and onions ranged from 10.8 to 65 mg nickel/kg dry weight. In beans and peas, nickel levels ranged from 42 to 65.1 mg/kg and 16.5 to 23.4 mg/kg dry weight, respectively. In root vegetables, nickel was accumulated to a lesser extent; concentrations of 7.95–26.9 mg nickel/kg dry weight were measured. Increasing nickel concentrations in the soil resulted in increasing nickel accumulation in the plants. This also held true for maize used as fodder for animals. However, in maize, nickel levels were generally lower ranging from 2.32 to 4.27 mg/kg dry weight in kernels, 6.69 to 10.7 mg/kg in leaves, and 4.33 to 5.53 mg/kg in stems. The nickel concentration in the soil was 745 mg/kg.

Reddy & Dunn (1984) grew soya beans on sewage sludge-treated soil in glass houses and found increasing nickel levels in plant tissues with increasing rates of sludge application. Concentrations were greater in leaves than in stems and ranged from 2.1 to 8.5 mg/kg dry weight in leaves and from 1.2 to 6.2 mg/kg dry weight in stems, following application of 0–8.4 kg nickel/ha.

Keefer et al. (1986) found accumulation of nickel in both the edible and inedible parts of vegetables grown on soils treated with different types of sewage sludge. Nickel loading of the soil was in the range of 2–2540 kg/ha. Nickel concentrations detected in cabbage heads were 2.04–22.1 mg/kg dry weight (control, 3.78 mg/kg). Radish roots and tops contained 1.28–12.3 mg nickel/kg (control, 1.64 mg/kg) and 3.12–18.3 mg/kg (control, 3.16 mg/kg), respectively. In green bean leaves and pods, nickel levels were 4.68–14.0 mg/kg (control, 4.00 mg/kg) and 5.0–11.0 mg/kg (control 5.04 mg/kg). The water-soluble nickel concentration in sewage sludges was related to nickel uptake in the species tested. Soil characteristics, such as texture, drainage status, and sorptive capacity, play a dominant role in nickel availability to plants. When vegetables were grown in greenhouse pots on sewage sludge-treated soils of a calcareous loam, a clay, and a sandy loam type, the highest nickel accumulation occurred in cabbage grown on clay and lettuce grown on sandy loam (Alloway & Morgan, 1986). Gignac & Beckett (1986) found a negative correlation between the nickel content of peat and the percentage organic content.

Increased acidity of soils resulting, e.g., from SO_2^- emission, enhanced nickel solubility and uptake by plants (Hutchinson & Whitby, 1977; Brown et al., 1983b; Sanders et al., 1986). Liming of soil can reduce nickel uptake by plants (Machelett & Podlesak, 1980; Francis et al., 1985).

Nickel accumulation in plants growing in the vicinity of a nickel smelter was investigated by Hutchinson & Whitby (1977). The nickel contents of foliage from *Comptonia peregrina, Deschampsia flexuosa, Acer rubrum,* and *Betula papyrifera*, growing at a distance of 1.6 km from the Coniston nickel smelter near Sudbury, Ontario, were 113 mg/kg, 902 mg/kg, 109 mg/kg, and 148 mg/kg dry weight, respectively. The corresponding nickel concentration in the upper soil surface was 2.679 mg/kg dry weight. The nickel contents of the soil and the plant leaves declined with increasing distance from the smelter. Similar observations were reported by Gignac & Beckett (1986) for vascular plant species, sphagnum species, and bryophytes growing near Sudbury.

Determination of nickel concentrations in plant species growing on a copper mine spoil heap demonstrated relative bioconcentration values (mg nickel per kg in plants/mg nickel per kg in EDTA soil fraction) of 2.7 (leaves) and 1.4 (branches) in *Thlaspi montanum*, and 2.0 (leaves) and 0.9 (branches) in *Phlox austromontana* (Hobbs & Streit, 1986).

Animals grazing on nickel-contaminated vegetation accumulated nickel in various organs. Wild ruminants grazing near nickel-emitting industrial sources accumulated nickel in the ribs and kidneys at levels of from 1.13–1.50 mg/kg dry weight and 0.47–0.86 mg/kg dry weight, respectively. The nickel contents of their winter grazing were determined to be in the range of 3.12–14.49 mg/kg dry weight (Groppel et al., 1980). Highly elevated nickel levels (27 times the control value) were detected in primary flight feathers of mallard and black duck in the Sudbury district (20–140 km from a nickel smelter) (Ranta et al., 1978).

4.2.2 Aquatic organisms

In general, aquatic organisms resorb metals over their entire surface. They also incorporate metals from their food. In rooted aquatic plants, metals can be absorbed not only by the roots but also by submerged stems and leaves (Mortimer, 1985). Most groups of aquatic organisms include some species capable of accumulating nickel (Jenkins, 1980a). The highest levels have been found in aquatic organisms near sources of pollution, especially nickel smelters.

Clark et al. (1981) studied the accumulation and depuration of nickel by the duckweed *Lemna perpusilla*. Plants collected from a fly-ash pond were allowed to depurate in dechlorinated tap water at 20 °C for 14 days. Accumulation was then examined over a 10-day period and depuration over the following 8 days. During the 14-day depuration period, nickel concentrations fell from 160 mg/kg dry weight to less than 40 mg/kg dry weight and, in clean water, remained at this level. When exposed to 0.1 mg nickel/litre in the water, duckweed accumulated nickel to levels of about 800 mg/kg dry weight at the end of a 10-day exposure period. The peak bioaccumulation concentration of nickel occurred 2 days after the depuration period began, with most nickel elimination occurring in the 2 succeeding days. After 8 days, the nickel level was down to the original value of < 160 mg/kg. Because the concentration of nickel in the water of the fly-ash pond was also about 0.1 mg/litre, greater accumulation occurred in the laboratory than in the field.

Cowgill (1976) found that *Euglena gracilis* accumulated nickel to a concentration of 1.8 mg/kg dry weight when exposed to 8.9×10^{-4} mg nickel/litre in spring water. A biological concentration factor of about 2000 was calculated.

In a study by Hutchinson & Czyrska (1975), *Lemna minor* was collected from 23 ponds and lakes in Southern Ontario in which the mean content of nickel in the water was 0.027 mg/litre. The plants contained 5.4–35.1 mg nickel/kg dry weight, equivalent to concentration factors of 200–1300. The authors also cultured *Lemna minor* in a growth medium containing 0.01–1.00 mg nickel/litre at a temperature of 24 ± 52 °C and a pH of 6.8, for 3 weeks. Nickel accumulation ranged from 4000 (0.01 mg nickel/litre in the growth medium) to 6134 (0.5 mg nickel/litre in the growth medium). There was a correlation between levels of nickel in the plant and levels in water. Nickel accumulation was greater in the presence of copper.

Elodea densa, cultivated at 21–25 °C in a flowing water system with a constant nickel concentration in the medium of 0.01 mg/litre, showed an accumulation factor of 200 after 12 days (Mortimer, 1985).

The highest concentration factor reported, approximately 20 000, was found in an aquatic ecosystem study, conducted by Hutchinson et al. (1975), in periphyton algae sampled from a section of the metal-contaminated Wanapitei river in the Sudbury area. Analysis of aquatic macrophytes, which were collected from metal-contaminated rivers in this area, indicated a species specificity for uptake and a significant correlation between total nickel content in the sediment and water and in rooted macrophytes.

Watras et al. (1985) studied the accumulation of nickel in two levels of a simple aquatic food chain using *Scenedesmus obliquus* and *Daphnia magna*. The algae accumulated nickel to concentrations 30–300 times the ambient concentration. In *Daphnia*, concentration factors were only 2–12. There was little difference in accumulation from incubation in ^{63}Ni-labelled medium without algae or from incubation in labelled medium with labelled algae. The data indicated that direct uptake from the medium rather than uptake from ingested algae was the primary accumulation mechanism. These results confirmed earlier studies by Hall (1982), who described nickel accumulation in *Daphnia magna* as the sum of five processes occurring in the various body components, namely, adsorption to, and desorption from, body and tissue surfaces, absorption, retention or storage, and excretion.

In the course of the aquatic ecosystem study performed by Hutchinson et al. (1975), nickel levels were determined in aquatic animals. Accumulation factors in animals were lower than in aquatic vegetation and were found to be 643 in zooplankton, 929 in crayfish, and 262 in clams. In fish species caught in the Wanapitei river (42 mg nickel/litre), nickel levels in muscle tissue were lower than those in the liver, kidneys, and gills. The predatory yellow pickerell exhibited the highest nickel levels with 51.6 mg/kg wet weight in kidney tissue, giving a concentration value of 229.

Calamari et al. (1982) reported nickel levels of 2.9 mg/kg wet weight in liver, 4.0 mg/kg in kidneys, and 0.8 mg/kg in muscle in *Salmo gairdneri*, after 180 days exposure to 1 mg nickel/litre in the water. Nickel levels at the start of the study were 1.5, 1.5, and 0.5 mg/kg, in liver, kidneys, and muscle, respectively. The authors also found, by means of a toxicokinetic model, that theoretical asymptotic values for liver, kidney, and muscle should be reached in 397, 313, and 460 days, respectively, yielding bioconcentration values of 3.1, 4.2, and 1.0, respectively. Laboratory studies showed that nickel had little capacity for accumulation in all the fish species studied. However, it was also demonstrated that this relatively low concentration of nickel in tissues could cause biochemical damage.

The range of concentrations reported in whole fish in uncontaminated waters, on a wet-weight basis, is 0.2–2 mg/kg. This value could be increased by a factor of ten in contaminated areas (Calamari et al. 1984).

White et al. (1986) investigated nickel levels in coots (*Fulica americana*) resting and feeding by a pond that was used for the disposal of fly ash from a nearby coal-fired power plant. Though the nickel concentration in the pond sediment was much higher than the concentration in the water (which was below detection limit, except at one collection period), accumulation of nickel in coot livers was not observed in 2 years of plant operation.

4.3 Biomagnification

Accumulation factors in different trophic levels of aquatic food chains suggest that biomagnification of nickel along the food chain, at least in aquatic ecosystems, does not occur.

Hutchinson et al. (1975), in their investigations on nickel compartmentation in an aquatic ecosystem, found large concentration factors in the vegetation and decreasing factors in the higher trophic levels. In a small food chain consisting of an alga (*Scenedesmus obliquus*) and a zooplankton species (*Daphnia magna*), there was no biomagnification (Watras et al., 1985). Because nickel in aquatic ecosystems decreases in concentration with increasing levels of the food chain, biomagnification does not occur.

5. ENVIRONMENTAL LEVELS AND HUMAN EXPOSURE

5.1 Environmental levels

5.1.1 Air

Owing to the large number of sources releasing nickel into the atmosphere and their uneven distribution over the globe, ambient nickel concentrations may vary over several orders of magnitude.

Urban and rural areas usually exhibit air nickel levels ranging from 5 to 35 ng/m^3 (Bennett, 1984). Higher values were recorded in heavily industrialized areas and larger cities. Nickel concentrations, monitored continuously over one year in 4 American cities, were found to be in the range of 18–42 ng/m^3 (Saltzman et al., 1985). In the vicinity of a nickel smelter in the Sudbury area, Ontario, levels of 124 ng/m^3 were measured (Chan & Lusis, 1986). Atmospheric concentrations at Spitsbergen, measured during two months, ranged between approximately 1 and 2 ng nickel/m^3 (Pacyna et al., 1985). In the Canadian Arctic, the annual mean concentration was 0.38 ng/m^3. The winter mean was 0.62 ng/m^3, indicating a seasonal cycle (Hoff & Barrie, 1986). Assuming a ventilation rate of 20 m^3 and air concentrations of 5–35 ng/m^3, the amount of nickel entering the human respiratory tract is in the range of 0.1–0.7 µg/day.

The distribution of nickel among suspended particulates in the air will determine the fraction that is inhalable. Data on the size distribution of nickel particulates are limited. Lee & von Lehmden (1973) summarized data on the size of nickel particulates in urban air and found mass median diameters of 0.83–1.67 µm, 28–55% of the particles being < 1 µm. A more recent summary of size distribution of trace elements in different areas yielded a mass median diameter for nickel particulates of 0.98 µm (Milford & Davidson, 1985). Particles of less than about 1 µm are deposited predominantly in the alveolar regions of the lung (Stern et al., 1984). Nickel was found to be most concentrated in the smallest particles emitted from coal-fired power plants (Natusch et al., 1974).

Particles of a mass median diameter of 0.65–1.1 μm contained 1600 mg nickel/kg while 4.7–11 μm particles contained about 400 mg nickel/kg.

Another important route of nickel exposure is tobacco smoking. Cigarette tobacco may contain approximately 1.3–4.0 mg nickel/kg. About 0.04-0.58 μg nickel is released with the main stream smoke (gas phase and particle phase) of one cigarette (Szadowski et al., 1969a; Menden et al., 1972; Gutenmann et al., 1982). Smoking 40 cigarettes per day may result in the inhalation of 2–23 μg nickel per day. The formation of nickel carbonyl in the main stream smoke is suspected (Sunderman & Sunderman, 1961a; Stahley, 1973), but it could not be detected using Fourier-transform infrared spectrometry (detection limit 0.1 μg/litre) (Alexander et al., 1983).

5.1.2 Drinking-water

On the basis of determinations of nickel concentrations in 969 water supplies in the USA during 1965–70, the average concentration of nickel in water samples taken at the consumer's tap was 4.8 μg/litre (NAS, 1975).

In Italy, nickel levels in drinking-water were mostly below 10 μg/litre (Clemente et al., 1980). Schumann (1980) measured levels of 6.8–10.9 μg/litre in the German Democratic Republic. Leaching processes from water taps and fixtures contribute to nickel levels already present in drinking-water. Between 18 and 900 mg of nickel were leached from 10 used water taps, which had been filled, in an inverted position, with 15 ml deionized water, and left overnight for 16 h (Strain et al., 1980). In Denmark, levels of up to 490 μg/litre were observed, when water was left standing overnight in nickel-containing plumbing fittings (Andersen et al., 1983).

In areas where nickel is mined, as much as 200 μg nickel/litre has been recorded in drinking-water (McNeely et al., 1972).

Assuming a daily intake of 1.5 litres water and a level of 5–10 μg nickel/litre, the mean daily intake of nickel from water for adults would be between 7.5 and 15 μg.

5.1.3 Food

Nickel is ingested by human beings through the consumption of plants and animals that contain nickel. Nickel levels were determined, using EAAS, in foods in the Netherlands by Ellen et al. (1978). They were found to be less than 0.5 mg/kg fresh weight in most products, except for cacao products and nuts, which contained nickel levels of up to 9.8 and 5.1 mg/kg, respectively. Smart & Sherlock (1987) reported that nickel levels, determined using FAAS, in English meat, fruit, and vegetables were of the order of ≥ 0.2 mg/kg fresh weight. Aquatic organisms, e.g., molluscs and fish, may contain relatively large amounts of nickel, if the nickel concentration in the water is high (Table 11).

Nickel levels were determined in 234 foods from a 1984 US FDA Total Diet Study, using ICP-AES (Pennington & Jones, 1987); 91% of the foods contained nickel levels of less than 0.4 mg/kg and 66.2% contained levels of less than 0.1 mg/kg. Only seven foods had values exceeding 1 mg/kg. Foods highest in nickel included nuts, legumes, items containing chocolate, canned foods, and grain products.

Food processing and storage methods apparently add to the nickel levels already present in foodstuffs through: leaching from nickel-containing alloys in food-processing equipment made from stainless steel; the milling of flour; and the catalytic hydrogenation of fats and oils using nickel catalysts (NAS, 1975). Grandjean (1984) estimated that leaching from cooking ware, kitchen utensils, and water piping could occasionally add 1 mg to the daily intake of nickel, i.e., much more than the intake resulting from nickel in food and beverages.

Schroeder et al. (1962) calculated the average oral intake of nickel by American adults to be about 300–600 µg/day, but more recent estimates are lower. Myron et al. (1978) determined the nickel content of 9 typical diets in the USA and calculated an average intake of 165 µg/day. Clemente et al. (1980) evaluated the available data and derived a mean intake value of 200–300 µg/day for the normal adult in the Western countries. However, because of the wide variation in the nickel contents of single food items, the daily intake may vary considerably (100–800 µg/day) as a function of dietary habits. Levels of 250–270 µg/day, determined by Smart & Sherlock (1987)

Table 11. Nickel concentrations in aquatic organisms

Species	Tissue or organ	Locality	Concentration [a] (mg/kg)		Reference
Plants					
Lemna minor (duckweed)	whole	Southern Ontario; ponds and lakes	5.4-35.1	D	Hutchinson & Czyrska (1975)
Eichhornia crassipes	whole	Yamuna river, India; near big cities	4.4-83.0	D	Ajmal et al. (1985)
Fontiralis antipyretrica		Augraben river, Italy; near motorway	5.5-10.6	D	Dallinger & Kautzky (1985)
		Leiferer Graben river, Italy; metal industry area	17.7-29.2	D	
Ranunculus fluitans		Augraben river, Italy; near motorway	69	D	Dallinger & Kautzky (1985)
Animals					
Penaeus semisulcatus [b] (shrimp)	body tissue	AD-Damman, Saudi Arabia; sewage outfall area	0.54	W	Sadiq et al (1982)
Paphia undulata [b] (clam)	whole soft parts	Gulf of Thailand	1.30-2.00	W	Phillips & Muttarasin (1985)
Anadara granosa [b] (cockle)			0.65-2.31	W	

Table 11 (continued)

Species	Tissue or organ	Locality	Concentration[a] (mg/kg)		Reference
Perna viridis[b] (green mussel)	whole soft parts	Gulf of Thailand	0.26-4.74	W	Phillips & Muttarasin (1985)
Crassostrea commercialis[b] (rock oyster)			0.60-2.52	W	
Crassostrea virginica[b] (oyster)	body tissue	Estuary of Mississippi; pollutants from rivers, bayous, municipal and agricultural run-off	2.1	D	Byrne & Deleon (1986)
Rangia cuneata[b] (clam)			28.2-28.7	D	
Elliptio complanata (bivalve)	shell tissue	Great Lakes, Ontario	6.56-7.6 6.79-10.7	D D	Dermott & Lum (1986)
Mytilus edulis[b] (mussel)	tissue	Eastern Scheldt, Netherlands, 1977-1980	0.24-1.0	W	Vos et al. (1986)
Crangon crangon[b] (shrimp)			0.13-0.70	W	
Salmo trutta[b] (brook trout)	muscle liver body tissue	Leine river, Germany, municipal discharge area	0.220-0.220 0.327-0.469 0.359-0.477	W W W	Abo-Rady (1979b)

Table 11 (continued)

Species	Tissue	Location	Value	W/D	Reference
Leuciscus cephalus [a] (chub) Vimba vimba (vimba bream) Abramis brama (bream) Esox lucius (pike)	muscle	River Danube, Federal Republic of Germany	0.5-1.2	W	Wachs (1982)
Psetta maxima [b] (turbot)	muscle	Southern Baltic, Poland	0.24	W	Falandysz (1985)
Pleuronectes platessa [b] (plaice)			0.19	W	
Platichthys flesus [b] (flounder)			0.14	W	
Heteropnuestes fossilis [b]	muscle	Yamura river, India; near big cities	1.9-32.7	D	Ajmal et al. (1985)
Basilichthys bonariensis [b] (pejerry)	whole carcass	Lago Poopo, Bolivia; tin mining and refining area	1.37	D	Beveridge et al. (1985)
Orestias luteus [b] (killifish)			2.85	D	

Table 11 (continued)

Species	Tissue or organ	Locality	Concentration[a] (mg/kg)		Reference
Salmo gairdneri[b] (rainbow trout)	gills	Augraben river, Italy; near motorway	14.5	D	Dallinger & Kautzky (1985)
	liver		5.8	D	
	kidney		7.9	D	
	muscle		5.9	D	
	gonads		3.6	D	
	gills	Leiferer Graben river, Italy; metal industry area	21.8	D	
	liver		8.4	D	
	kidney		11.5	D	
	muscle		7.8	D	
	gonads		9.9	D	
Salvelinus namaycush[b] (lake trout)	kidney	Lakes near Sudbury, Ontario	2.0–5.1	D	Bradley & Morris (1986)
Perca flavescens[b] (yellow perch)	muscle		2.8; 3.4	D	
	liver		2.2; 2.9	D	
white sucker[b]	liver		16.4	D	
Gadus morhua (cod)	muscle	Southern Baltic Sea	0.081	W	Falandysz (1986a)
Clupea harengus (herring)	muscle	Southern Baltic Sea	0.10	W	Falandysz (1986b)

Table 11 (continued)

Solea solea (sole)	edible parts	Coast of Netherlands, 1977-80	0.01-0.05	W	Vos et al. (1986)
Gadus morhua (cod)			0.03-0.22	W	
Clupea harengus (herring)			0.02-0.12	W	
Anguilla anguilla (eel)			0.02-0.31	W	
Stenella coeruleoalba (stripped dolphin)	cartilage skull vertebrae ribs	Kii peninsula, Japan	0.029-0.009 0.025-0.083 0.024-0.057 0.036-0.44	D D D D	Honda et al. (1984)

^a D = dry weight.
   W = wet weight.
^b Intended for human consumption.

for United Kingdom diets, are within this range. This includes an estimated 100 μg/day for nickel migrating from metal cookware. An actual measured value using glass cookware for food preparation was only 140–150 μg/day. The maximum daily nickel intake from an average Danish diet was calculated to be 900 μg/day (Nielsen & Flyvholm, 1984).

5.1.4 Terrestrial and aquatic organisms

Levels in terrestrial and aquatic organisms may vary over several orders of magnitude, according to the species and environmental factors (Jenkins, 1980 a,b). Tables 11 and 12 include data on nickel levels in tissue samples of different organisms, including plants and animals relevant to human nutrition.

5.2 General population exposure

The general population is exposed to nickel species through the oral, inhalation, and dermal routes.

5.2.1 Oral

Nickel is ingested through the consumption of foodstuffs and beverages that contain nickel and through the ingestion of inhaled material returned to the pharynx by mucociliary clearance. Nickel levels in European and USA foodstuffs generally ranged from less than 0.1 mg/kg to 0.5 mg/kg, though a few foodstuffs, e.g., nuts and legumes, contained levels of more than 1 mg/kg (section 5.1.3). The results of dietary studies in the United Kingdom and the USA showed average intakes from food of 200–300 μg nickel/day. In the United Kingdom, it has been estimated that 100 μg nickel/day is contributed by nickel-containing cooking utensils and that the level in food ranges from 100 to 200 μg/day. Drinking-water may contain nickel derived from its source, from environmental contamination, and from nickel-containing plumbing fittings. Concentrations in uncontaminated water may range from 5 to 11 μg/litre (section 5.1.2). Assuming a nickel level of 5–10 μg/litre and a daily intake of 1.5 litres water, a mean daily intake of nickel from water would be 7.5–15 μg.

Table 12. Nickel concentration in terrestrial organisms

Species	Tissue or organ	Locality	Concentration [a] (mg/kg)		Reference
Plants					
Aster tripolium [b]	shoots	Salt marsh in Netherlands; discharge from urban and industrial areas	0.3-4.9	D	Beeftink & Nieuwenhuize (1982)
Halimione portulacoides			0.1-3.3	D	
Limonium vulgare			0.8-5.4	D	
Plantago maritima			0.9-4.5	D	
Puccinellia maritima			0.5-5.0	D	
Salicornia europaea [b]			0.4-4.2	D	
Spartina anglica			0.2-1.8	D	
Suaeda maritima			0.3-4.6	D	
Triglochin maritim			0.5-6.2	D	
Peanuts	kernels		<0.14-14	W	Wolnik et al. (1983)
Soybeans	beans (without pod)		0.35-29	W	
Sweet corn	kernels		<0.026-0.35	W	
Field corn	kernels		<0.2-1.1	W	Wolnik et al. (1985)
Spinach	leaves		<0.02-0.3	W	
Onions	peeled bulb		<0.02-0.16	W	
Tomatoes	fruit		<0.002-0.255	W	
Rice	kernels		<0.2-1.2	W	
Carrots	root		<0.02-0.46	W	

Table 12 (continued)

Species	Tissue or organ	Locality	Concentration [a] (mg/kg)		Reference
Thlaspi montanum	leaves	Grand Canyon, Arizona; copper mine spoil	54	D	Hobbs & Streit (1986)
	shoots		27	D	
Phlox austromontana	leaves		39	D	
	shoots		<17	D	
Juniperus osteosperma	leaves		8	D	
	shoots		17	D	
Animals					
Lagopus lagopus (willow ptarmigan)	plumage	remote areas in Ontario	0.91	D	Parker (1985)
Bonasa umbellus (ruffed grouse)			0.74	D	
Canachites canadensis (spruce grouse)			1.09	D	
Egretta alba (great white egret)	muscle	Central Korea	0.03-0.04	W	Honda et al. (1985)
	bone		0.03-0.19	W	
	feather		0.14-1.59	W	
Sterna kirundo (common tern)	liver	Providence, Rhode Island; former electroplating industry centre	ND-1.0	D	Custer et al. (1986)
Fulica americana (coot)	liver	Texas; fly ash pond control site	0.06-0.24	W[c]	White et al. (1986)
			0.07-0.22	W	

Table 12 (continued)

Microtus pensylvannicus		Sudbury, Ontario; nickel and copper mine tailings control site			Cloutier et al. (1986)
	liver		N.D.-9.9	D[d]	
	kidney		N.D.-18.4	D	
	muscle		5.9	D	
	liver		N.D.-8.7	D	
	kidney		N.D.-25.3	D	
	muscle		5.5	D	

[a] D = dry weight.
W = wet weight.
ND = not detected.
[b] Intended for human consumption.
[c] No statistically significant difference between sites.
[d] Nickel levels in most samples were below detection limit; site effects were not observed.

5.2.2 Inhalation

Measured atmospheric nickel concentrations have been in the ranges of 18–42 μg/m^3 in industrial areas and 5–35 μg/m^3 in non-industrial urban and rural areas. Taking a concentration range of 5–35 μg/m^3 and a daily ventilation rate of 20 m^3, 0.1–0.7 mg nickel/day will enter the respiratory tract; absorption will then be a function of the nickel species and the physical state. Cigarette smoking also contributes to the inhalation of nickel; for example, smoking 40 cigarettes daily may result in the inhalation of 2–23 μg nickel (section 5.1.1).

5.2.3 Dermal

This route of exposure is of particular significance for nickel contact hypersensitivity. Because of the ubiquity of nickel-containing commodities in an industrial society, dermal exposure to nickel is almost continuous in the general population. This is reflected by the constantly increasing number of patients with nickel contact dermatitis (Dooms-Goossens et al., 1980). Sources of environmental exposure include jewellery, coinage, clothing, fasteners, tools, cooking utensils, and stainless steel kitchen utensils (NAS, 1975). Even jewellery made of white gold containing 2–15% of nickel may cause eczema (Fischer, 1984).

A number of common sources of nickel dermatitis are listed in Table 13. The amount of nickel released from these items depends on the corrosion resistance of the object and the presence of sweat and other fluids in the environment; because of its chloride content and relative low pH, sweat can dissolve nickel. The solution of nickel in synthetic sweat was examined for 15 different nickel alloys and surfaces; a high release of nickel ions was documented for nickel-plated items (both electrolytically and chemically plated), nickel-iron alloy (65% nickel), German silver (10–20% nickel), Monel 400® alloy (66% nickel), Nicrobraze LM® alloy (83% nickel), and coin alloy (25% nickel). These nickel objects were also tested for their ability to provoke nickel allergy in sensitized persons. Nine of the materials tested caused a medium to severe degree of allergic skin reactions, and the amount of nickel dissolved was related to the degree of the allergic reaction (Kato & Samitz, 1975).

Table 13. Non-occupational exposure to nickel [a]

Source causing dermatitis	Location
Earrings	Earlobes
Garter clasps, metal chairs	Thighs
Thimbles, crochet and knitting needles, scissors, nickel coins, cover of fountain pen	Fingers
Handles of car doors, baby carriage, umbrella, refrigerator, and handbags	Palms
Clasp of necklace, zipper	Neck
Watch band, bracelet	Wrists
Wire support of brassiere cup	Breast
Handbags	Antecubital area
Bobby pins	Side of face
Spectacle frames	Back of ears
Nickel coin (patient had rubbing habit)	Side of nose
Eyelash curler	Eyelids
Zipper	Axilla
Safety pin	Pubic area
Eyelets of shoe	Dorsum of foot
Metal arch support	Plantar aspect of foot
Bobby pins held in mouth, metal lipstick cases	Lips
Hair grips [b]	Scalp
Jeans buttons [c]	Belly

[a] Adapted from: Fisher & Shapiro (1956).
[b] From: Cronin (1980).
[c] From: Fisher (1985).

Fisher's dimethyglyoxime spot test can be used, in most cases, to identify nickel alloys and nickel-plated surfaces releasing nickel ions. Important exceptions are Inconnel 600®, which has a low nickel release to synthetic sweat, but was positive when tested on nickel sensitive patients, and Nicrobraze®, which has a high nickel release and was positive when tested on patients (Menné et al. 1987b).

Stainless steel, which contains about 10% nickel, is very resistant because of the protective effect of chromium oxides on the surface. It did not release noticeable amounts of nickel ions into synthetic sweat and did not produce skin reactions in nickel-sensitive patients

(Kato & Samitz, 1975; Menné et al., 1987b). However, nickel can be released from stainless steel in a very corrosive environment or by prolonged treatment in a sweat solution.

5.3 Iatrogenic exposure

Iatrogenic exposures to nickel arise from: (*a*) nickel-containing implants (joint replacements, intraosseous pins, cardiac valve replacements, cardiac pacemaker wires, and dental prostheses); (*b*) intravenous fluids and medications that are contaminated with nickel; and (*c*) haemodialysis with nickel contamination of the dialysate fluid (Sunderman, 1986; Hopfer et al., 1989). Marek & Treharne (1982) investigated nickel release from surgical implant alloys (14 and 36% nickel) into Ringer's solution. Nickel release from the alloy shavings declined after a few days and reached a constant value of approximately 0.3 ng/cm^2 per day. Thus, a prosthesis with a surface area of 200 cm^2 would release about 60 ng nickel per day or 22 µg/year. The significance of nickel release from implants has been questioned (Fisher, 1977; Burrows et al., 1981). Hypernickelaemia (serum-nickel > 1.1 µg/litre) occurred in only one out of 13 patients with stainless steel hip prostheses. This patient suffered from mild renal insufficiency, suggesting an impaired ability to excrete nickel absorbed from the prosthesis (Linden et al., 1985; Sunderman, 1986).

Brune (1986) compiled data on the amounts of different metals released from dental alloys in natural or artificial saliva in *in vitro* and *in vivo* tests. The quantities released were calculated on the basis of a standard man with a specified number or area of dental restorations. Nickel was released *in vitro* from a metal alloy with a high nickel content (ca 80%), at the same level as from food and drink. Nickel release from cobalt-based alloys (0.1–0.2% nickel) was less than 2 µg per day.

Contamination of dialysate fluids may produce parenteral exposure to nickel. Hopfer et al. (1985) demonstrated that serum-nickel concentrations in haemodialysis patients may, on average, be 11–22 times the mean concentration in serum from healthy subjects. During normal dialysis, the average intravenous nickel uptake has been estimated to be 100 µg per treatment (Sunderman, 1983a).

Hypernickelaemia has also been observed in patients following intravenous administration of meglumine diatrizoate (Renografin-76®, a radiographic contrast medium) for coronary arteriography. Nickel analysis revealed levels of 144 ± 44 µg nickel/litre contrast medium. Approximately half an hour after arteriography, the incremental serum-nickel concentration in patients averaged 1.81 ± 0.39 µg/litre. The authors recommended that nickel concentrations in radiographic contrast media should not exceed 10 µg/litre (Leach & Sunderman, 1987).

5.4 Occupational exposure

Nickel concentrations may be significantly higher in the working environment than normal atmospheric air levels. US NIOSH (1977b) estimated that about 250 000 individuals in the USA were occupationally exposed to inorganic nickel. A list of occupations, identified as involving exposure to nickel, is presented in Table 14 (US NIOSH, 1977b).

Table 14. Occupations with potential exposure to nickel [a]

Battery makers, storage	Nickel-alloy makers
Cashiers	Nickel miners
Catalyst workers	Nickel refiners
Cemented-carbide makers	Nickel smelters
Ceramic makers	Oil hydrogenators
Disinfectant makers	Organic-chemical synthesizers
Dyers	Paint makers
Electroplaters	Penpoint makers
Enamellers	Petroleum-refinery workers
Gas-mask makers	Spark-plug makers
Glass makers[b]	Stainless-steel makers
Ink makers	Textile dyers
Jewellers	Vacuum-tube makers
Magnet makers	Varnish makers
Metallizers	Welders
Mond-process workers	

[a] Adapted from: US NIOSH (1977b).
[b] From: Raithel et al. (1981).

Representative exposure data are difficult to obtain, and most published values of occupational exposure were gained during biological monitoring studies. US NIOSH (1977b) compiled data on the concentrations of nickel in air, in smelting and refining operations, the alloy industry, welding operations, and other processes; the concentrations ranged from a few $\mu g/m^3$ to several mg/m^3. Recent exposure data from the primary and secondary nickel industries are summarized in Table 15. Levels may vary considerably, according to the individual operations, or areas of a manufacturing process. For example, in alloy production an average concentration of airborne nickel during pickling and handling was determined to be 0.008 mg/m^3, whereas, during grinding, the average airborne nickel level was 0.298 mg/m^3 (Warner, 1984). Generally, the concentration of nickel in the material being handled and the operation being performed affect the concentration of nickel in air. Levels of airborne nickel are expected to be higher in dusty operations involving fine, dry particulates, e.g., metal powder or salts. Welding operations can create airborne nickel levels of 0.004–0.24 mg/m^3. A higher exposure, especially to metallic nickel, exists in the user industries. Improvements in operational techniques and ventilation reduce nickel concentrations in the air of work-places (Boysen et al., 1982; Coenen et al., 1986).

In the occupational environment, nickel dust may also enter the body through oral exposure, because of poor personal hygiene or inadequate work practices. This route of exposure was reported in a battery factory where high faecal levels of nickel were related to dusty working conditions (Adamsson et al., 1980).

The dermal route of exposure is of significance to workers sensitized to nickel. Occupational dermal exposure to nickel may occur in battery makers, nickel-catalyst makers, ceramics makers, duplicating machine workers, dyers, electronics workers, electroplaters, ink-makers, jewellers, spark-plug makers, and rubber workers (NAS, 1975). In some occupations, the skin may be directly exposed to dissolved nickel, e.g., in the electroplating and electroforming industry (Wall & Calnan, 1980). Nickel contact dermatitis was observed in hairdressers and was attributed to the handling of nickel-bearing tools and contact with liquids throughout the day (Wahlberg, 1975). Nickel contact dermatitis has been reported in hospital cleaning personnel (Gawkrodger, 1986a). The level of nickel in the water they used for washing surfaces increased during the cleaning process by transfer from cleaned areas on wash cloths. In water from used cloths, a mean level of 90 µg nickel/litre was found (Clemmensen et al., 1981).

Table 15. Occupational exposure to nickel

Process/Operations	Average nickel concentration in air (mg/m³) [a]		Number of samples [b]	Chemical form [b]	Reference
Nickel mining and refining					
Mining	0.025	A	ns	mineral form	Warner (1984)
Concentrating	0.03–0.13	P	ns	mineral form	Warner (1984)
Roasting/smelting	<0.1	A	ns	nickel subsulfide, nickel oxide	Boysen et al. (1982)
Roasting/smelting	0.048; 0.075	A	ns	mineral form, nickel subsulfide, nickel oxide combined with iron oxide, nickel sulfate	Warner (1984)
Converting	0.033; 0.037	A	ns	nickel subsulfide	Warner (1984)
Smelting to ferronickel	0.0022–0.274 0.0047–0.029 0.005–0.193	P A P	>79	nickel oxide combined with iron oxide, nickel subsulfide	Warner (1984)
Hydrometallurgical refining	0.029–0.336	A,P	577	nickel subsulfide, nickel oxide, metallic nickel, nickel sulfate, nickel chloride	Warner (1984)

Table 15 (continued)

Electrolytic refining	<0.1	A	ns	ns	Boysen et al. (1982)
	0.34; 0.19	A	ns	as under Hydrometallurgical refining	Warner (1984)
Use of primary nickel products					
Stainless steel welding	0.004	A	35	ns	Wilson et al. (1981)
	0.313	P	7	ns	
	0.07–0.24[c]	P	182	ns	Coenen et al. (1986)
	0.02–0.22[c]	A	280	ns	
Stainless steel production	0.014–0.134	P	40[d]	nickel oxide, metallic nickel	Warner (1984)
High nickel alloy production	0.008–0.298	P	1530	nickel oxide, metallic nickel	Warner (1984)
Foundry operations (melting, casting, grinding)	<0.3	P	ns	nickel oxide, nickel alloys	Bernacki et al. (1978b)
	0.013–0.310	P	217	nickel oxide, nickel alloys	Warner (1984)
Electroplating	0.03–0.16	P	25	nickel sulfate	Tola et al. (1979)
	0.005–0.016	P	15	ns	Bernacki et al. (1980)
	<0.003–<0.011	A,P	48	various nickel salts	Warner (1984)

Table 15 (continued)

Process/Operations	Average nickel concentration in air (mg/m³)[a]		Number of samples[b]	Chemical form[b]	Reference
Metal sintering					
Furnace maintenance	0.001-0.168	P	20	ns	Lichty & Zey (1985)
Metal powder mixing	0.006-1.28	P	6	ns	
Nickel-cadmium battery manufacturing	0.012-0.033	P	213	ns	Adamsson et al. (1980)
	0.02-1.91	P	36	ns	Warner (1984)
Glass production	0.03-3.8	A	ns	ns	Raithel et al. (1981)
	0.07-2.622	P			

[a] A = area samples; P = personal samples.
[b] ns = not specified.
[c] 90% value.
[d] Number of companies reporting exposure.

6. KINETICS AND METABOLISM

Health hazards associated with exposure to nickel in the occupational environment have resulted primarily from inhalation. For this reason, deposition, retention, and clearance of nickel from the human respiratory tract are of special importance. However, in addition to this main exposure through inhaled air (ambient and at the work-place), human beings are also exposed to nickel in drinking-water and food, and through skin contact, which is of special concern in view of resulting adverse effects, namely, nickel contact dermatitis.

6.1 Absorption

Human exposure to nickel originates from a variety of sources and is highly variable. Nickel and its inorganic compounds can be absorbed via the gastrointestinal tract as well as the respiratory passages. Under certain circumstances, the skin is a qualitatively important route by which nickel enters the body. However, percutaneous absorption is less important for the systemic effects of nickel than for the allergenic responses to it. Placental transfer is of importance because of the effects on the fetus. Knowledge of this route of absorption makes it possible to estimate the contribution to the body burden at birth. The diverse routes of parenteral administration of nickel compounds are mainly of interest in toxicity studies on animals and are particularly useful in assessing the kinetics of nickel transport, distribution, and excretion.

The relative amounts of nickel absorbed by an organism are determined, not only by the quantities inhaled, ingested, or administered, but also by the physical and chemical characteristics of the nickel compound. Solubility is an important factor in all routes of absorption. Soluble salts of nickel dissociate readily in the aqueous environment of biological membranes, thus facilitating their transport as metal ions. Conversely, insoluble nickel compounds are relatively poorly absorbed.

Kuehn & Sunderman (1982) incubated 17 nickel compounds in water, rat serum, and renal cytosol for 72 h at 37 °C. Concentrations of dissolved nickel were determined by electrothermal atomic absorption and dissolution half-times were calculated. Eleven of the nickel compounds (Ni, β-NiS, amorphous NiS, α-Ni$_3$S$_2$, NiSe, Ni$_3$Se$_2$, NiTe, NiAs, Ni$_{11}$As$_8$, Ni$_5$As$_2$, and NiFeS$_4$) dissolved more rapidly in serum or cytosol than in water. Dissolution of 4 of the compounds (NiO, NiSb, NiFe alloy, and NiTiO$_3$) was not detectable in any of the media (half-time, > 11 years). One compound (NiAsS) had approximately equal dissolution half-times in the 3 media. Because of precipitation, the half-time value for NiS$_2$ could not be determined. These findings were in close agreement with the elimination half-time (24 days) obtained from elimination of ^{63}Ni in the urine and faeces of rats, after intramuscular injection of α-^{63}Ni$_3$S$_2$. The authors suggested that the *in vitro* dissolution half-times of nickel compounds might be used to predict their *in vivo* elimination half-times, since the dissolution process is rate limiting for the metabolism and elimination of the compounds.

It is possible that other factors, such as host, nutritional and physiological status, or stage of development, also play a role, but these have not been studied.

Several studies on the dissolution kinetics of nickel subsulfide have been performed. Autoradiographic observations (Kasprzak, 1974) showed that extracellular particles of ^{63}Ni subsulfide or Ni$_3$^{35}S$_2$ could persist at the site of injection for many months, without detectable alterations, and could eventually become surrounded by neoplastic tissue. Intracellular localization of ^{63}Ni or ^{35}S was not detected within muscle or tumour cells. This finding was confirmed quantitatively by Sunderman et al. (1976b), who measured the elimination of ^{63}Ni following ^{63}Ni subsulfide administration to rats, and by Oskarsson (1979), who carried out whole-body autoradiography in mice. After 20 weeks, as much as 19% of the ^{63}Ni dose was found at the site of injection, while the retention of nickel in organs distant from the injection site was less than 0.1% (Sunderman et al., 1976b). Addition of manganese to the administered nickel subsulfide, which significantly decreased the carcinogenicity of the latter, did not affect the gross elimination of ^{63}Ni. The role of manganese was to effect the subcellular partition of soluble ^{63}Ni derived from ^{63}Ni subsulfide. Ultrafiltered homogenates of muscle tissue injected with ^{63}Ni derived from ^{63}Ni subsulfide + Mn contained less nickel than those injected with ^{63}Ni subsulfide alone (Sunderman et al., 1976b). Nickel dissolution kinetics, similar to those *in vivo*, were obtained when leaching of ^{63}Ni was measured during two weeks of *in vitro* incubation of ^{63}Ni subsulfide with rat serum or aqueous triethylenetetramine, but not with water (Kasprzak & Sunderman, 1977). This finding suggested that the interaction of nickel subsulfide with body fluids could be of the same nature under both *in vivo* and *in vitro* conditions. Evaluation of the kinetic curves and the X-ray diffractometry of the sediments following incubation of nickel subsulfide in the three media (Kasprzak & Sunderman, 1977) revealed that solubilization of nickel required the presence of oxygen and involved two reactions:

(a) $2\ \alpha\text{-Ni}_3\text{S}_2 + \text{O}_2 + 2\text{H}_2\text{O} \leftrightarrow 4\ \beta\text{-NiS} + 2\text{Ni(OH)}_2$

(b) $\beta\text{-NiS} + 2\text{O}_2 \leftrightarrow \text{Ni}^{2+} + \text{SO}_2^-$

When complexing agents, i.e., proteins, amino-acids, etc., are present, Ni^{2+} and $Ni(OH)_2$ form water-soluble Ni(II)-complexes and undergo fast mobilization and elimination. The remaining β-Ni monosulfide constitutes a surface coating on the nickel subsulfide particles and requires more oxygen for the further dissolution of nickel.

6.1.1 Absorption via the respiratory tract

Respiratory absorption of nickel is normally the principal route for its entry into the human body, under conditions of occupational exposure. It usually involves the inhalation of one of the following substances: dust of relatively insoluble nickel compounds, aerosols derived from nickel solutions (soluble nickel), and gaseous forms containing nickel (usually nickel carbonyl). The inhalation route is also of importance in exposure from the general environment, including tobacco smoke. It has been reported that cigarette smoke may contain nickel carbonyl (section 5.1.1).

The relative amounts of inhaled nickel absorbed from various compartments of the pulmonary tract are a function of both the chemical and physical forms.

6.1.1.1 Particulate nickel

The Task Group on Lung Dynamics (1966) considered that the respiratory absorption of nickel compounds in particulate form was influenced by three processes in the lung, namely deposition, mucociliary clearance, and alveolar clearance. This Group developed deposition and clearance models for man for inhaled particulate matter of whatever chemical origin, as a function of particulate size, chemical category, and compartmentalization within the respiratory tract. Nickel oxide and nickel halides are classified as compounds having moderate retention in the lungs and a clearance time of weeks. Although this model approach has its limitations, it can be of some value in assessing deposition and clearance rates for nickel compounds of known particle size and chemical composition.

Removal of material deposited in the lung depends on its solubility characteristics and is slow for metallic nickel or nickel oxide dust, faster for soluble nickel salts, and most rapid for the volatile and lipid-soluble nickel carbonyl.

Nickel has a tendency to accumulate in lung tissue and in the regional lymph nodes. Thus, only part of the nickel retained will be transferred to the blood, depending on the solubility of the nickel compound.

Absorption from the pulmonary tract of nickel in particulate matter is considerably less than that of nickel carbonyl. Smaller particles penetrate deeper in the respiratory tract than larger particles and the relative absorption is greater. Soluble nickel compounds are absorbed quickly, making them less available for mucociliary clearance. A solubility model may be the most accurate means of evaluating the rate of absorption of the dust retained in the alveoli (Mercer, 1967; Morrow, 1970).

(a) Experimental animals

There are few animal studies dealing with respiratory absorption, and data on the pulmonary uptake of nickel in particulate form are limited.

Wehner & Craig (1972) exposed Syrian golden hamsters to nickel oxide (nickel oxide not specified) particles with a mass mean aerodynamic diameter (MMAD) of 1.0–2.5 μm, and observed that inhalation for 2 days (7 h/day) at a concentration of 10–190 mg/m^3 air resulted in a deposition of 20% of the inhaled amount. On the 10th day after exposure, more than 75% of the nickel oxide was still present in the lungs, and, even after 45 days, approximately 50% of the total amount inhaled still remained. As no significant quantities of nickel oxide were found in the liver and kidney at any time after exposure, absorption seemed to be negligible during this period. In ancillary studies, Wehner et al. (1975) exposed hamsters for 61 days to nickel oxide aerosol (nickel oxide not specified) for 7 h/day and cigarette smoke (nose-only exposure of approximately 10 min duration, twice before, and once after, the daily 7-h dust exposure). It was found that the inhalation of cigarette smoke did not change either the deposition or the clearance pattern of the nickel oxide.

In a later study, Wehner et al. (1979) exposed Syrian golden hamsters to a highly respirable aerosol (MMAD = 2.8 μm) of nickel-enriched fly ash (NEFA; nickel content 9%), at concentrations of 220 mg/m^3 (one 6-h exposure) and 190 mg/m^3 (for 60 days, 6 h/day). In the acute exposure, approximately 95% (6.8 μg nickel = 75 μg NEFA) of the total deposited amount (7 μg nickel = 78 μg NEFA) was found in the deep lung, 1 month after exposure, indicating a very slow clearance. The findings also show, that nickel in the NEFA is retained and does not leach appreciably from the NEFA into tissue fluids. This assumption is supported by the observation that the average nickel content remained practically the same from 7 days after exposure (7 μg) to 30 days after exposure (6.8 μg), which would not be the case if the nickel were to leach from NEFA. In the 2-month study, the deposition was 5.7 mg NEFA or 510 mg nickel on the third day after exposure.

In another study, Wehner et al. (1981) exposed Syrian golden hamsters to a high (70 mg/m^3) and a low (17 mg/m^3) respirable NEFA concentration, for up to 20 months. The NEFA contained approximately 6% nickel. Exposure resulted in heavy deposits of NEFA in the lungs of 731 ± 507 μg and 91 ± 65 μg nickel/lung in the 70 mg/m^3 and 17 mg/m^3 exposure groups, respectively).

The short- and long-term NEFA studies showed that the time from the end of exposure to sacrifice (64 h–7 days) was not long enough to allow for mucociliary clearance of the nickel deposited in the ciliated part of the respiratory tract. In addition, clearance mechanisms may be disturbed by the repeated 6-h exposures to high aerosol concentrations, making them decreasingly efficient and resulting in the retention of larger quantities of material.

However, trace elements are not homogeneously distributed, even in particles of similar size, and the mean content of a given trace element in fly ash is determined by relatively few particles with a very high content of that element. This means that cells in the respiratory tract or the lung would not come into contact with fly ash particles containing 0.03% nickel, but instead with particles containing many times that quantity. It should be noted that respirable NEFA is emitted from coal-fired power plants (Natusch et al., 1974).

Leslie et al. (1976) exposed mice for 4 h to nickel-containing welding fume aerosols. Particle size and nickel content were determined. The nickel content was highest (8.4 $\mu g/m^3$) with particles 0.5–1.0 μm in diameter. It was reported that no clearance of lung-deposited nickel had occurred by 24 h, nor was there any elevation in blood-nickel levels, indicating that there had been no absorption into the bloodstream. When rats were exposed through inhalation to nickel oxide aerosol (0.4–70 mg/m^3) for 6–7 h, 5 days/week, for a maximum of 3 months, the fraction deposited in the lung significantly decreased with increasing mass median diameter and slightly decreased with increasing exposure concentration (Kodama et al., 1985).

The data obtained by Wehner et al., showing poor absorption and retention of the greatest portion of inhaled nickel oxide in the lungs, are supported by the study of Rittmann et al. (1981), who exposed Wistar rats via inhalation to nickel oxide, produced by the pyrolysis of nickel acetate at 500–600 °C (50 $\mu g/m^3$ for 15 weeks). They found a half-life for the clearance of nickel from the deep tract of 36 days. The half-life for clearance from the tracheobronchiolar compartment was less than one day.

Valentine & Fisher (1984) administered $^{63}Ni_3S_2$ intratracheally to mice (11.7 μg $^{63}Ni_3S_2$/animal). During the initial phase of clearance, 38% of the instilled dose was cleared, with a biological half-time of 1.2 days in the final phase. Four hours after instillation, the total lung burden was 85% of the administered dose. Thirty-five days after exposure, 10% of the administered dose was still retained by the lung.

In a study by Graham et al. (1978), mice were exposed, through inhalation, to nickel chloride (particle diameter = 3 μm, 644 μg nickel/m^3) for 2 h. Clearance of 70% (5.77 mg nickel/kg dry weight) of the deposited fraction (8.06 ± 0.506 mg nickel/kg dry weight) was found in the lung on the fourth day after exposure. In rats that had received 1 mg nickel, administered intratracheally as a single dose of ^{63}Ni chloride, most of the administered dose was found in the kidney (53%) and the lung (30%), the rest being distributed among the adrenals, liver, pancreas, spleen, heart, and testes (Clary, 1975). As clearance by 3 days was faster in the kidney, the lungs became the organ with the highest ^{63}Ni level (64% of the total amount deposited; kidney: 19%). Lung clearance within 6 h was 27%, which means that 70% of the material originally deposited had been absorbed.

Appreciable amounts of radioactive nickel, administered to male rats intratracheally as the chloride (1.27 μg nickel), were absorbed (Carvalho & Ziemer, 1982). Twenty-one days after exposure, the only measurable activity was in the lungs and kidneys. For example, 1 day after exposure, 29% of the initial burden was retained in the lungs, decreasing to 0.1% on day 21.

Following intratracheal instillation of nickel carbonate in mice (0.05 mg/animal), most of the nickel was eliminated after 12 days (Furst & Al-Mahrouq, 1981).

Medinsky et al (1987) administered nickel sulfate solution intratracheally to rats at doses of 1 μg, 11.2 μg or 105.7 μg nickel/rat. After 4 h, 49%, 21%, or 8%, respectively, of the instilled dose/g tissue was found in the lungs.

An important factor for retention in the lung is the solubility of the nickel compounds. Insoluble forms, such as nickel oxide and metallic nickel, seem to be retained in the lung for a longer time, whereas the more soluble nickel salts are absorbed. They are also solubilized in the fluids and mucus cleared from the lung by the mucociliary mechanisms into the alimentary tract.

(b) Human beings

Particulate nickel can be taken up from ambient air and from cigarette smoke. Respiratory absorption of nickel in particulate form is the major route of entry, under conditions of occupational exposure.

The amount of nickel absorbed from the air is expected to vary according to ambient atmospheric levels. Schroeder (1970) calculated that 75% of respiratory nickel intake is retained in the body and 25% is expired, depending on the particle size distribution. About 50% of the inhaled nickel would be deposited on the bronchial mucosa (and swept upward by mucociliary transport to be swallowed), and 25% in the pulmonary parenchyma.

6.1.1.2 Nickel carbonyl

In the toxicology of nickel, a special position is occupied by nickel carbonyl, a volatile, liquid compound. After nickel carbonyl inhalation, removal of nickel deposited in the lung is the most rapid, compared with the clearance of all other compounds, indicating an extensive absorption and clearance. Since the alveolar cells are covered by a phospolipid layer, the lipid solubility of nickel carbonyl vapours is of importance for their penetration of the alveolar membrane. This explains why nickel carbonyl is the only one of the nickel compounds to cause acute symptoms of poisoning, when inhaled.

(a) Experimental animals

Because of its industrial importance, nickel carbonyl absorption through inhalation has been studied extensively in experimental animal species including the dog, cat, rabbit (Armit, 1908; Tedeschi & Sunderman, 1957; Sunderman et al., 1961; Mikheyev, 1971), rat (Barnes & Denz, 1951; Sunderman et al., 1957, 1961; Ghiringhelli & Agamennone, 1957; Sunderman & Selin, 1968; Sunderman et al., 1968) and mouse (Oskarsson & Tjälve, 1979a). Animals received single doses ranging from 200 to 3050 mg nickel/m^3 air for periods ranging from 5 to 240 min. Sunderman et al. (1957) administered concentrations of 30 or 60 mg nickel/m^3 air for 30-min periods, 3 times/weeks, for 3 or 52 weeks. In all the studies, nickel was found in the respiratory tissues, brain, liver, kidneys, urinary bladder, adrenals, renal cortex, heart, diaphragm, and blood, from where it was rapidly mobilized after exposure (within 2 days). Sunderman & Selin (1968) reported that, 24 h after inhalation of ^{63}Ni carbonyl, the partition of the body burden of ^{63}Ni was: viscera 50%, muscle and gut 30%; bone and connective tissue 16%, and nervous tissue 4%. During 2-4 h following exposure of rats or rabbits to nickel carbonyl, the lung was the major excretory organ for the compound (Sunderman & Selin, 1968; Mikheyev, 1971). Elimination of nickel has been reported to be mainly via the urine: 62% (after 3 days) in rabbit (Mikheyev, 1971); 75% in dog, cat, rabbit (Armit, 1908); 90% in rat and dog (Tedeschi & Sunderman, 1957). Sunderman & Selin (1968) indicated that 26% of the inhaled amount was excreted via the urine in 4 days. Since the same amount or more might have been exhaled during the same period, the authors speculated that at least 50% of the inhaled dose could have been absorbed.

(b) Human beings

The extensive absorption of nickel carbonyl by human beings after respiratory exposure has been demonstrated by the measurements of enhanced nickel concentrations in organs of workers who died from nickel carbonyl poisoning (Brandes, 1934; Bayer, 1939; Sunderman & Kincaid, 1954; Ludewigs & Thiess, 1970; Sunderman, 1971; National Academy of Sciences, 1975) and of increased levels in the blood, serum, plasma, and urine in nickel-refining workers (Kincaid et al., 1956; Sorinson et al.; 1958; Hagedorn-Götz et al., 1977). In general, the highest tissue concentrations after inhalation of nickel carbonyl have been found in the lungs; lower concentrations have been measured in the kidneys, liver, and brain. However, there are no firm data on the dose levels of nickel carbonyl that are toxic for human beings. Experience suggests that, not only is there considerable interpersonal variation, but also that a certain degree of resistance can develop. Exposure to concentrations estimated to have been of the order of 0.5 mg $NiCO_4/m^3$ for half an hour have caused severe illness (Morgan, 1989, personal communication).

6.1.2 Absorption via the gastrointestinal tract

Absorption of nickel from the gastrointestinal tract occurs after ingestion of food, beverages, or drinking-water. In the occupational environment, an appreciable amount of nickel dust may be swallowed via the mucociliary clearance mechanisms; insufficient personal hygiene, poor work practices, and inadequate work-place conditions may also increase this uptake. Gastrointestinal intake of nickel leaching from nickel-containing dental alloys is of limited importance.

The rate of nickel absorption from the gastrointestinal tract is dependent on its chemical form. While soluble nickel compounds (e.g., $NiSO_4$) are better absorbed than relatively insoluble ones, the contribution of the poorly soluble compounds to the total nickel absorption may be more significant, since they are more soluble in the acidic gastric fluids.

6.1.2.1 Experimental animals

Nickel is poorly absorbed from ordinary diets and is eliminated mainly in the faeces. This has been shown in a nickel balance study on dogs (Tedeschi & Sunderman, 1957), in which the nickel intake in the food was equal to the output in the urine and faeces. The results also indicated that an average of 90% (1.01 ± 0.44 mg) of the amount of nickel ingested (1.12 ± 0.16 mg) was eliminated in the faeces. In rats, even at very high intakes of nickel (approx. 14.5 mg) from different sources, over periods of 4, 8, 12, and 16 days, nickel was absorbed poorly (0.15 mg, i.e., 1%) and was eliminated mainly in the faeces (13.9 mg = 96%). Levels of retained nickel averaged 0.5 mg (range: 0.29–0.8 mg), i.e., 3.5% (Phatak & Patwardhan, 1950, 1952).

Schroeder et al. (1969, 1974) did not find any measurable absorption of nickel in mice that were given 5 mg nickel/litre in the drinking-water, throughout their lives.

It was reported by Elakhovskaya (1972) that nickel, given orally to rats as the chloride in the drinking-water (0.005, 0.5, or 5 mg/litre), was eliminated mainly in the faeces. Ho & Furst (1973) reported that intubation in rats of ^{63}Ni (as the chloride) in 0.1 N HCl led to 3–6% absorption of the labelled nickel, regardless of the administered dose (1.8 μg/animal, or 4, 16, and 64 mg nickel/kg body weight). From these two studies (Elakhovskaya, 1972; Ho & Furst, 1973), it can be concluded that very little nickel in water or beverages is bioavailable. Phatak & Patwardhan (1950) showed that large doses are required to overcome the intestinal absorption-limiting mechanism.

The mechanism of nickel absorption in the perfused rat jejunum was studied by Foulkes & McMullen (1986). In step 1, the absorption process (uptake from lumen of perfused jejunum) proceeded at a rate linearly dependent on the concentration up to about 20 µmol nickel/litre perfusate. At higher levels, it approached apparent saturation. In step 2 of nickel absorption (movement of nickel from mucosa into the body), nickel was not appreciably retained in the mucosa.

6.1.2.2 Human beings

The intake of nickel via the gastrointestinal tract in human beings can be high, compared with that of other trace elements. Although the daily dietary intake may range up to 900 µg nickel, average values have been estimated to be around 200 µg (section 5.1.3).

Nodiya (1972) performed nickel balance studies on 10 male volunteers, aged 17 years, who ingested a mean of 289 ± 23 mg nickel/day (range 251–309 mg) and found that faecal elimination of nickel averaged around 89% (258 ± 23 mg/day).

In human volunteers who ingested nickel sulfate in the drinking-water or food, at doses of between 12 and 50 µg/kg body weight (one treatment), the amount of nickel absorbed averaged 27 ± 17% of the dose ingested in water compared with 0.7 ± 0.4% of the same dose ingested in food (Sunderman et al., 1989a).

6.1.2.3 Factors influencing gastrointestinal absorption

In assessing toxicity from ingested nickel, it is important to keep in mind possible factors that might change the absorption rate.

(a) Bioavailability

The experimental data obtained by Ho & Furst (1973) and Elakhovskaya (1972) did not indicate any change in absorption efficiency in rats, whether the nickel was taken up from liquids or drinking-water (section 6.1.2.1). However, Cronin et al. (1980)

reported that ingestion of a soluble nickel compound during fasting resulted in high urinary elimination rates of 4–20% of the dose. Fifteen female volunteers, who received single oral doses of 2.5, 1.25, or 0.6 mg nickel (as the sulfate, in gelatine-lactose capsules, together with 100 ml water), excreted 95–206 µg, 62–253 µg, and 48–89 µg nickel, respectively, (normal value 9 µg) in the urine. Nickel naturally occurring in food items also increases urinary nickel elimination; a diet containing 850 µg nickel resulted in an increased nickel elimination corresponding to about 1% of the amount in the diet (Nielsen et al., 1987a).

Solomons et al. (1982) estimated the bioavailability of nickel in human subjects by the serial determination of the changes in plasma-nickel concentrations following a standard dose of 22.4 mg of nickel sulfate hexahydrate (5 mg nickel), given in each of two standard meals, as well as in the drinking-water and 5 beverages (cow's milk, coffee, tea, orange juice, and Coca Cola®). The plasma-nickel concentration was stable in the fasting state and after an unlabelled test meal, but was elevated after the standard dose of nickel in water. It did not rise above fasting levels in the two labelled standard meals. When 5 mg of nickel was added to each of the 5 beverages, the rise in the plasma concentration was significantly suppressed with all but Coca Cola®. These results indicate that nickel absorption may be suppressed by binding or chelating substances, competitive inhibitors, or redox reagents; on the other hand absorption is often enhanced by substances that increase pH, solubility, or oxidation, or by chelating agents that are actively absorbed. Such compounds, which were constituents of the meals and beverages studied by Solomons et al. (1982), include: ascorbic acid, citric acid, pectins (from orange juice), which affect trace mineral absorption; tannins (in tea and coffee), which inhibit absorption of iron and zinc; ascorbic acid, which suppresses nickel absorption; and complexing agents, such as NaFeEDTA and EDTA, which depress plasma-nickel levels.

Sunderman et al. (1989a) studied the kinetics of nickel absorption, distribution, and elimination in healthy human volunteers who ingested $NiSO_4$ in the drinking-water or added to food (section 6.1.2.2). Nickel levels were determined, using electrothermal atomic absorption spectrophotometry, in samples of serum, urine, and faeces collected 2 days before, and 4 days after, a specified

NiSO$_4$ dose (12 μg Ni/kg, N = 4; 18 μg Ni/kg, N = 4, or 50 μg Ni/kg, N = 1). Absorbed nickel averaged 27 ± 17% (mean ± SD) of the dose ingested in water versus 0.7 ± 0.4% of the same dose ingested in food (40-fold difference). The results of this study confirmed that dietary constituents profoundly reduce the bioavailability of the Ni^{2+} for gastrointestinal absorption. Approximately one-quarter of the nickel ingested in drinking-water after an overnight fast was absorbed from the human intestine and excreted in urine, compared to only 1% of nickel ingested in food. The kinetic parameters provided by this study reduce the uncertainty of toxicological risk assessments of human exposures to nickel in the drinking-water and food.

Nickel absorption and distribution can be influenced by other factors. An example is disulfiram, which is used for alcohol aversion therapy, and which is immediately metabolized into two molecules of DDC (diethyldithiocarbamate). Following oral exposure of mice to ^{57}Ni (3 μg/kg), the residual body burdens of nickel after 22 h and 48 h were increased several fold in groups receiving clinically effective doses of DDC, either orally or intraperitoneally, compared with controls. The organ distribution was considerably changed compared with the control values: at 48 h, the amount deposited in the brain was at least 100 times greater than the control value; deposition in the kidneys, liver, and lungs was also increased (Nielsen et al., 1987b).

(b) Nickel/iron interaction

Becker et al. (1980) suggested that iron might affect nickel absorption. Using isolated intestinal segments of rats in an *in vitro* test system, they found that nickel ions had their own transport system located in the proximal part of the small intestine, thus making it likely that iron nutrition could affect nickel absorption. Forth & Rummel (1971) found that the transfer of nickel from the mucosal to the serosal side was elevated in iron-deficient intestinal segments.

6.1.3 Absorption through the skin

Percutaneous absorption is of negligible significance, quantitatively, but is clinically important in the pathogenesis of contact dermatitis. Because of the ubiquity of nickel-containing objects in industrialized society, nickel-sensitive patients frequently face considerable problems in the work-place in a wide range of jobs, and also because of contact with nickel-containing material in household and everyday items.

6.1.3.1 Experimental animals

The majority of studies on the dermal uptake of nickel do not permit the calculation of absorption data. Norgaard (1957), using guinea-pigs and rabbits (two of each species), applied 10 μl of a 5% solution of ^{57}Ni (as the sulfate heptahydrate) to a shaved area of 5 × 5 cm on the animal's back and measured the radioactivity in the organs and body fluids, 24 h after the application. The relative distribution levels in the urine, blood, kidney, and liver, measured as impulses/min using a Geiger-Müller counter, are shown in Table 16. The findings demonstrate that nickel absorption took place through the skin of the two animal species examined.

Mathur et al. (1977) found that nickel (as the sulfate) was absorbed systemically in male albino rats. Nickel sulfate in saline solution was applied daily, for 15 or 30 days, at doses equivalent to 40, 60, or 100 mg nickel/kg body weight. There were no clinical symptoms of toxicity in any of the animals and no gross changes were noted at autopsy. However, in rats sacrificed at 15 days, the livers of those

Table 16. Radioactivity (impulses/min) in organs, blood, and urine of rabbits and guinea-pigs, 24 h after application of ^{57}Ni to the skin [a]

Organ or body fluid	Rabbit 1	Rabbit 2	Guinea-pig 1	Guinea-pig 2
Urine, 5 ml	72	15		
Blood, 10 ml	25	9	4	4
Kidney	26	22	2.6	2.6
Liver	5	4	2	1.4

[a] From: Norgaard (1957)

that had received 60 or 100 mg nickel/kg body weight showed microscopic changes consisting of swollen hepatocytes and feathery degeneration; the testes were normal. In rats sacrificed at 30 days, the liver changes were more marked, with focal necrosis in those that had received 60 or 100 mg nickel/kg body weight; the testes showed tubular damage and degeneration.

In a study on dermal absorption, Lloyd (1980) applied ^{63}Ni chloride (40 μCi) to the shaven flanks of guinea-pigs and reported that a small amount of the applied dose passed through the skin and appeared in the plasma. After 4, 12, and 24 h exposure, 0.005, 0.07, and 0.05%, respectively, of the total ^{63}Ni dose were found in the plasma, and 0.009, 0.21, and 0.51%, respectively, of the absorbed nickel, were measured in urine. In excised skin, levels of 1.94, 7.30, and 5.33%, respectively, were found. Using micro-autoradiography of ^{63}Ni chloride exposed skin (2 μCi; periods of 1/2–48 h, shaved flanks of guinea-pigs), Lloyd (1980) also found that the radioactive nickel accumulated within 1 h in the highly keratinized areas, the stratum corneum, and hair shafts, showing a route of entry via the hair follicles and sweat glands. Increased radioactivity was also measured in the serum and urine. Wells (1956) had also reported that nickel ions can penetrate via the sweat ducts and hair follicle ostia and that they have a special affinity for keratin. Based on histochemical evidence, it was suggested that nickel is bound by the carboxyl groups of keratin (Samitz & Katz, 1976).

The greatest accumulation of nickel was found in the Malpighian layer, the sweat glands, and the walls of the blood vessels. Wells (1956) reported that nickel sulfate did not penetrate the skin, because the stratum corneum was a barrier to its penetration. Samitz & Pomerantz (1958) found that the extent of penetration (nickel sulfate, 0.5% in aqueous solution) was enhanced by sweat and sodium lauryl sulfate (1% aqueous solution) in animals (species not indicated). However, there was no evidence of the actual absorption of nickel sulfate.

6.1.3.2 Human beings

Nickel/epidermal interactions were studied *in vitro* in diffusion cells by Samitz & Katz (1976), who found that the diffusion of ^{63}Ni (as the sulfate, specific activity 1 μCi/ml; 0.1, 0.01, or 0.001 mol/litre in physiological salt solution) through the human epidermis was only slight after 17, 24, and 90 h, respectively. Diffusion did not take place within the first 5 h. Sweat or surfactants (0.2% in physiological saline solution) slightly enhanced the diffusion of nickel (0.002 mol nickel sulfate/litre, containing 0.2 8 μCi ^{63}Ni/ml).

Spruit et al. (1965) using human cadaver skin reported that nickel ions (from nickel chloride) penetrate, and are bound by, the dermis. He suggested that nickel bound by the dermis can serve as a reservoir for the subsequent release of nickel ions. In a study (using ^{57}Ni as an indicator) on normal and nickel-hypersensitive persons, it was shown that when 10 μlitre of a 5, 2.5, 1.25, or 0.68% solution of nickel sulfate was applied to the skin, about two thirds of the nickel was absorbed in 24 h (as measured with a Geiger-Müller counter) (Norgaard, 1955). The absolute amount absorbed was highest during the first few hours following application. Absorption was the same in normal and in hypersensitive patients. In hypersensitive patients, the eczematous reaction appeared at the time when only about 10% of the quantity of nickel absorbed was left on the skin. However, the findings have not been verified by examining nickel levels in the skin, other organs, and in body fluids. Kolpakov (1963), using skin from persons who had died suddenly from accidental injury, found that the Malphigian layer of the epidermis, the dermis, and the hypodermis were readily permeable to nickel sulfate. Permeation of nickel from nickel chloride and nickel sulfate solution through the human skin was determined *in vitro* in diffusion cells (Fullerton et al., 1986). The permeation process was slow with a lag time of around 50 h. Without occlusion, the amount of nickel permeation was negligible. Permeation of nickel ions was faster from a nickel chloride solution than from a nickel sulfate solution. After about 200 h, 13–43% of the nickel from a nickel chloride solution and about 4.7% from a nickel sulfate solution were present in the skin matrix.

6.1.4 Other routes of absorption

Parenterally injected nickel is only of practical interest in toxicity studies, where it is particularly useful in assessing the kinetics of nickel transport, distribution, and elimination.

6.1.4.1 Experimental animals

Bergman et al. (1980a) implanted specimens of non-precious dental casting alloys containing 70–75% nickel (by weight) subcutaneously in the neck region of mice. After 5 months of exposure, most of the nickel, released from the implants through electrochemical corrosion, had accumulated in the soft tissue capsule at a concentration of 123 mg/kg (wet weight), whereas only 0.31 mg/kg (wet weight) appeared in the kidney; other tissues contained amounts that were more than 10 times less.

Samitz & Katz (1975) found that nickel was leached from stainless steel spheres (nickel content not indicated) implanted in incisions in both hind legs of 3 rabbits. Taking biopsies, from both legs, at the site of implant and at distances of 1, 2, 3, 4, and 5 cm around it, nickel was found only in the tissue near the implant (1 cm distance); levels ranged from 34.3 to 39.8 mg/kg (wet weight) after 3 weeks, and from 46.2 to 53.3 mg/kg (wet weight) after 6 weeks. Whether nickel was translocated to organs and the blood was not examined.

6.1.4.2 Human beings

In human beings, absorption may occur from a variety of implanted nickel-containing medications, metallic de-vices, and prostheses, which release nickel by leaching (section 5.3). However, leaching from implanted metals is difficult to assess in human beings, because of the few and conflicting experimental data. Leaching of nickel from nickel-coated containers may contaminate intravenous fluids (section 5.3).

6.1.5 Transplacental transfer

Transplacental transfer provides an initial body burden that will be augmented by later environmental exposures. For this reason, and, in view of the possible adverse effects associated with the exposure of pregnant women to nickel during early pregnancy, transplacental transfer is important. Placental transfer is influenced by gestational age and the availability of nickel in the maternal blood. Species differences in placental structure and implantation, which may possibly influence nickel transfer, must also be considered.

6.1.5.1 Experimental animals

Several reports indicate that transplacental transfer of nickel occurs in animals. In a study by Phatak & Padwardhan (1950), the newborn offspring of rats, fed nickel in various chemical forms (nickel carbonate, nickel catalyst, nickel soap[a]) at dietary concentrations of 250–1000 mg/kg, showed whole-body levels of 22–30 mg/kg and 12–17 mg/kg body weight, when dams received 1000 mg nickel/kg diet, and 500 mg/kg, respectively.

Lu et al. (1981) injected pregnant mice (ICR strain, 12–14 weeks old) intraperitoneally, on day 16 of gestation, with a single 0.1 ml dose of nickel chloride solution (equivalent to 4.6 mg nickel/kg). The kinetics of nickel chloride in the fetal tissues showed a different pattern from that in maternal tissues. The concentration of nickel in the maternal blood and the placenta were found to be at a maximum (19.8 and 3.9 mg/kg, respectively) 2 h after injection. The maximum concentration in fetal tissues (1.1 mg/kg) was reached 8 h after injection, and only a slight and gradual decrease in concentration was observed up to 24 h. The concentration began to decrease rapidly between 24 and 48 h. The biological half-life was calculated to be 8.9 h in the rapid phase and 33 h in the slow phase. In this study, the mean concentration of nickel in the placenta was less than those in the maternal kidneys and blood, but higher than those in the maternal liver and spleen after 24–48 h.

In a study by Lu et al. (1979), 10 pregnant ICR mice given an intraperitoneal injection of nickel chloride solution, equivalent to 4.6 mg nickel/kg body weight, on day 8 of gestation, were sacrificed after 4 h and the embryos removed. The concentration of nickel retained in embryonic tissues was 800 times higher in the exposed animals than in the controls.

When pregnant mice were given a single intraperitoneal injection of ^{63}Ni chloride (50 μCi; 0.14 mg/kg body weight) on day 18 of gestation, passage of ^{63}Ni from mother to fetus was rapid and concentrations in fetal tissues were generally higher than those in the dam (Jacobsen et al., 1978).

---

[a] Nickel soap was prepared by neutralizing mixed fatty acids (obtained by saponification of refined groundnut oil) with nickel carbonate.

Olsen & Jonsen (1979a) investigated nickel uptake and retention after intraperitoneal injection of 0.5 ml ^{63}Ni chloride in a 16-day pregnant mouse. They observed that placental transfer of nickel occurred throughout gestation (in the visceral yolk sac during early gestation 10–11 days, and in the visceral yolk sac and chorioallantoic placenta during late gestation). A significant uptake of nickel was seen on days 5–6 of gestation. Fetal accumulation of nickel took place up to day 16 of gestation. Nickel was distributed throughout the tissues in the early embryo; distribution became more differentiated with increasing gestation and became similar to that in the dam.

Sunderman et al. (1978a) administered ^{63}Ni intramuscularly to groups of pregnant Fischer 344 rats on days 8 or 18 of gestation and determined maternal and fetal tissue concentrations by autoradiography. In the fetuses of dams injected on day 8 of gestation, the mean ^{63}Ni concentration in the embryos and membranes, after 24 h, was equivalent to the ^{63}Ni concentrations in the maternal lungs, adrenals, and ovaries. In dams injected on day 18 of gestation, there was localization of ^{63}Ni in the placentas and ^{63}Ni was present in the yolk sacs and fetuses, after 24 h. The fetal organ with the highest concentration of ^{63}Ni was the urinary bladder, suggesting that there was renal elimination of the ^{63}Ni that had entered the fetuses on day 18 of gestation.

Nadeenko et al. (1979) administered nickel to rats in the drinking-water for 7 months before, and during, pregnancy. The nickel contents increased in the placentas, but not in the fetuses.

Dostal et al. (1989) studied the effects of nickel on lactating rats, their suckling pups, and the transfer of nickel via the milk. Dose-dependent increases were observed in the concentrations of nickel in the milk and plasma, 4 h after a single subcutaneous injection of nickel chloride at 10, 50, or 100 μmol/kg body weight to lactating dams giving a milk/plasma nickel ratio of 0.02. Peak plasma nickel concentrations in the dams occurred 4 h after the injection, while the peak concentration in the milk was observed at 12 h and remained elevated at 24 h. Daily subcutaneous injections of dams with 50–100 μmol/kg body weight for 4 days increased the milk/plasma nickel ratio to 0.10. Significant alterations in milk composition included increased solids and lipids (42% and 110%, respectively) and decreased milk protein and lactose (29% and 62%, respectively). In multiple dose studies where 50 or 100 μmol nickel chloride/kg body weight were given subcutaneously, once daily, on days 12–15 of lactation, the plasma-nickel concentrations in suckling pups, sacrificed 4–6 h after the third daily injection of 50 or 100 μmol/kg body weight to the dams, were 24 and 48 μg/litre, respectively. Liver weights were decreased in the pups whose dams received 100 μmol/kg body weight, but no changes in hepatic lipid peroxidation or thymus weight were reported.

6.1.5.2 Human beings

Nickel has been shown to cross the human placenta; it has been found in both the fetal tissue (Schroeder et al., 1962) and the umbilical cord serum, where the average concentration from 12 newborn babies was 3 ± 1.2 μg/litre (range 1.7–4.9 μg/litre) and was identical with that in the mother's serum, immediately after delivery (McNeely et al., 1971b). Measurable concentrations have been found in various fetal tissues. Stack et al. (1976) found a mean nickel concentration of 23 mg/kg (SD = 7.2) in developing teeth from 26 cases of stillbirth and neonatal death, while enamel and dentine of developing teeth from 4 fetuses showed levels ranging from 11 to 19 mg/kg for dentine and 12 to 20 mg/kg for enamel. In tissues such as liver, kidney, brain, heart, lung, skeletal muscle, and bone, nickel was found in mean concentrations similar to those in adults ranging from 0.24 to 0.69 mg/kg dry matter (SD ranging from 0.16 to 0.6) (Casey & Robinson, 1978). The passage of nickel across the human placental barrier is of relevance because of the presence of female workers in industry.

Appreciable amounts of nickel have been found in breast milk (Stovbun et al., 1962; Medvedeva, 1965).

6.1.6 Nickel carbonyl

Nickel carbonyl absorption and toxicity is primarily of concern in occupational inhalation exposure (section 6.1.1.2). Other routes of absorption are not of practical significance. Absorption by the gastrointestinal route could be of importance in case of accidental intake, but no data are available. Because of its lipid-soluble properties, nickel carbonyl may be absorbed dermally, but this has not been demonstrated. Parenteral absorption is only of experimental significance in studying nickel metabolism.

6.2 Distribution, retention, and elimination

The distribution of nickel in the body and its mode of elimination are relevant in view of the occupational and non-occupational exposures to nickel resulting from its wide industrial applications. Studies on the distribution of nickel in the tissues of animals are useful for the understanding of the interaction between nickel and biological materials and, consequently, of its toxic and carcinogenic effects.

Nickel is concentrated in the kidneys, liver, and lungs; it is excreted primarily in the urine. Nickel can also be found in the urine of non-occupationally exposed persons. Since the bioavailability of nickel and the rate of elimination depend very much on the nature of the nickel compound, urinary excretion may not always be an appropriate measure of exposure.

6.2.1 Transport

A few nickel-binding serum proteins have been identified in *in vivo* and *in vitro* studies using labelled nickel chloride. Nickel has been found in three major fractions of human and rabbit serum:

(*a*) macroglobulin-bound nickel;

(*b*) albumin-bound nickel; and

(*c*) nickel bound to ultrafiltrable ligands (e.g., amino acids) (Nomoto et al., 1971; Hendel & Sunderman, 1972; van Soestbergen & Sunderman, 1972; Callan & Sunderman, 1973).

Albumin is the principal transport protein for nickel in human, bovine, rabbit, and rat sera (van Soestbergen & Sunderman, 1972; Callan & Sunderman, 1973).

A metalloprotein, designated nickeloplasmin, has been isolated from the sera of rabbits and man (Nomoto et al., 1971; Decsy & Sunderman, 1974). It is a macroglobulin with an estimated relative molecular mass of 7×10^5 and contains approximately 0.8 g atomic nickel/mole. Disc gel- and immunoelectrophoresis have shown that purified nickeloplasmin is an α-2-macroglobulin in rabbit serum and a 9.5 S α-glycoprotein in human serum. These results have been confirmed, using more refined and sensitive techniques (Nomoto & Sunderman, 1988).

Ultrafiltrable, nickel-binding ligands play an important role in extracellular transport and in the elimination of nickel in urine (Hendel & Sunderman, 1972; van Soestbergen & Sunderman, 1972; Asato et al., 1975). The low-relative-molecular-mass, nickel-binding constituent of human serum has been identified as the amino acid, L-histidine (Lucassen & Sarkar, 1979; Glennon & Sarker, 1982). In an *in vitro* system, L-histidine was found to have a greater affinity for nickel than serum-albumin. Nickel-binding to human albumin became evident only when no more L-histidine was available. *In vivo*, the concentration of albumin was much higher than the concentration of L-histidine and most of the nickel was associated with albumin. The equilibrium between L-histidine-nickel and serum-albumin-nickel may be biologically significant. The L-histidine nickel complex, which has a much smaller molecular size than the albumin-nickel complex, may mediate the transport through a biological membrane by virtue of the equilibrium between these two molecular species of nickel. The equilibrium in favour of the L-histidine-nickel complex may be an explanation for the rapid urinary excretion of nickel observed by Onkelinx et al. (1973) and Sarkar (1980). The exchange and transfer of nickel between L-histidine and albumin appear to be mediated by a ternary complex in the form of albumin-nickel-L-histidine. The amount of nickel in each compartment varies from species to species and this may be due, in part, to species variation in the affinity of albumin for nickel (Hendel & Sunderman, 1972; Callan & Sunderman, 1973).

6.2.2 Tissue distribution

Parameters, such as the nickel compound administered, the dose, the number of administrations, the length of time between exposure and sacrifice, the strain of animal, and the route of absorption, may strongly influence the organ distribution pattern. Thus, the uptake of nickel by organs has differed in the reports of various investigators.

6.2.2.1 Experimental animals

Levels of nickel in the tissues of experimental animals have been determined following exposure to various nickel compounds under different experimental conditions (Table 17). The studies (in most cases with radioactive nickel, ^{63}Ni) indicate that nickel is widely distributed and rapidly eliminated. Administration of divalent nickel salts intravenously, intraperitoneally, or subcutaneously, as single or repeated injections, led to the highest accumulation in the kidney, endocrine glands, lung, and liver. Concentrations in nerve tissue were low; this is consistent with the observed low neurotoxic potential of divalent nickel salts. There was also slight uptake by bone, consistent with the rapid and extensive elimination of nickel from the organism. Low concentrations are retained in soft and mineral tissue.

A large amount of nickel was found in the guinea-pig kidney and pituitary gland, after daily subcutaneous injections of 1 mg nickel chloride/kg body weight for 5 days (Clary, 1975). In a study by Parker & Sunderman (1974), substantial concentrations of ^{63}Ni in the pituitary gland of the rabbit were found following single or repeated intravenous injections of ^{63}Ni chloride. The ^{63}Ni level in the pituitary gland was second only to that in the kidneys. The findng that nickel is particularly localized in the pituitary gland may have physiological significance. LaBella et al. (1973a) suggested that nickel may exert a direct, specific inhibitory action on prolactin-secreting cells in the anterior pituitary gland.

Studies by Herlant-Peers et al. (1982) and Kasprzak & Poirier (1983) showed that the number of administrations and the time interval between nickel injection and sacrifice of the animals influenced the distribution pattern of nickel chloride injected intraperitoneally (Table 17, Fig. 1).

Table 17. Relative tissue distribution of nickel in experimental animals [a]

Species (Number)	Dosage and stp [b]	Relative distribution [c]	Reference
Rat (4)	617 µg/kg (single iv injection) 2 h	kidney > lung > adrenal > ovary > heart > gastrointestinal tract > skin > eye > pancreas > spleen = liver > muscle teeth > bone > brain = fat	Smith & Hackley (1968)
Rabbit (3)	240 µg/kg (single iv injection) 2 h	kidney > pituitary > serum > whole blood > skin > lung > heart > testis > pancreas > adrenal > duodenum > bone > spleen > liver > muscle > spinal cord > cerebellum > medulla oblongata = hypothalamus	Parker & Sunderman (1974)
Rabbit (4)	4.5 µg/kg (34-38 daily consecutive injections) 24 h	kidney > pituitary > spleen > lung > skin > testis > serum = pancreas = adrenal > sclearae > duodenum = liver > whole blood > heart > bone > iris > muscle > cornea = cerebellum = hypothalamus > medulla oblongata > spinal cord > retina > lens > vitreous humor	Parker & Sunderman (1974)
Rat (4-11)	3.3 or 6.5 mg/kg (single iv injection) 2 h	kidney > adrenal > lung > heart > pancreas > small intestine > eye > thymus > muscle > epididymis > ovary > liver > spleen > bone > brain > incisor > fat	Chausmer (1976)

Table 17 (continued)

Mouse (12)	38.3 µg or 76.6 µg/kg (single iv injection) 1 h	kidney > lung > sternal cartillage > liver > pancreas	Oskarsson & Tjälve (1979a)
Mouse (3)	0.5 mg/kg (single iv injection) 2 h	kidney > urinary bladder (urine) > lung > skin > cartilage > eye (retina) > hair follicle > blood > oral epithelium > gastric epithelium > tooth enamel > tooth dentine > salivary glands > liver > gastric mucosa > pancreas > spleen > brown fat > bone > muscle	Bergman et al. (1980a)
Mouse (8)	6.2 mg/kg (one ip injection) 2 h	kidney > lung > plasma > liver > erythrocytes > spleen > bladder > heart > brain > carcass (muscle, bone, and fat)	Wase et al. (1954)
Rat (18)	3 or 6 mg/kg (7 or 14 injections as $NiSO_4 \cdot 6H_2O$) 48 h following last injection	7 or 14 × 3 mg/kg: heart > spleen > kidney > testis > bone > liver 7 × 6 mg/kg: heart > spleen > kidney > testis > liver > bone 14 × 6 mg/kg: heart > spleen > kidney > bone > testis > liver	Mathur et al. (1978)

Table 17 (*continued*)

Species (Number)	Dosage and stp [b]	Relative distribution [c]	Reference
Rat	82 µg/kg (ip) 6 h	kidney > spleen > lung > heart > liver > muscle	Sarkar (1980)
Rat (5)	6 mg/kg (single ip injection) 24 h	kidney > serum > heart > liver	Tandon (1982)
Mouse	50 mCi (single ip injection)	*6 days after injection:* lung > kidney > skin > small intestine > spleen > liver > uterus > brain > stomach > heart > whole blood > muscle > serum	Herlant-Peers et al. (1982)
Mouse	100 µCi (7 successive ip injections at 24 h intervals	*24 h after last injection:* lung > kidney > uterus > skin > stomach > spleen > small intestine > liver > brain > heart > serum > muscle > whole blood	Herlant-Peers et al. (1982)
Mouse (5)	1 or 24 (3 times/week) ip injections of 42.5 µmol ^{63}Ni acetate/kg 48 h	*1 injection:* kidney > lung > pancreas > spleen > liver > brain > heart > blood *24 injections:* pancreas > lung > heart > kidney > spleen > liver > brain > blood	Kasprzak & Poirier (1983)

Table 17 (continued)

Mouse (10)	specimen of non-precious dental casting alloy implanted subcutaneously in the neck region (10 × 5 × 1 mm) containing 71% nickel, exposure time 5 months	soft tissue capsule around implants > kidney > lung > spleen > liver > pancreas > heart > blood	Bergman et al. (1980b)
Guinea-pig (6)	1 mg/kg (subcutaneously for 5 days) 6 h	kidney > pituitary > lung liver > spleen > heart > adrenal > testis > pancreas > medulla oblongata = cerebrum = cerebellum	Clary (1975)
Rat (5)	7.5 µg/kg (one intratracheal injection) 1 h	lung > kidney > blood > skin > bone = spleen = testes = liver > heart	Carvalho & Ziemer (1982)
Rat (30)	1 mg/animal (one intratracheal injection) 6 h	kidney > lung > adrenal > liver > pancreas = spleen = heart > testes	Clary (1975)

Table 17 (continued)

Species (Number)	Dosage and stp [b]	Relative distribution [c]	Reference
Hamster (51)	53.2 % 11.1 mg nickel oxide/m^3, (inhalation) life-span exposure	lung > liver > kidney	Wehner et al. (1975)
Rat (12)	25, 50, or 100 mg NiCO$_3$/100 g diet week 9	*100 mg*: bones > heart > kidney > blood > spleen > intestine > testes > skin > liver *50 mg*: bones > testes > spleen > intestine > heart > liver > kidney > blood > skin *25 mg*: bones > intestine > testes > kidney > heart > spleen > blood > skin > liver	Phatak & Padwardhan (1950)
Rat (72)	100 nmol NiCl$_2$ (single intratracheal injection) 0.5 h	lung > mediastinal lymph nodes > kidney > ovaries > blood > femur > heart = adrenals > skin = pancreas > duodenum > pituitary > liver > spleen	English et al. (1981)
	100 nmol NiO (single intratracheal injection) 0.5 h	lung > mediastinal lymph nodes > kidney > heart > femur > duodenum > kidney > pancreas > ovaries > spleen > blood > adrenals > skin > pituitary > liver	

Table 17 (continued)

Calf (12)	62.5, 250, 1000 mg NiCO$_3$/kg dietary supplementation for 8 weeks	*1000 mg/kg:* serum > kidney > vitreous humor > lung > testis > bile > tongue > pancreas > rib > spleen > brain > liver > heart *250 mg/kg:* lung > serum > kidney *62.5 mg/kg:* lung > kidney > liver > testis	O'Dell et al. (1971)
Rat (24)	100, 500, 1000 mg/kg diet as nickel acetate for 6 weeks	liver > heart > kidney > testis	Whanger (1973)
Rat (64)	5.4 mg/kg in food and water for life	spleen > heart > kidney > lung liver	Schroeder et al. (1974)
Lamb (12)	65 µg/kg or 5 mg nickel/kg in diet for 97 days; on day 94: single oral dose of 40 µCi ^{63}Ni/kg	*65 mg/kg:* kidney > lung > spleen > heart > liver > brain > testis > *5 mg/g:* kidney > spleen > lung > liver > testis > heart > brain	Spears et al. (1978)
Rat (45)	1, 11.2 or 105.7 µg sulfate/rat (single intratracheal injection) 4, 24, or 96 h	lung > trachea > larynx > kidney > urinary bladder > adrenal glands > blood > large intestine > thyroid	Medinsky et al. (1987)

Table 17 (*continued*)

Species (Number)	Dosage and stp [b]	Relative distribution [c]	Reference
Rat (15)	2.5 µg/animal orally for 30 days as nickel sulfate	trachea > nasopharynx > skull bone > oesophagus > spleen > kidneys > lungs > heart	Huang et al. (1986)

[a] Partially adapted from: NAS (1975).
[b] Nickel given as ^{63}Ni chloride unless other compound indicated; stp = sacrifice time postexposure.
[c] Distribution in decreasing nickel concentration.

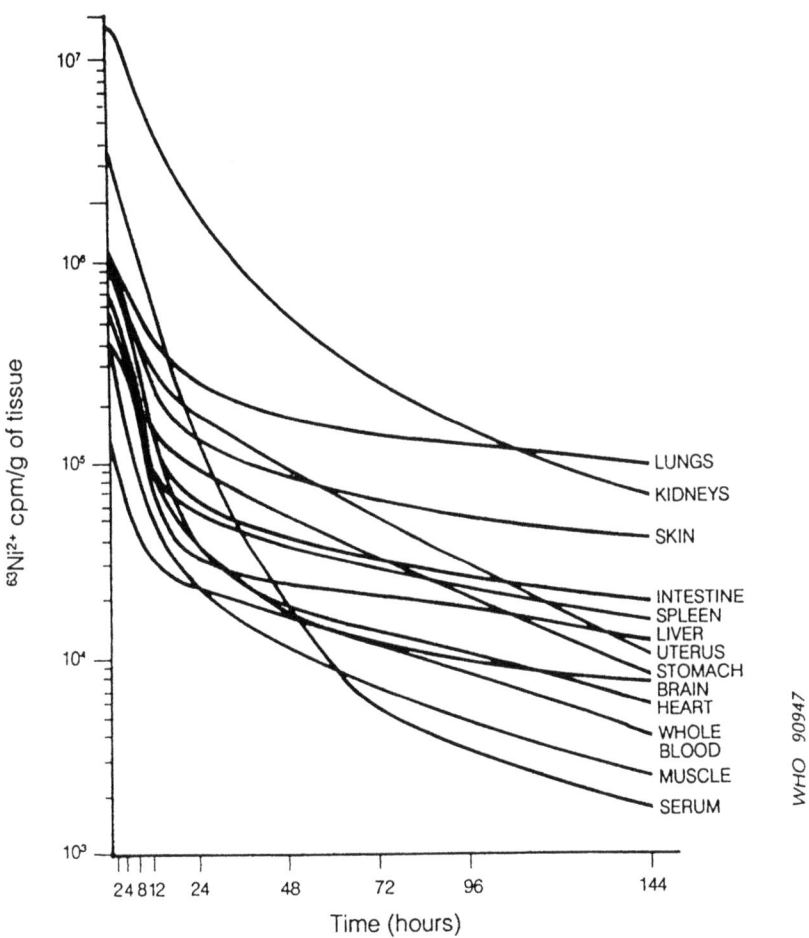

Fig. 1. *In vivo* kinetic study of $^{63}Ni^{2+}$ incorporation after a single intraperitoneal injection of 50 mCi of a ^{63}Ni chloride solution (spec. act. = 4.5 mCi/mg Ni) in Balb/C mice.
From: Herlandt-Peers et al. (1982).

The most striking findings of the study by Kasprzak et al. (1983) were the high accumulation of nickel in the pancreas and the decreasing nickel accumulation in the kidney and heart, following multiple intraperitoneal injections of nickel acetate. The first finding would link nickel with zinc and insulin metabolism and relate the commonly observed nickel-induced elevation of serum glucose to this interaction (Clary, 1975; Sunderman et al., 1976a). The second finding suggests the development of some detoxifying mechanisms in the kidney and heart during prolonged exposure to nickel. Huang et al. (1986) administered 2.5 µg nickel sulfate/animal to rats, orally, for 30 days. The nickel contents in the trachea, nasopharynx, oesophagus, lungs, skull, bone, heart, spleen, and kidneys of rats fed with nickel were significantly higher than those in the control animals.

After exposing rats to nickel at a concentration of 5 mg/litre in the drinking-water for their lifetime, Schroeder et al. (1974) did not find any measurable accumulation of nickel in tissues. When rats were fed nickel carbonate, nickel soaps, or metallic nickel catalyst, tissue accumulation was significant only in the case of the carbonate (Phatak & Padwardhan, 1950). O'Dell et al. (1971) fed calves supplemental dietary nickel at levels of 62.5, 250, or 1000 mg/kg and found pronounced increases in nickel levels in the pancreas, testes, and bone at the highest dietary level. While comparison of data for monogastric and ruminant animals may not be valid, these data indicate that the skeleton is the main storage depot for nickel, even though the nickel concentration in bone differs greatly between the studies by Phatak & Padwardhan (1950) and O'Dell et al. (1971). The data agree on the limited capacity of the liver to store nickel (which is in contrast to most of the trace elements). A major difference between the data for rats and calves is in the nickel level in the heart muscle. Nickel concentrations in this tissue were somewhat elevated in rats (250 mg/kg), but not in calves.

Similar studies on weanling rats (Whanger, 1973) and lambs (Spears et al., 1978) given soluble nickel salts (acetate or ^{63}Ni chloride) in the diet at various levels up to 1000 mg nickel/kg showed the highest accumulation in the kidney. As the nickel dose increased, the nickel contents of the tissues (kidney, liver, heart, lung, and testes) also increased. In rats, treated intratracheally, the distribution was virtually analogous (except for the pituitary gland), though, as expected, the lung (rather than the kidney) showed the highest accumulation (Clary, 1975; Carvalho & Ziemer, 1982).

6.2.2.2 Kinetics of metabolism

Following inhalation of high concentrations of nickel oxide (10–190 mg/m^3) by hamsters (7 h daily, repeated exposures for up to 3 months), 20% of the inhaled amount of nickel oxide was still present in the lungs 3–4 days after exposure. Complete clearance of this oxide was estimated to take weeks to months; 75% of the nickel oxide was still present in the lungs 10 days after exposure and 40% was still present 100 days after exposure (Wehner & Craig, 1972; Wehner et al., 1975). The lungs retained more than 99% of the nickel oxide deposited there. The liver and kidney retained small amounts of 0.21 and 0.04%, respectively. Kodama et al (1985) exposed rats to nickel oxide by inhalation at concentrations of 0.4–70 mg/m^3 for 6–7 h/day, 5 days/week, for a maximum of 3 months. Deposition of nickel oxide in the lungs ranged from 2.3 ± 0.9 to 23.4 ± 1%. The deposition fraction significantly decreased with increase in the mass median diameter of the particles, and slightly decreased with increasing exposure concentration. The clearance rate was estimated to be approximately 100 µg/year.

In contrast to the prolonged retention of nickel oxide in the lung after inhalation, the more soluble nickel chloride was rapidly cleared after a single intratracheal injection (1 mg/kg body weight) in rats (Clary, 1975). Six h after exposure, the kidneys showed the greatest amount of nickel followed, in order, by the lung and adrenals, with decreasing amounts in the pancreas, spleen, heart, and testes. Other tissues, such as whole brain, thymus, eyes, and femur showed only trace amounts. By 3 days, 90% of the injected nickel had been excreted, mainly in the urine (75%). Carvalho & Ziemer (1982) studied the deposition, clearance, and distribution of ^{63}Ni after intratracheal instillation in rats of very low dose levels (1.27 µg ^{63}Ni per animal); the highest concentrations of ^{63}Ni were retained in the lungs and kidneys. These were the only organs containing measurable amounts of ^{63}Ni, 21 days after exposure. Urinary excretion was the main route of elimination (78.5% of the initial deposition within 3 days). On day 21, almost all the ^{63}Ni

(96.5% of the initial body burden) had been excreted in the urine. The lungs retained 29% of their initial deposition (35 min after exposure), decreasing to less than 1% on day 21. Medinsky et al. (1987) gave 1, 11.2, or 105.7 μg nickel sulfate/animal, by intratracheal instillation, to Fischer 344 rats. Urinary excretion accounted for 50% of the dose, at doses of 1 and 11.2 μg/rat, and 80% at a dose of 105.7 μg/rat. The half-time for urinary excretion of nickel increased from 4.6 h at the highest dose to 23 h at the lowest dose. Faecal elimination of the initial dose was 30% (1 and 11.2 μg doses) or 13% (105.7 μg dose). Over 50% of the nickel remaining in the body at the end of 96 h was in the lungs. The half-time for lung clearance of nickel sulfate ranged from 21 h (highest dose) to 36 h (lowest dose). The results of this study indicate that the differences in the lung clearance of soluble nickel compounds reported by Carvalho & Ziemer (1982) and Clary (1975) can be explained by differences in instilled doses.

Tanaka et al. (1985, 1988) estimated the biological half-time of nickel monosulfide (amorphous) aerosol and of green nickel oxide in rats exposed through inhalation. The biological half-time of NiS (A) in the rat lung was 20-h, while that of the oxide was 21 months (particle size 4.0 μm) or 11.5 months (particle size 1.2 μm).

In several studies, ^{63}Ni-labelled nickel salts have been used to study the distribution and elimination of nickel after parenteral injection (Wase et al., 1954; Smith & Hackley, 1968; Parker & Sunderman, 1974; Clary, 1975; Mathur et al., 1978; Oskarsson & Tjälve, 1979b; Bergman et al., 1980b; Tandon, 1982). Using this route of administration, most nickel was excreted in the urine, causing a high labelling of the kidneys, which may be related to the role of the kidneys in nickel clearance. However, there have been different observations on its localization in other organs. Smith & Hackley (1968) found a good correlation between the blood volume of each tissue and the amount of nickel retained by that tissue. Mathur et al. (1978) investigated the effects of dose and duration of exposure on the relative distribution and found that, while a single exposure to nickel may not have a lasting effect on the body tissues, regular exposures could have a cumulative effect, particularly on the kidneys and the heart. An autoradiographic distribution study by Bergman et al. (1980b) on the albino mouse showed that, between 30 min and 24 h, there were high concentrations of ^{63}Ni in the urogenital, circulatory, and respiratory organs. Accumulation was also found in cartilage, lacrimal glands, and the skin. After one day, the distribution pattern changed, so that the highest concentrations were in the lungs, kidneys, central nervous system, skin, and the epithelia of the oral cavity and the oesophageal part of the stomach, with the long residence time of 3 weeks; accumulation was highest in the lung, central nervous system, kidneys, hard tissues (teeth, cartilage, bone), and the skin. A distribution study by Bergman et al. (1980a), who implanted specimens of non-precious dental casting alloys subcutaneously in the neck region of mice, did not yield any information about the dynamic pattern of the release of nickel, the uptake in various tissues and organs, or its elimination.

Metabolic data from nickel-balance studies carried out by Phatak & Padwardhan (1950) who fed rats, nickel carbonate, nickel soaps, or nickel catalyst (250, 500, or 1000 mg/kg in the diet for 2 months) demonstrated that appreciable quantities of nickel from all the nickel-containing diets were retained. Retention from the nickel carbonate diet was greater than that from the other two nickel preparations. This can be attributed to the ready solubility of the compound in the stomach and the easier absorption from the intestine. The proportion of the ingested nickel found in the faeces was lowest in the carbonate group. The amount of nickel excreted in the urine was only slightly higher in this group than in the other two.

O'Dell et al. (1971) fed calves a basal diet supplemented with nickel (as the carbonate) at levels similar to those used by Phatak & Padwardhan (1950) and for the same period of time. The results of this study showed that the absorption and tissue retention of dietary nickel can be increased and that the increase is related to the rate of nickel intake as well as to the total nickel intake.

(a) Kinetic modelling

The whole-body kinetics of nickel chloride (or other soluble metal salts) can be studied by injecting animals with suitable radiotracers, following the metal concentration in plasma as a function of time after injection and measuring the amounts eliminated in urinary and faecal collections. In general, data obtained in this fashion can be analysed mathematically. This allows the formulation of compartment models describing the metabolism of nickel in terms of distribution volumes, clearances by elimination and clearances by exchange. Nickel metabolism is characterized by a typical distribution and elimination pattern (Fig. 2, Table 18). Onkelinx et al. (1973) injected ^{63}NiCl$_2$ intravenously in male and female Wistar rats (17 μg/animal) and New Zealand albino rabbits (816 μg/animal) and measured the radioactive label in the urine and faeces, 3 days after injection, and, in the blood, at intervals ranging from 1 h to 7 or 9 days. In the rats, during the first day, 68% ^{63}Ni was excreted in the urine and, after 3 days, 78%. In the rabbit (2 animals), 9% of the

Table 18. Parameters of the two-compartment model of Ni(II) metabolism[a]

Parameters	Symbol	Units	Wistar rat[b]	Fischer rat[c]	Rabbit[b]
Compartment I	V_1	ml	75.1	59.2	697
Compartment II	V_2	ml	8.3	31.6	265
Exchange I-II	f_e	ml/h	0.12	0.92	2.31
Total excretory clearance	f_t	ml/h	8.14	7.73	61.2
Urinary clearance	f_u	ml/h	6.39	6.70	54.1
Faecal clearance	f_d	ml/h	1.28	0.75	-[d]
Clearance into sink	f_s	ml/h	0.47	0.28	-[d]
Average body weight	wt	g	208	165	3400
Injected dose (μg Ni/animal)		μg	17 (iv)	2173 (ip)	816 (iv)

[a] From: Onkelinx & Sunderman (1980).
[b] From: Onkelinx et al. (1973).
[c] From: Sunderman et al. (1976a).
[d] Measurements of faecal ^{63}Ni(II) were not performed in rabbits; hence f_d and f_s could not be calculated.

Kinetics and metabolism

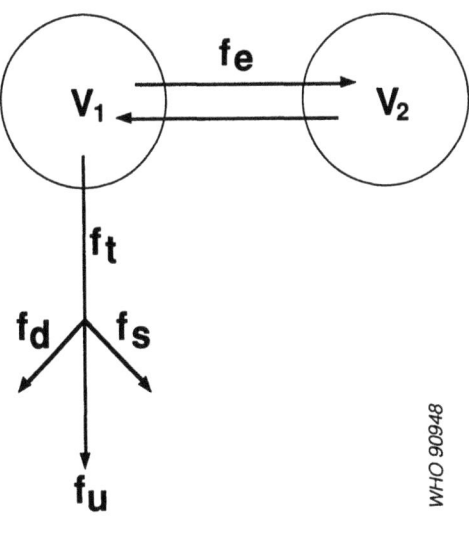

V_1/V_2	= volumes of the compartments
f_t, f_u, f_d, f_s	= total excretory clearance, urinary clearance, faecal clearance, and clearance into the body "sink"
f_e	= clearance by exchange between the two compartments

Fig. 2. Diagram of the 2-compartment model for Ni^{2+} metabolism in rats and rabbits. From: Onkelinx & Sunderman (1980).

administered dose was excreted in the urine, 5 h after injection, and 78% during the first day (Fig. 3). Faecal elimination of ^{63}Ni in the rat was 15% of the administered dose during the first 3 days following injection; faecal elimination was not determined in the rabbits. Sunderman et al. (1976a) administered 2173 μg ^{63}Ni chloride/animal to Fischer rats by intraperitoneal injection. Blood samples were taken at intervals between 10 min and 24 h after injection. Urine and faeces were collected at intervals between 6 h and 5 days after injection. Both studies showed that absorption and

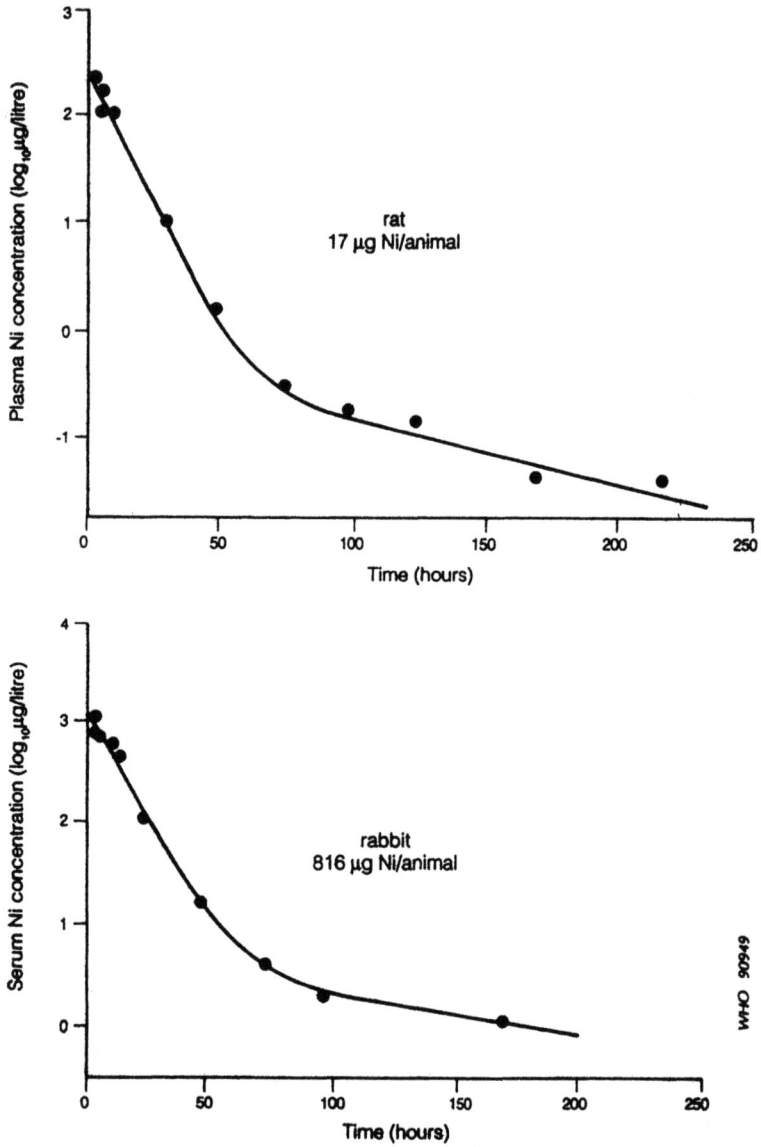

Fig. 3. Time course of $^{63}Ni^{2+}$ concentrations in rat plasma and rabbit serum following a single intravenous injection of ^{63}Ni chloride. From: Onkelinx & Sunderman (1980).

elimination fitted a 2-compartment model[a] in both species, comprising a rapid clearance phase from plasma or serum during the first 2 days and a much slower phase between the third and seventh days.

Chausmer (1976) determined tissue exchangeable pools (in rats injected intravenously with ^{63}Ni chloride) at a number of intervals following injection, and found (after performing a compartmental analysis of tissue exchangeable pools by computer evaluation of the percentage retained radioactive nickel) a rapid intracellular compartment having a half-life of several h in most tissues. The slower compartment had a half-life of several days. The kidney (followed by the lung, liver, and spleen) was found to have the largest rapidly exchangeable pool, 16 h after injection, with a two-compartment distribution, whereas bone had the best fit with a single compartment.

The distribution and elimination of nickel, given to animals as ^{63}Ni chloride, has been studied extensively. Most of the introduced nickel is rapidly excreted in the urine (65–87% in 24 h), the rest undergoing much slower elimination (76–90% in 5 days) (Sunderman & Selin, 1968; Sunderman et al., 1976a).

The 2-compartment model described by Onkelinx & Sunderman (1980) also provides a satisfactory fit for experimental data from human volunteers who ingested nickel sulfate in the drinking-water or food at doses of 12, 18, or 50 µg nickel/kg body weight (Sunderman et al., 1989a). Faecal elimination of nickel during 4 days following treatment averaged 76 ± 19% of the dose ingested in water versus 102 ± 20% of the dose ingested in food. The elimination half-time for absorbed nickel averaged 28 ± 9 h (range 17–48 h). Renal clearance was determined to be 8.3 ± 2.0 ml/min per 1.73 m^2 in human beings who had ingested nickel sulfate in water, and 5.8 ± 4.3 ml/min per 1.73 m^2 in those who had received nickel sulfate in food. The difference was not statistically significant.

---

[a] Compartment I: central compartment including the plasma and from which elimination takes place.
Compartment II: hypothetical volume that is connected to compartment I by reversible exchange.

6.2.2.3 Nickel carbonyl

There are few studies on the fate of nickel carbonyl in experimental animals. After the administration of nickel carbonyl, deposition occurs in the lung and in tissues, such as the brain, liver, and adrenals, and part of the administered dose of nickel is recovered in the urine (Armit, 1908; Barnes & Denz, 1951; Sunderman & Selin, 1968). It was assumed earlier that nickel carbonyl was rapidly dissociated in the lung and that the nickel was then transported to other tissues. However, the results of several studies (Sunderman & Selin, 1968; Sunderman et al., 1968; Kasprzak & Sunderman, 1969) have indicated that unchanged nickel carbonyl is present in the blood several hours after administration and can pass across the pulmonary alveoli in either direction without decomposition. It was suggested by Kasprzak & Sunderman (1969) that the nickel carbonyl that was not exhaled underwent a slow intracellular decomposition to NiO and CO. The released NiO was then oxidized to Ni^{2+}, which might become bound to nucleic acids or proteins, or to albumin in the plasma and, ultimately, would be excreted in the urine; the released CO would become bound to haemoglobin and finally exhaled.

Oskarsson & Tjälve (1979a) studied the distribution of intravenously administered ^{63}Ni and ^{14}C-labelled nickel carbonyl ($^{63}Ni(CO_4)$ and $Ni(^{14}CO_4)$) in mice by whole-body autoradiography and liquid scintillation counting. Radioactivity in the animals given ^{14}C-labelled carbonyl was mainly confined to the blood, indicating the formation of ^{14}CO-haemoglobin. This confirms the findings of Kasprzak & Sunderman (1969). After the administration of ^{63}Ni-labelled carbonyl, the highest level of ^{63}Ni was found in the lung, followed by the brain, spinal cord, heart, diaphragm, brown fat, adrenals, and corpora lutea. Additional studies showed that nickel was present in the lung, brain, heart, and blood as the cation.

6.2.2.4 Nickel levels in human beings

In human beings, wide variations have been reported in body nickel levels. This makes it difficult to appraise and compare the results obtained by various investigators. In addition to variations in the geographical origin of data and individual dietary and smoking habits, major differences can be attributed to the analytical methods employed. Only limited comparisons can be made using variants of spectrography, atomic absorption spectrometry, photometric methods, and special analytical techniques. Furthermore, no uniform reference samples have been used. Nickel values from tissue analyses have been related to ash or dry weight as well as to wet weight. The normal ranges of nickel concentrations in body fluids or tissues (serum, blood, lung, kidney) are not significantly influenced by age, sex, or pregnancy (McNeely et al., 1971; Turhan et al., 1983; Zober et al., 1984).

(a) Body fluids, hair, and nasal mucosa

The levels of nickel in biological fluids, hair, and some other materials increase remarkably in persons with increased occupational or environmental exposure and decline rapidly when exposure is reduced or stopped (Tables 19, 20, 21, and 22). Thus, measurements of nickel, particularly in the urine, serum, or hair, may serve as indices of exposure.

Data for normal nickel values in urine, blood, plasma, and serum, published in the last three decades, vary widely. Lower levels have been obtained by later investigators, because of the use of more sensitive analytical methods. Reference values in specimens from healthy, non-exposed persons are listed in Table 19. Because of doubts about the reliability of older studies, only recent data have been included.

A large number of workers exposed to various nickel compounds have been found to have elevated levels of nickel in the urine. These include those working in: nickel refineries (Morgan, 1960; Kemka, 1971; Norseth, 1975; Hogetveit & Barton, 1976, 1977; Bernacki et al., 1978a; Hogetveit et al., 1978; Morgan & Rouge, 1979; Torjussen & Andersen, 1979; Hogetveit et al., 1980; Boysen et al., 1982), the welding of nickel alloy steels (Norseth, 1975; Bernacki et al., 1978a; Grandjean et al., 1980; Polednak, 1981; Kalliomäki et al., 1981), nickel electroplating plants (Tandon et al., 1977; Tola et al., 1979; Bernacki et al., 1980; Tossavainen et al., 1980), nickel battery factories (Bernacki et al., 1978a; Adamsson et al., 1980), different occupations in shipyards (Grandjean et al., 1980), the pigment industry (Tandon et al., 1977), the glass industry (Raithel et al., 1981), nickel carbonyl processing in nickel refining (Kincaid et al., 1956;

Sorinson et al., 1958; Morgan, 1960; Nomoto & Sunderman, 1970; Hagedorn-Götz et al., 1977), aircraft mechanics and metal spraying (Bernacki et al., 1978a). The results of the investigations of Bernacki et al. (1978a), who analysed urine samples from nickel-exposed workers in 10 occupational groups, are listed in Table 20.

Serum or plasma nickel levels have been determined in workers in the following occupations: nickel refining (Hogetveit & Barton, 1976, 1977; Hogetveit et al., 1978, 1980; Torjussen & Andersen, 1979; Boysen et al., 1982), welding (Grandjean et al., 1980), electroplating (Tola et al., 1979; Tossavainen et al., 1980), battery manufacture (Adamsson et al., 1980) and shipyards (Grandjean et al., 1980), and the Mond process in nickel refining (Sorinson et al., 1958; Nomoto & Sunderman, 1970).

The highest nickel concentrations were found in the body fluids of nickel refinery workers. Concentrations in workers in electroplating shops, battery factories, and aircraft engineering works were lower (Table 10). After occupational exposure to nickel in electroplating processes, biological half-times ranging from 13 to 39 h for nickel in urine and from 20 to 34 h for nickel in plasma have been reported (Tossavainen et al., 1980).

Serum specimens of 22 residents of Sudbury, Ontario, who had been environmentally exposed to nickel (including air and tap water), contained nickel concentrations ranging from 0.2 to 1.3 µg/litre (mean 0.6 ± 0.3 µg/litre). These were significantly higher than the serum levels of residents without environmental exposure (Hopfer et al., 1989).

Nickel determinations in blood and urine, are widely used and accepted methods for monitoring nickel exposure. Although more data are available for urine, no clear-cut choice can be made between the use of blood or urine.

Grandjean et al. (1980, 1988) reported that analyses of nickel concentrations in both urine and plasma samples should be obtained to assess worker exposure. There were significantly higher ratios of plasma/urine nickel levels in painters and lower ratios in welders, compared with other workers in the shipyard. These differences probably reflected the different toxicokinetic characteristics of the nickel compounds to which the workers were exposed.

Table 19. Normal nickel concentrations in specimens from healthy non-exposed adults

Specimen	No. of subjects (m/f)	Nickel concentrations mean ±SD	Nickel concentrations range	Units	References
Whole blood	30 (15, 15)	0.34 ±0.28	< 0.05 -1.05	µg/litre	Linden et al. (1985)
Serum	10 (6, 4)	0.32 ±0.17	0.1 -0.6	µg/litre	Sunderman et al. (1989a)
Serum	43 (22, 21)	0.2 ±0.2	< 005 -1.0	µg/litre	Hopfer et al. (1989)
Serum	30 (15, 15)	0.28 ±0.24	< 0.05 -1.08	µg/litre	Linden et al. (1985)
Lymphocytes	10 (4, 6)	0.72 ±0.75	< 0.05 -1.10	µg/10^{10} cells	Wills et al (1985)
Urine (spot collection)	34 (18, 16)	2.0 ±1.5 2.8 ±1.9	0.5 -6.1 0.5 -8.8	µg/litre µg/litre (1.024 sp.gr.)[a]	Sunderman et al. (1986a)
Urine (24-h collection)	50 (24, 26)	2.2 ±1.2 2.6 ±1.4	0.7 -5.2 0.5 -6.4	µg/litre µg/day	Sunderman (1977)
Faeces (3-day collection)	10 (6, 4)	14.2 ±2.7 258 ±126	10.8 -18.7 80 -540	mg/kg (dry weight) µg/day	Horak & Sunderman (1973)
Faeces	10 (6, 4)	1.5 ±0.5	1.0 -2.2	mg/kg (wet weight)	Sunderman et al. (1989a)
Sweat	14 (6, 8)	51 ±38	8 -158	µg/litre	Christensen et al. (1979)

Table 19 (continued)

Specimen	No. of subjects (m/f)	Nickel concentrations		Units	References
		mean ±SD	range		
Bile	5[b]	2.3 ±0.8	1.5 -3.3	µg/litre	Rezuke et al.(1987)
Saliva	38 (32, 6)	1.9 ±1.0	0.8 -4.5	µg/litre	Catalanatto et al. (1977)
Hair	102 ([b])	0.29	0.0 -13.0	mg/kg (net weight)	Bencko et al. (1986)
Hair	22 (15, 7)	0.22 ±0.08	0.13 -0.51	mg/kg (dry weight)	Nechay & Sunderman (1973)
Hair	905 (437, 468)	—[c]	0.26 -2.70	mg/kg (wet weight)	Tagaki et al. (1986)
Nasal mucosa	57 (57m)	0.13 ±0.20	< 0.53 [d]	mg/kg (wet weight)	Torjussen & Andersen (1979)

[a] Urine nickel concentrations, factored to specific gravity = 1.024.
[b] Sex not indicated.
[c] Range of means of samples from five countries.
[d] Upper 95th percentile of nickel concentrations in nasal biopsies from non-exposed subjects.

Table 20. Nickel concentrations in urine of workers in various occupational groups [a]

Occupation	No.	Description	Atmospheric nickel ($\mu g/m^3$)	Concentration Urine [b] (μg/litre)	Creatinine [b] (μg/litre)
External grinders	9	abrasive wheel grinding of exteriors of articles made of nickel alloys	1.6 ± 3.0 (0.02-9.5)	5.4 ± 2.4 (2.1-8.8)	3.5 ± 1.6 (1.7-6.1)
Arc welders	10	DC arc welding of aircraft made of nickel alloys	6.0 ± 14.3 (0.2-46)	6.3 ± 4.1 [c] (1.6-14)	5.6 ± 6.2 (1.1-17)
Bench mechanics	8	assembling, fitting, and finishing parts made of nickel alloys	52 ± 94 (0.01-252)	12.2 ± 13.6 [c] (1.4-41)	7.2 ± 6.8 [c] (0.7-20)
Nickel battery workers	6	fabricating nickel-cadmium or nickel-zinc electrical storage batteries	Not measured	11.7 ± 7.75 [d] (3.4-25)	10.2 ± 6.4 [d] (7.2-23)
Metal sprayers	5	flame-spraying nickel-containing powders in phase on to aircraft parts	2.4 ± 2.6 (0.04-2.1)	17.2 ± 9.8 [d] (1.4-26)	16.0 ± 21.9 (1.4-54)

Table 20 (continued)

Occupation	No.	Description	Concentration		
			Atmospheric nickel ($\mu g/m^3$)	Urine[b] (μg/litre)	Creatinine[b] (μg/litre)
Electroplaters	11	intermittent exposure to nickel in combined electro-deposition operations involving silver, cadmium, chromium plating, as well as nickel	0.8 ± 0.9 (0.04-2.1)	10.5 ± 8.1[d] (1.3-30)	5.9 ± 5.0[c] (1.0-20)
Nickel platers	21	full-time work in nickel plating operations	Not measured	27.5 ± 21.2[e] (3.6-65)	19.0 ± 14.7[e] (2.4-47)
Nickel refinery workers	15	workers in a nickel refinery using electrolytic processes	489 ± 560 (20-2200)	222 ± 226[e] (8.6-8.3)	124 ± 109[e] (6.1-287)

[a] From: Bernacki et al. (1978a).
[b] Mean SD with range in parantheses.
[c] $P < 0.05$ versus controls, calculated by t-test.
[d] $P < 0.01$ versus controls.
[e] $P < 0.001$ versus controls.

The correlation between exposure levels and nickel concentrations in body fluids is poor in most studies (Table 21). The closest positive relationships of nickel concentrations in body fluids with ambient air levels were found by Norseth (1975) and Rahkonen et al. (1983) in welders, and by Tola et al. (1979) and Bernacki et al. (1980) in electroplaters.

There seem to be at least three reasons for the inconsistencies in the correlation between exposure levels and biological measurements of nickel:

(a) exposure is not to a single chemical species, but to a variety of nickel compounds of very different solubility, absorption, transportation, and elimination rates. The same workers may be simultaneously exposed to both insoluble and readily soluble compounds, having half-lives ranging from days (Tola et al., 1979) to years (Torjussen & Andersen, 1979). The influence of the nickel species in ambient air on the concentration in body fluids has been shown in a study by Bernacki et al. (1978a), who found widely varying ratios of air/urine levels in different occupational groups (Table 22);

(b) differences in personal working habits and hygiene;

(c) failure to standardize sampling methods.

The nickel contents of hair and nasal mucosa have been determined in occupationally-exposed persons. Theoretically, the nasal mucosa is one of the target tissues for nickel carcinogenicity. However, practical problems of sampling and standardization preclude the routine use of these measurements. Torjussen & Andersen (1979) analysed biopsy specimens of nasal mucosa from 318 nickel workers, 15 retired nickel workers, and 57 unexposed controls. The results showed that nickel exposure led to significantly raised nickel concentrations in the nasal mucosa in both active and retired nickel workers (2.74 ± 4.12 mg/kg wet weight, and 1.14 ± 1.78 mg/kg wet weight, respectively, versus 0.13 ± 0.2 mg/kg wet weight in the controls). The average nickel concentration in the nasal mucosa was highest in workers exposed to the highest atmospheric nickel concentration, inhaled as nickel subsulfide and nickel oxide dust. Workers exposed to aerosols of soluble nickel components, such as the chloride and sulfate, at a lower atmospheric nickel concentration, had the highest mean nickel

Table 21. Studies on the correlation between nickel concentrations in the air and in biological fluids in occupational exposure to nickel compounds[a]

Exposure	Biological matrix	Correlation coefficient (r)	Reference
welding	urine	0.85	Norseth (1975)
roasting-smelting	urine	none	
welding	urine	none	Bernacki et al. (1978a)
bench mechanics	urine	none	
electroplating	urine	none	
metal spraying	urine	none	
refinery	urine	none	
roasting-smelting	plasma	-0.11	Hogetveit et al. (1978)
	urine	0.14	
electrolysis	plasma	0.21	
	urine	0.31	
refinery (other)	plasma	0.67	
	urine	0.47	
refinery (nickel salts)	urine	0.49; 0.55	Morgan & Rouge (1979)
refinery:			
Mond process	urine	0.01	
calciner	urine	0.22	
powder plant	urine	-0.05	
electroplating	plasma	0.83	Tola et al. (1979)
	urine	0.82[b]	
	urine	0.96[c]	
electroplating	urine	0.70[d]	Bernacki et al. (1980)
battery manufacture	urine	significant[e]	Adamsson et al. (1980)
welding	blood	0.56	Rahkonen et al. (1983)
	urine	0.95	

[a] From: Aitio (1984).
[b] Afternoon.
[c] Next morning.
[d] After shift.
[e] $P < 0.01$.

Table 22. Nickel concentrations in serum, urine, nasal mucosa, and personal air samples from workers at the Falconbridge Nickel Refinery in Kristiansand, Norway [a]

Category of subjects/work	No. of subjects	Plasma nickel (µg/litre) mean ± SD	Urine nickel (µg/litre) mean ± SD	Nasal mucosal nickel (µg/100g) wet weight)	Air nickel (mg/m^3) mean ± SD
First study [b]					
Roasting/smelting	24	7.2 ± 2.8	65 ± 58		0.86 ± 1.20
Electrolysis	90	11.9 ± 8.0	129 ± 106		0.23 ± 0.40
Other process	13	6.4 ± 1.9	45 ± 27		0.42 ± 0.49
Second study [c]					
Controls	57	1.9 ± 1.4	4.9 ± 4.2	13 ± 20	
Roasting/smelting	97	5.2 ± 2.7	34 ± 35	467 ± 595	
Electrolysis	144	8.1 ± 6.0	73 ± 85	178 ± 235	
Other process	77	4.3 ± 2.2	22 ± 18	211 ± 301	

[a] From: Sunderman et al. (1986b).
[b] From: Høgetveit et al. (1978).
[c] From: Torjussen & Andersen (1979).

concentrations in the plasma and urine and the lowest in the nasal mucosa. The mucosal concentration was significantly correlated with the duration of nickel exposure.

For hair, normal values ranged between 0.13 mg/kg (dry weight) and 2.7 mg/kg (wet weight). In hair samples from 45 occupationally-exposed adults, Bencko et al. (1986) found nickel concentrations ranging from 1.6 to 3.5 mg/kg (mean 2.39 mg/kg) in welders and from 42.7 to 2140 mg/kg (mean 222.5 mg/kg) in nickel smelter workers. Control values ranged from 0 to 13 mg/kg (mean 0.29 mg/kg). In an accidental case of exposure to nickel carbonyl, nickel concentrations in hair samples from 5 workers ranged from 4 to 48.1 mg/kg (Hagedorn-Götz et al., 1977).

Although hair has been studied as a rapid, non-invasive measurement of exposure/absorption relationships, conflicting values have been obtained. Furthermore, the use of hair as an internal exposure index is controversial because of various factors, such as the external contamination of the hair surface, different sampling methods, and non-standardized cleaning methods.

(b) Tissues

There are few data on human tissue concentrations of nickel. Spectrographic analyses indicate that the retained nickel is widely distributed in very low concentrations in the body. Information on normal nickel levels in organs and tissues is presented in Tables 23 and 24.

Generally, there is no significant influence of sex or age on human organ levels of nickel (McNeely et al., 1971a; Turhan et al., 1983; Zober et al., 1984). The ribs, liver, and kidneys in babies up to the age of 3 months, were found to accumulate significantly more nickel than those in persons between 1 and 90 years of age (Schneider et al., 1980). Few data exist on the organ levels of nickel in occupationally exposed persons. In lung autopsy samples from 4 deceased persons living in the vicinity of a nickel-processing industry in the German Democratic Republic, the mean nickel concentration was 2135 ± 1867 mg/kg dry weight (Schneider et al., 1980).

Table 23. Reference values for nickel concentrations in human autopsy tissues[a]

Tissue	No. of subjects	Wet weight (μg/kg) Mean ±SD	Range	Dry weight (μg/kg) Mean ±SD	Range	Reference
Lung	4	16 ± 8	(8-24)	86 ± 56	(33-146)	Sunderman et al. (1971)
	9			132 ± 99	(50-290)	Chen et al. (1977)
	41	7 ± 10	(<1-70)			Zober et al. (1984)
	15			180 ± 105	(43-361)	Seemann et al. (1985)
	9	18 ± 12	(7-46)	173 ± 94	(71-371)	Rezuke et al. (1987)
	15	44 ± 56	(16-242)			Raithel (1987)
Kidney	6			125 ± 54	(50-210)	Chen et al. (1977)
	36	14 ± 27	(<1-165)			Zober et al. (1984)
	18			34 ± 22	(<5-84)	Seemann et al. (1985)
	10	9 ± 6	(3-25)	62 ± 43	(19-171)	Rezuke et al. (1987)
Liver	4	9 ± 3	(5-13)	32 ± 12	(21-48)	Sunderman et al. (1971)
	23			18 ± 21	(<5-86)	Seemann et al. (1985)
	10	10 ± 7	(8-21)	50 ± 31	(11-102)	Rezuke et al. (1987)
Heart	4	6 ± 2	(4-8)	23 ± 6	(16-30)	Sunderman et al. (1971)
	9	8 ± 5	(1-14)	54 ± 40	(10-110)	Rezuke et al. (1987)
Spleen	22			23 ± 20	(<5-85)	Seemann et al. (1985)
	10	7 ± 5	(1-15)	37 ± 31	(9-95)	Rezuke et al. (1987)

[a] Adapted from: Rezuke et al. (1987).

Table 24. Normal nickel concentration in Japanese human tissues (mg/kg wet weight) [a]

Tissue	Sex/number	Median	Average	Range	Mean ± SD
Rib	M/6	0.230	0.19	0.13-0.35	0.23 ± 0.07
	F/6		0.27		
Lung	M/15	0.160	0.21	0.04-0.44	0.16 ± 0.09
	F/15		<0.10		
Small intestine	M/5	0.120	0.11	0.05-0.29	0.13 ± 0.07
	F/5		0.15		
Large intestine	M/5	0.111	0.14	0.04-0.30	0.14 ± 0.10
	F/5		0.15		
Trachea	M/3	0.098	0.09	0.06-0.11	0.09 ± 0.02
	F/1		0.11		
Kidney	M/14	0.081	0.10	0.01-0.30	0.10 ± 0.07
	F/14		0.10		
Skin	M/4	0.072	0.09	0.02-0.22	0.10 ± 0.08
	F/2		0.14		
Muscle	M/5	0.070	0.11	0.02-0.27	0.10 ± 0.08
	F/5		0.09		
Liver	M/14	0.068	0.10	0.03-0.22	0.08 ± 0.05
	F/13		0.05		
Cerebrum	M/2	0.025	0.06	0.02-0.11	0.05 ± 0.11
	F/1		0.03		
Cerebellum	M/1	NM[b]	<0.03	NI[c]	NM
	F/1				

Table 24 (continued)

Tissue	Sex/Number	Median	Average	Range	Mean ± SD
Heart		NM	NC[d]	NC	NM
Pancreas	M/6	NM	<0.10	NM	NM
	F/2				
Spleen	M/1		<0.30		
		NM		NM	NM
Adrenal glands	M/1	M/1	<0.10		
		NM		NM	NM
Testis	M/1		0.05		
Ovary		NM	-	NM	NM
Fat	F/3	NM	<0.01	NI	NM

[a] From: Sumino et al. (1975).
[b] NM = not measured.
[c] NI = not indicated.
[d] NC = not calculated because there were less than 5 samples available or there was no mean.

c) *Body burden*

One assessment of nickel metabolism in human beings indicated that the body burden of nickel in normal adults averaged 0.5 mg (7 µg/kg for a 70-kg adult person) (Bennett, 1984). However, in a study on 30 Japanese subjects, Sumino et al. (1975) calculated a total body burden of about 5.7 mg (for a body weight of 55 kg), and Schroeder et al. (1962) indicated a body burden of 10 mg nickel for an adult person. Bennett (1984) concluded that the oral intake of nickel averaged 170 µg/day, of which approximately 5% would be absorbed (8.5 µg/day). Inhalation of nickel averaged 0.4 µg/day for urban dwellers, of which 35% was retained (0.07–0.14 µg/day); this involves the assumption that 70% of the nickel absorbed into the blood is promptly excreted by the kidneys and that the remaining 30% is deposited in the tissues, with a mean retention time of 200 days (Bennett, 1984).

6.2.2.5 Pathological states influencing nickel levels

Nickel metabolism is known to be altered in several common diseases, as well as in some physiological states. Alonzo & Pell (1963) observed increased nickel concentrations in the serum of 19 out of 20 patients with acute myocardial infarction, sampled within 24 h of admission to the hospital. Sunderman et al. (1970, 1971) reported increased nickel concentrations in the serum of 25 out of 35 patients with acute myocardial infarction, sampled 12–36 h after onset of symptoms. The frequent occurrence of hypernickelaemia after acute myocardial infarction has been confirmed by studies in the Federal Republic of Germany (Völlkopf et al., 1981), Pakistan (Khan et al., 1984), the United Kingdom (Howard, 1980), the USA (Leach et al., 1985), and the USSR (Nozdryukhina, 1978). McNeely et al. (1971a) showed that hypernickelaemia is not specific for myocardial infarction, because nickel concentrations in serum are also increased in patients with cerebral stroke and thermal burns, as well as in patients with myocardial ischaemia without infarction. Hypernickelaemia has been observed in patients with unstable angina pectoris, without infarction, and in patients suffering from coronary atherosclerosis, who developed cardiac ischaemia during treadmill exercise (Leach et al., 1985).

Volini et al. (1968) observed increased nickel concentrations in the liver in both the early and advanced stages of hepatic cirrhosis. In a patient suffering from aspartylglycosaminurea, a 10-fold increased concentration of nickel in the hepatic tissue was reported by Palo & Savolainen (1973).

Significantly decreased serum-nickel levels have been measured in steel-mill workers exposed to extreme heat (Szadkowski et al., 1969b).

In an investigation by Rubanyi et al. (1982a), serum-nickel levels in postpartum mothers were found to be reduced by 60%. However, a significant 20-fold elevation in the concentration of nickel was observed immediately after delivery of the infant, but before delivery of the placenta. Nomoto et al. (1983) did not confirm the occurrence of hypernickelaemia. Post-operative hypernickelaemia and nickeluresis were observed in patients following total knee and hip arthroplasty with porous coated nickel alloy prostheses (Sunderman et al., 1989b).

Nickel concentrations in the serum, whole blood, and urine from 61 patients with chronic alcoholism were elevated 17-, 15-, and 39-fold, respectively, after 4 months to 4 years of disulfiram treatment. Disulfiram (tetraethyl-thiuram-disulfide), a nickel-chelating agent, is used in alcoholism therapy (Hopfer et al., 1987).

6.3 Elimination and excretion

The elimination routes for nickel in human beings and animals depend, in part, on the chemical form of the compound and the mode of intake. In general, relatively low gastrointestinal absorption explains the elimination of dietary nickel in the faeces. In human beings and animals, urinary excretion is usually the major clearance route for absorbed nickel. Other routes of elimination are of minor importance. All body secretions appear to have the ability to excrete nickel; it has been found in saliva, sweat, tears, and milk. Biliary excretion is minimal in animals, but may be significant in human beings. Hair is an excretory tissue for nickel.

6.3.1 Experimental animals

In experimental animals, urinary excretion is the main clearance route for nickel compounds, introduced parenterally. Only a small portion of an injected dose is excreted via the gastrointestinal tract. Wase et al. (1954) studied the distribution and elimination of ^{63}Ni in mice using a high dose (102 μg ^{63}Ni/animal), administered intraperitoneally, and found faecal and urinary elimination in the ratio of 30:70%. The primary route of elimination of supplemental dietary nickel (carbonate) fed to calves was faecal (Tedeschi & Sunderman, 1957; O'Dell et al., 1971).

Biliary excretion of nickel was minimal following subcutaneous injection of 0.1 mg ^{63}Ni, as nickel chloride, in rats (Marzouk & Sunderman, 1985).

In rats or rabbits, after inhalation of nickel carbonyl, Sunderman & Selin (1968) and Mikheyev (1971) found the lungs were the major excretory organ besides excretion via the urine, 2–4 h after exposure. Other studies indicated that up to 90% nickel was excreted in the urine (Armit, 1908; Tedeschi & Sunderman, 1957; Sunderman & Selin, 1968; Mikheyev, 1971) and 38% was exhaled via the lungs as nickel carbonyl (Sunderman & Selin, 1968).

After intravenous injection of ^{14}C-nickel carbonyl in rats, Kasprzak & Sunderman (1969) found that 30% of the ^{14}C was excreted in the expired air as ^{14}C-nickel carbonyl and 50% as ^{14}C-carbon monoxide.

Nickel is excreted in the urine, not as the free metal, but bound to a protein that is similar to, or a fragment of, the soluble low relative molecular mass glycoprotein associated with nickel in renal tissue (Verma et al., 1980; Abdulwajid & Sarkar, 1983).

6.3.2 Human beings

As human beings take up most nickel via ingestion, it is eliminated unabsorbed, mainly in the faeces (Drinker et al., 1924; Tedeschi & Sunderman, 1957; Sunderman et al., 1963; Nodiya, 1972). Horak & Sunderman (1973) found that the faecal elimination of nickel in 10 healthy adults averaged 258 µg/day (SD 126) or 14.2 mg/kg, (dry weight) (SD 2.7), thus, the normal faecal elimination of nickel was approximately 100 times greater than the normal urinary excretion (2.6 µg/day (SD 1.4) or 2.2 µg/litre (SD 1.2)).

The urinary nickel levels of persons occupationally exposed to appreciable nickel concentrations, via inhalation at the work-place, are raised significantly. Positive correlations have been reported between air and urinary nickel concentrations in workers in the nickel industry (section 6.2.2.4). Urinary nickel concentrations of normal and exposed persons are given in Tables 19 and 20. The large amounts of nickel also found in the faeces, in some cases, indicated that, either retrograde loss from the lung into the oesophagus, or considerable oral exposure via contaminated surfaces, had occurred.

Nickel concentrations in samples of human bile (section 6.2.2.4) suggest that biliary excretion of nickel may be quantitatively significant in human beings (Rezuke et al., 1987). Sweat may constitute an excretory route of significance under conditions of physical exertion. Hohnadel et al. (1973) demonstrated that, in sauna bathers, the mean concentrations of nickel in the sweat from healthy men and women were significantly higher than the mean concentrations in the urine (men: 52 µg/litre, SD = 36; women: 131 µg/litre, SD = 65). Under conditions of profuse sweating, appreciable losses of nickel occurred. This may account for the diminished concentrations of serum nickel that were reported by Szadkowski et al. (1969b) in blast-furnace workers who were exposed to extreme heat over a long period.

The role of nickel deposition in human hair as an excretory mechanism has been studied(section 6.2.2.4).

Measurements of salivary nickel were performed by Catalanatto et al. (1977) on specimens of parotid saliva from 38 healthy adults. The concentrations of nickel in saliva averaged 1.9 ± 1.0 µg/litre (range 0.8–4.5 µg/litre). There was no significant correlation between the concentrations of salivary nickel and protein. No significant differences were observed between the mean concentrations of nickel in saliva samples from men and women.

7. EFFECTS ON ORGANISMS IN THE ENVIRONMENT

7.1 Microorganisms

Nickel is considered essential for certain metabolic processes in bacteria. Bartha & Ordal (1965) demonstrated a nickel requirement in the "Knallgas" bacterium *Alcalignes entrophus*. A nickel requirement was reported by van Baalen & O'Donnell (1978) for the blue green algae *Oscillatoria* sp.

Fungi and microorganisms demonstrate a fairly wide variety of sensitivity to nickel, but are generally more tolerant than the higher organisms. The toxic effects of nickel on microorganisms, including eubacteria (non-marine and marine), actinomycetes, yeasts, and filamentous fungi, were studied by Babich & Stotzky (1982a,b; 1983). Filamentous fungi varied considerably in their response to nickel, growth of *Achyla* sp. being inhibited at 5 mg nickel/litre whereas *Aspergillus niger* and *Gliocladium* sp. were only affected at concentrations as high as 1000 mg nickel/litre. With actinomycetes and eubacteria, there was less variability in toxicity. Concentrations of nickel inhibiting growth ranged from 5 to 30 mg nickel/litre, with the exception of *Caulobacter leidyi*, which exhibited some growth at 100 mg nickel/litre. Growth inhibition in yeasts occurred at 1–40 mg nickel/litre. In all microorganisms, toxicity increased as the pH decreased.

Babich & Stotzky (1983) investigated the influence of various factors on the toxicity of nickel for eubacteria, an actinomycete, and yeasts. Reductions in cell number occurred at 5 or 10 mg/litre and viable cells were eliminated at 10–50 mg/litre, though some species were unaffected after 24 h at 100 mg/litre. Reductions in pH from 6.8 to 5.3 enhanced the toxicity of 75 mg nickel/litre in some species, but not in others. The toxicity of nickel (100 mg/litre) for marine microbes was reduced by increasing the salinity and decreasing the temperature. Addition of a simulated sediment (a mixture of organic and inorganic particles from soil) reduced toxic effects after exposure to 100 mg nickel/litre . In freshwater microbes, addition of a clay mineral (50 mg/litre) provided protection against the toxicity of 10 mg nickel/litre. This effect was probably because of the adsorption of nickel on the particulates. Increasing the hardness of the water, by adding calcium carbonate at 200 or 400 mg/litre, reduced the toxicity of 10 mg nickel/litre. Long-term studies indicated that microbial survival was greater in marine than in fresh water. Bringmann et al. (1980) reported that, in fairly hard water (approximately 150 mg $CaCO_3$/litre) and a pH of 6.9, a nickel concentration of 0.82 mg/litre reduced the numbers of the saprozoic flagellate *Chilomonas paramecium*.

Thus, fungi and microorganisms have a wide range of sensitivities to nickel, but are usually more tolerant than higher organisms.

7.2 Aquatic algae and plants

Nickel at concentrations of 0.05–0.1 mg/litre inhibited the growth of algae, though some species may be more tolerant (Spencer, 1980). Upitis et al. (1980) reported growth inhibition in blue-green algae at concentrations of 1–5 mg nickel/litre. The chlorophyll content was found to be significantly reduced leading to discoloration of the cells. Concentrations of 10–30 mg nickel/litre were lethal for *Chlorella* sp. The same authors investigated the influence of various environmental factors on the toxicity of nickel for *Chlorella*. Nickel inhibition could be overcome by the addition of ethylenediaminetetramine (EDTA) (40 mg/litre) and also by the addition of zinc (10 mg/litre) to a medium containing 5 mg nickel/litre. A synergistic effect of copper and nickel was demonstrated by Hutchinson (1973) for *Chlorella vulgaris, Scenedesmus acuminata, Haematococcus capensis,* and *Chlamydomonas eugametos*.

A nickel concentration of 0.1 mg/litre at 20 °C inhibited the growth of 4 species of green algae, *Pediastrum tetras, Ankistrodesmus falcatus, Scenedesmus quadricauda,* and *S. dimorpha*. However, a concentration of 0.6 mg/litre did not affect the blue-green alga *Anabaena cylindrica*, though it reduced the rate of growth of *Anabaena flos-aquae* (Spencer & Greene, 1981). Stokes (1975) studied *Scenedesmus acutiformis* var. *alternans*, from a lake in a nickel-mining and smelting area. The lake water contained about 2.5 mg nickel/litre. At a nickel concentration of 1.9 mg/litre, the

Scenedesmus grew at 53% of the rate of controls grown in clean water and, at 3.0 mg/litre, growth was still 18% of the control rate.

A nickel concentration of 0.125 mg/litre inhibited the growth of *Anabaena inequalis*, but a concentration of 10 mg/litre was required to inhibit photosynthesis, and 20 mg/litre, to inhibit nitrogenase activity (Stratton & Corke, 1979).

Chiaudani & Vighi (1978) exposed *Selenastrum capricornutum* to nickel in a standard medium with, and without, EDTA. At 24 °C and a pH of 6.9–6.3 the 7-day EC_{50} (inhibition of growth to 50% of control values) was 0.9925 mg nickel/litre. When 0.3 mg EDTA/litre was added to medium, the 7-day EC_{50} of nickel was increased to 0.013 mg/litre. In a further study, the addition of 0.04 mg nickel/litre to the water samples did not inhibit growth as much as was predicted from the laboratory studies.

When the diatom *Navicula pelliculosa* was exposed to 0.1 mg nickel/litre (of which all but 0.2% was said to be Ni^{2+}) for 14 days, growth was retarded (50% of control value) (Fezy et al., 1979).

Hutchinson & Czyrska (1975) exposed *Lemna minor* (Valdiviana), for 3 weeks, to nickel concentrations ranging from 0.01 to 1.0 mg/litre in an artificial medium at pH 6.8, and a temperature of 24 ± 2 °C, with 16 h of light per 24 h. They found that a concentration of 0.05 mg/litre stimulated growth, and that concentrations greater than 0.1 mg/litre inhibited growth. At 1 mg nickel/litre, growth was prevented. Nickel uptake and toxicity were enhanced by the presence of copper.

The same authors examined *Lemna minor* from 23 sites, where the mean concentration of nickel in the water was 0.027 mg/litre. The plants contained from 5.4 to 35.1 mg nickel/kg (dry weight), equivalent to bioaccumulation factors (BCFs) of 200 and 1300. *Lemna*, grown on a culture medium containing 0.01–1 mg nickel/litre at pH 6.8 and a temperature of 24 ± 2 °C for 3 weeks, accumulated nickel concentrations ranging from 40 mg/kg dry weight, at 0.01 mg/litre, to 3067 mg/kg, at 0.5 mg/litre.

Clark et al., (1981) studied the accumulation and depuration of nickel by *Lemna perpusilla*. Plants collected from a fly ash basin (nickel concentration, 0.1 mg/litre) were allowed to depurate in dechlorinated tap water at 20 °C for a 14-day depuration period.

Concentrations of nickel fell from about 160 mg/kg dry weight to about half of this value. In the accumulation studies, *Lemna* accumulated nickel readily, particularly at the lowest ambient concentration of 0.1 mg/litre, and levels reached 500–600 mg/kg in 10 days. After a return to depuration conditions, the nickel concentration in the plants fell to 160 mg/kg, in 8 days.

Euglena gracilis, exposed to 8.9×10^{-4} mg nickel/litre in springwater, accumulated the metal to a concentration of 1.8 mg/kg, a BCF of about 2000 (Cowgill, 1976).

Ipomea aquatica took up 200 mg nickel/kg dry plant in 48 h, mostly in the roots, from water containing 5 mg nickel/litre (BCF = 40) (Low & Lee, 1981).

It is noted that, under laboratory conditions, the growth of a macrophyte (*Lemna*) was inhibited at a concentration of 0.1 mg/litre, but the growth of algae was inhibited at concentrations as low as 0.04 mg/litre. However, in natural waters, a nickel concentration of 0.04 mg/litre had a less inhibiting effect.

7.3 Aquatic invertebrates

Timourian & Watchmaker (1972) investigated the uptake of nickel chloride and its effects on the development of sea urchin embryos. After fertilization, sea urchin eggs exhibited increased rates of nickel uptake that appeared to be a result of an active transport mechanism. When exposed to 59–590 mg nickel/litre, gastrulation of embryos was prevented. Embryos grown in 0.59–5.9 mg nickel/litre were able to gastrulate, but failed to develop dorsoventral symmetry.

Acute and long-term toxicity studies performed by Powlesland & George (1986) revealed a different sensitivity to nickel in different developmental stages of *Chironomus riparis* larvae. First instar larvae were found to be significantly more sensitive to nickel than second instars with 48-h LC_{50} values of 79.5 mg/litre and 169 mg/litre, respectively. Longer term toxicity tests (30 days), in which larvae were allowed to develop from eggs until just prior to pupation, indicated that nickel concentrations up to 25 mg/litre appeared to have little effect on the percentage hatch. However, the growth of larvae was significantly reduced at 2.5 mg/litre. A threshold concentration for the effect of nickel on growth was estimated to be 1.1 mg nickel/litre.

Bryant et al., (1985) investigated the effects of temperature and salinity on the toxicity of nickel in two estuarine invertebrates. In the amphipod *Corophium volutator*, and the bivalve *Macoma baltia*, 96-h LC_{50} values varied from 5 to 54 mg nickel/litre and from 95 to 1100 mg nickel/litre, respectively. A decrease in salinity from 35 to 5 mg/litre resulted in greater toxicity in both species. Toxicity also increased in *Corophium volutator* with an increase in temperature from 5 to 15 °C.

Mathis & Cummings (1973) measured nickel levels in sediments, water, and biota in a river. The water contained the lowest concentrations of nickel (< 0.01 mg/litre) and the sediments, the highest (3–124 mg/kg). Two species of tubificid worms (*Limnodrilus hoffmeisteri* and *Tubifex tubifex*) contained 4–18 mg nickel/kg wet weight. Three species of clam were examined: in order of increasing nickel content (mg/kg on a wet-weight basis) they were *Quadrula quadrula* (0.4–1.6), *Amblema plicata* (0.4–2.3) and *Fusconaia flava* (0.7–3.0). Neither worms nor clams were starved before being examined, and nickel may also have been present in their gut contents. Brkovic-Popovic & Popovic (1977) studied the effects of nickel and other heavy metals on the survival of *Tubifex tubifex* in water of different pH and hardness. At a hardness of 0.1 mg $CaCO_3$/litre, the 48-h LC_{50} was 0.082 mg nickel/litre. Increases in hardness to 34.2 mg $CaCO_3$/litre and 261 mg $CaCO_3$/litre increased the 48-h LC_{50} to 8.7 mg nickel/litre and 61.4 mg nickel/litre, respectively, thus decreasing the toxic effects.

A 64-h LC_{50} was determined for *Daphnia magna* of 0.32 mg nickel/litre, at a temperature of 25 °C and a hardness of around 100 mg $CaCO_3$/litre (Anderson, 1950). Baudouin & Scoppa (1974), using *Daphnia hyalina*, estimated a 48-h LC_{50} value of 1.9 mg nickel/litre at a temperature of 10 °C, pH 6.2, and hardness of 58 mg $CaCO_3$/litre.

Exposure of *Daphnia magna* to nickel sulfate at concentrations ranging from 5 to 10 μg nickel/litre for 3 generations resulted in extermination (Lazareva, 1985).

Hall (1982) exposed *Daphnia magna* to 0.25 mg nickel/litre, including ^{63}Ni, at pH 6.9, and a temperature of 18–21 °C in water with a hardness of 60 mg CaCO$_3$/litre. The uptake of nickel was initially rapid (about 12 mg in 80 h). Depuration also occurred, and 25–33% of the nickel was lost from the animal in the exuviae, shed on moulting. Gut tissue did not accumulate nickel until after the first 5 h of exposure, suggesting that the oral route was not important for nickel.

Daphnia, exposed for 3 weeks to 0.125 mg nickel/litre in water, at a temperature of 18 ± 1 °C, hardness of 42.3–45.3 mg CaCO$_3$/litre, and pH 7.74, had 43% lower weights than control *Daphnia*, 9% less proteins, and the glutamic oxalacetic transaminase activity was reduced by 26%. A 16% impairment of reproduction occurred at 0.03 mg nickel/litre with a 50% impairment at 0.095 mg nickel/litre (Biesinger & Christensen, 1972).

Cowgill (1976) reared *Daphnia magna* and *Daphnia pulex* for 3 months on *Euglena gracilis*, which had been cultured in spring water containing 8.9×10^{-4} mg nickel/litre. The algal cells contained 1.8 mg nickel/kg, the *Daphnia magna*, 3.6 mg nickel/kg, and the *D. pulex*, 4.2 mg nickel/kg, giving BCF values of 2020 and 4050.

The acute effects of nickel on the freshwater snails *Juga plicifera* and *Physa gyrina* were studied by Nebeker et al., (1986). The 96-h LC$_{50}$ values were 0.237 mg nickel/litre and 0.239 mg/litre, respectively. A no-observed-effect-level of 0.124 mg nickel/litre was determined for *Juga plicifera*. Data published for other species of snails did not indicate a pronounced effect of hardness on nickel toxicity (Nebeker et al., 1986). The eggs and adults of the snail *Amnicola* were exposed to nickel in water at a temperature of 17 °C, pH 7.6, and a hardness of 50 mg/litre with 6.2 mg dissolved oxygen/litre. The 24-h LC$_{50}$s were 26.9 mg/litre for eggs and 21.1 mg/litre for adults, whereas at 96 h, the LC$_{50}$s were 11.4 and 14.3 mg/litre, respectively (Rehwoldt et al., 1973).

Nickel influenced the rate of filtration in the marine bivalve *Villorita cyprinoides* (Abraham et al., 1986). Rates of filtration decreased exponentially with increasing nickel levels. The EC$_{50}$, i.e., the concentration that reduced the rate of filtration by 50%, was 0.003 mg nickel/litre. A 96-h LC$_{50}$ was 0.061 mg nickel/litre.

It is concluded that the nickel concentrations causing mortality in acutely exposed invertebrates were generally similar to those for fish, but that *Daphnia* sp. appeared more sensitive, with LC_{50} values of less than 2 mg nickel/litre.

7.4 Fish

Sensitivity to nickel varies considerably among fish species. However, 96-h median lethal concentrations generally fall within the ranges of 4–14 and 24–44 mg nickel/litre for tests conducted in soft, and hard water, respectively. For example, in water of hardness 100, 125, and 174 mg $CaCO_3$/litre, the LC_{50}s for rainbow trout, exposed from fertilization to 4 days after hatching, were 0.05, 0.06, and 0.09 mg/litre, respectively (Birge & Black, 1980).

Pickering & Henderson (1966) compared the toxicity of nickel chloride in waters of 2 levels of hardness (total hardness 20 or 300 mg $CaCO_3$/litre). The 96-h LC_{50} was 4.9 mg/litre for fathead minnow (*Pimephales promelas*) and 5.3 mg/litre for bluegill sunfish (*Lepomis macrochirus*) in soft water, and 43.5 mg/litre and 39.6 mg/litre, respectively, in hard water. Rainbow trout (*Salmo gairdneri*) showed a 4-fold increase in sensitivity between hard and soft water, the 48-h LC_{50} changing from about 80 to about 20 mg/litre (Brown, 1968).

In rainbow trout, a 48-h LC_{50} for nickel sulfate of 263 mg/litre was determined in a static test (Osterreichisches Forschungs-Zentrum Seibersdorf, 1983). The water had a hardness of 402 mg $CaCO_3$/litre, a pH of 7.6, and a temperature of about 15 °C. The first signs of toxicity were observed at a concentration of 85 mg nickel sulfate/litre.

Using the same test method and corresponding test conditions, Butz (1984) demonstrated that a decrease in water hardness from 270 to 49 mg $CaCO_3$/litre resulted in a 2.6-fold increase in toxicity.

Rehwoldt et al. (1971, 1972) studied the effects of temperature on the toxicity of nickel for 6 warm-water species of fish: banded killifish (*Fundulus diaphanus*), striped bass (*Roccus saxatilis*), pumpkin seed (*Lepomis gibbosus*), white perch (*Roccus americanus*), American eel (*Anguilla rostrata*) and carp (*Cyprinus carpio*). There was a wide range of sensitivity among these species, with 96-h LC_{50}

values at 17 °C ranging from 6.2 to 46.2 mg nickel/litre. However, each species showed very little variation in sensitivity at temperatures of 17 and 28 °C. At 28 °C and a hardness of 55 mg CaCO$_3$/litre, *Roccus saxatilis* and *Lepomis gibbosus* were the most sensitive, having 96-h LC$_{50}$s of 6.3 and 8.0 mg nickel/litre, respectively, while *Fundulus diaphanus* was the most tolerant (46.1 mg nickel/litre). *Roccus americanus*, *Anguilla rostrata*, and *Cyprinus carpio* were intermediate in their response, but relatively sensitive, the 96-h LC$_{50}$s being 13.7, 13.9, and 10.4 mg nickel/litre, respectively (Rehwoldt et al., 1972).

In static tests in softer water (20 mg CaCO$_3$/litre), *Pimephales promelas*, *Lepomis macrochirus*, *Carassius auratus*, and *Lebistes reticulatus* showed similar levels of sensitivity in terms of 96-h LC$_{50}$s with LC$_{50}$ values of 4.9, 5.3, 9.8, and 4.5 mg nickel/litre, respectively (Pickering & Henderson, 1966).

In a flow-through test, rainbow trout (*Salmo gairdneri*) were less sensitive than other fish species with a 48-h LC$_{50}$ of 20 mg nickel/litre, in soft water (Brown, 1968). In field studies, Hale (1977) reported a 96-h LC$_{50}$ for rainbow trout of 35.5 mg nickel/litre in continuous-flow tests in water with a hardness of 82–132 mg CaCO$_3$/litre. Arillo et al. (1982) found that rainbow trout (*Salmo gairdneri*), exposed to nickel, showed a reduction in glucidic stores. This effect is consistent with direct metal interaction on both membranes and enzyme thiolic groups of the pancreatic cells. Other effects, similar to those found with other metals, such as damage to the secondary lamellae of gills (Hughes et al., 1979) and sialic acid depletion in the gills (Arillo et al., 1982) have been described.

A 96-h LC$_{50}$ of 118.3 mg nickel/litre was determined for the marine grey mullet (*Chelon labrosus*) (Taylor et al., 1985a).

The toxicity of nickel(II) chloride was studied in 2 estuarine fish species (US EPA, 1987). In tidewater silverside (*Menidia peninsulae*) larvae, a 96-h LC$_{50}$ was 38 mg/litre. For adult spotfish (*Leiostomus xanthums*), the 96-h LC$_{50}$ was 70 mg/litre.

In short-term tests in soft water, the most sensitive species of freshwater fish were killed by exposure to concentrations of about 4–20 mg nickel/litre. Higher LC$_{50}$ values of nickel have been found for different species of fish in harder waters, ranging from about 30 to 80 mg nickel/litre. From the limited data available, it appears that hardness has the greatest effect on toxicity, while other determinants have not been proved to have any significant effects. In acute tests, there are interspecies differences in sensitivity, but these are within a single order of magnitude.

Shaw & Brown (1971) did not observe any effects on rainbow trout eggs fertilized in water containing 1 mg nickel/litre and then maintained in clean water. In a life-cycle study on fathead minnow, in water with a hardness of 210 mg $CaCO_3$/litre, pH 7.8, and an average temperature of 18 °C, Pickering (1974) found that nickel concentrations of 0.38 mg/litre and less (0.18 and 0.08 mg/litre) did not have any adverse effects on survival, growth, and reproduction. However, a concentration of 0.73 mg nickel/litre had a statistically significant effect on the number of eggs produced per spawning and on the hatchability of these eggs, though it did not affect the survival and growth of the first generation of fish.

In carp (*Cyprinus carpio*) eggs and larvae, the 72-h LC_{50}s were 6.1 and 8.4 mg/litre, respectively, while the 257-h LC_{50} for larvae was 0.75 mg nickel/litre in water with a hardness of 128 mg $CaCO_3$/litre, pH of 7.4, and a temperature of 25 °C. A concentration of 3 mg nickel/litre caused an increased incidence of abnormal larvae (23% compared with 8.6% in the controls) and 32.7% of embryos failed to hatch, compared with 5.9% in the controls (Blaylock & Frank, 1979).

Birge et al. (1978) exposed rainbow trout and largemouth bass (*Micropterus salmoides*) from fertilization to 4 days after hatching, to 11 trace metals found in coal, at a temperature of 12–13 °C, equivalent to a period of exposure of 28 days for rainbow trout and 8 days for bass. The LC_{50} values for these periods were 0.05 mg nickel/litre for rainbow trout and 2.06 mg nickel/litre for bass. The water used in the tests had a hardness of about 100 mg $CaCO_3$/litre and a pH of 7.2–7.8. For rainbow trout and goldfish, teratic larvae were observed at exposure levels that did not significantly affect egg hatchability. In water from a natural source with a pH of 7.8 and 174 mg $CaCO_3$/litre hardness, an LC_{50} for rainbow trout was 0.09 mg nickel/litre whereas in dechlorinated tap water (pH 7.6, hardness 125 mg $CaCO_3$/litre) the LC_{50} was 0.06 mg nickel/litre.

Using water with a hardness of about 100 mg $CaCO_3$/litre and pH 7.2–7.8, Birge & Black (1980) found that the LC_{50} from fertilization to 4 days after hatching was 0.71 mg nickel/litre for channel catfish (*Ictalurus punctatus*) and 2.78 mg nickel/litre for goldfish (*Carassius auratus*).

Calamari et al. (1982) found that, during the long-term exposure of fish to 1 mg nickel/litre, continuous uptake of nickel occurred for 180 days. The concentrations found were: liver, about 2.9 mg nickel/kg wet weight, kidneys, 4.0 mg/kg, and muscle, 0.8 mg/kg, while, at the beginning of the study, the levels had been 1.5, 1.5, and 9.5 mg nickel/kg, respectively. Toxicokinetic modelling indicated that theoretical asymptotic values for the liver, kidney, and muscle should be reached in 397, 313, and 460 days, respectively, at which times the calculated bioconcentration factors (BCF tissue concentration/environmental concentration) were 3.1, 4.2, and 1.9, respectively. Release was slower in clean water and the proportions of nickel remaining after 90 days in the liver, kidney, and muscle were 25, 41, and 31%, respectively. These data suggest that nickel had little capacity for accumulation in the tissues examined. However, even these relatively low concentrations are toxic (Arillo et al., 1982).

7.5 Terrestrial organisms

7.5.1 Plants

Nickel is ubiquitous in plant tissues. There is evidence that nickel is a required nutrient in a number of plant species. The urease enzyme of jack bean (*Canavalia ensiformis*) has been shown to be a nickel metalloenzyme (Dixon et al., 1975).

Although nickel levels above 50 mg/kg in plants are usually toxic, a number of plant species may tolerate higher levels (section 4.2.1).

In general, the effects of long-term, low-level exposure to nickel are only manifested in growth decrements with no visible signs. Nickel toxicity in plants is characterized by chlorosis and necrosis of the leaves, stunting of the roots, deformation of various plant organs, and wilting (Brooks, 1980; Prokipcak & Ormrod, 1986).

Apart from the solubility of nickel ions or nickel complexes, other factors can affect nickel toxicity. Of special interest is the presence of other heavy metals, which can act synergistically, and the ameliorating effects of calcium.

A synergistic effect of nickel and copper on the growth of bush beans was demonstrated by Wallace & Berry (1983). When barley was grown on loam soil with an elevated level of each of the 6 trace elements, lithium (13 mg/kg soil, dry weight), zinc (200 mg/kg), copper (200 mg/kg), nickel (100 mg/kg), and cadmium (100 mg/kg) there was no reduction in yield when they were applied singly. However, when all 6 were applied together, at the concentrations applied singly, there was a 40% reduction in yield, probably because of depressed phosphorus levels (Wallace et al., 1980a).

Investigations by Prokipcak & Ormrod (1986) of the growth responses of tomato and soy bean to combinations of nickel, copper, and ozone, indicated that the nature of the joint action of these chemicals is very complex and depends on species, the concentrations of the metals and ozone, and, to a lesser extent, the duration of exposure. In the first study, nickel was added to the nutrient solution of tomato and soy bean plants at 1.5, 7.5, or 37.5 mg/litre for 6 days, beginning on day 14 after seeding. The plants were then exposed to 0.15 or 0.30 µlitre atmospheric ozone/litre. Growth variables were markedly reduced by nickel, but ozone response depended on the nickel level. In the second study, 0.3 or 1.5 mg nickel/litre was provided from the 5th or 14th day onwards. There was little effect of duration of nickel treatment on growth. Increasing nickel levels and increasing ozone levels decreased growth, but there was no interaction. In the third study, treatments with 1.5 or 3.0 mg nickel/litre were combined with 3.0 or 6.0 mg copper/litre, prior to treatment with 0.25 ml atmospheric ozone/litre. There were complex interactive effects of all 3 compounds on tomato plant growth, but not on soy bean plant growth.

Calcium can reduce nickel toxicity. For example, when soy beans were grown in nutrient solution containing 1.2 mg nickel/litre, leaf yield depression was 74% at 4 mg calcium/litre, but only 45% at 400 mg calcium/litre (Wallace et al., 1980b).

7.5.2 Animals

Few data are available on the effects of nickel on terrestrial animals. Most data are derived from laboratory animals and indicate that nickel is an essential element in some species.

As land application of wastes is a common method of fertilization, studies were performed to evaluate the impact of heavy metals on the soil ecosystem, using the earthworm *Eisenia foetida* as a test organism. Following a 14-day exposure to nickel nitrate in artificial soil, the LC_{50} was calculated to be 757 mg/kg (Neuhauser et al., 1985). Hartenstein et al., (1981) determined the level at which added concentrations of heavy metals would cause an activated sludge to induce toxic effects on *Eisenia foetida*. Nickel was found to inhibit growth and to induce death at concentrations of 1200–12000 mg/kg dry weight. These concentrations seemed very high and the authors concluded that nickel might have been accumulated by the large population of microorganisms in the rich organic matrix, part of which might not be ingestible or digestible by earthworms. This would enable earthworms to grow in the presence of high nickel concentrations.

7.5.3 Essentiality of nickel for bacteria and plants

Evidence for specific biochemical functions of nickel has come from studies of microbial systems. Nickel is involved, in some way, in the "Knallgas" reaction, which is mediated by a number of bacteria of different genera (Tabillion et al., 1980; Friedrich et al., 1981; Albracht et al., 1982). The reduction of carbon dioxide to acetate, carried out by acetogenic bacteria, is dependent on nickel, which is needed to activate the enzyme carbon monoxide dehydrogenase (Diekert & Thauer, 1980; Drake, 1982). Diekert & Ritter (1982) demonstrated that *Acetobacterium woodii* growth on fructose was stimulated by, but not dependent on, nickel, unlike CO_2 reduction. A number of studies have established that nickel is the core metal in the tetrapyrrole ("Factor F_{430}"), found in methanogenic bacteria, and is essential for the growth of these organisms.

In plants, Dixon et al., (1975) showed that nickel is essential at the active site of urease in jack beans (*Canavalia ensiformis*), for its enzymatic activity.

7.6 Population and ecosystem effects

Few data are available that identify nickel as a specific cause for effects at the population level, because nickel is generally associated with other, often more toxic, trace metals or pollutants that could be involved in the effects.

Gradual ecological changes have been observed near sources emitting nickel and other trace metals, resulting in a decrease in the number and diversity of species (Hutchinson & Whitby, 1977; Gignac & Beckett, 1986). Yan et al. (1985) investigated 39 lakes in Ontario and found that the tracheophyte richness of acidic lakes decreased with increasing nickel and copper levels.

In acidic copper-, and nickel-contaminated lakes near Sudbury, Ontario, species richness and community biomasses were reduced in Crustacean zooplankton communities (Yan & Strus, 1980).

DeCantazaro & Hutchinson (1985a,b) demonstrated that the addition of nickel to microecosystems and incubated soil samples from boreal jack pine forests could disrupt nitrogen cycling. Nickel additions of 100–500 mg/kg soil were shown to stimulate nitrification and nitrogen mineralization, resulting in loss of nitrogen by leaching. The authors concluded that loss of nitrogen, which is probably the nutrient most limiting to growth in a boreal forest ecosystem, could have serious ecological consequences for forests in the vicinity of nickel smelters.

8. EFFECTS ON EXPERIMENTAL ANIMALS AND *IN VITRO* AND OTHER TEST SYSTEMS

Various studies have indicated that nickel is an essential element in a number of experimental animal species and that it may also have a physiological role in human beings. However, nickel deficiency has not been demonstrated in human beings, and the possible nickel requirement is probably very low. While the elucidation of nickel essentiality is in progress, it has not reached the stage where it can be quantified in relation to nickel deficiency.

8.1 Animals

8.1.1 Essentiality

Earlier studies in trace-element nutritional research did not demonstrate any consistent effects of nickel deficiency (Schroeder, 1968; Smith, 1969; Nielsen & Säuberlich, 1970; Wellenreiter et al., 1970; Nielsen & Higgs, 1971; Schroeder et al., 1974), in part, because of the technical difficulties of controlling nickel intake due to its ubiquity. Since 1975, diets and environments have been devised for adequately controlled studies on nickel metabolism and nutrition, and the effects of deprivation have been described for 17 animal species, including: chicken, cow, goat, mini-pig, pig, rat, and sheep.

Nickel is a component of several enzyme systems (certainly urease and some hydrogenases) and it seems essential for the well-being of several animal species (Spears et al., 1978).

8.1.1.1 Nickel deficiency symptoms

(a) Growth

In goats (Anke et al., 1977, 1978, 1980, 1986), pigs (Anke et al., 1977, 1986; Spears, 1984; Spears et al., 1984), and rats (Nielsen et al., 1975b; Schnegg & Kirchgessner, 1975a, 1980), a nickel-deficient diet resulted in significantly decreased growth. The growth depression depended on the nickel level and the duration of administration and only became evident after intrauterine nickel deficiency, i.e., in the second or later generations. In addition, species-specific differences seemed to exist (Anke et al., 1977).

(b) Reproduction and mortality

In goats, mini-pigs, and rats, reproduction was decreased only insignificantly by intra-uterine nickel deficiency (Anke et al., 1977; Schnegg & Kirchgessner, 1975a, 1980). Conception and abortion rates as well as the number of offspring were not influenced by nickel deficiency, but kidding in nickel-deficient goats and farrowing in nickel-deficient sows occurred later (Anke et al., 1974). Furthermore, at the end of the lactation period, significantly fewer offspring of nickel-deficient goats were still alive compared with control animals. Schnegg & Kirchgessner (1980) did not find any increase in mortality in intrauterinely nickel-deficient rats, whereas Nielsen et al., (1975b) found it to a remarkable extent. Smaller litter size has been observed in both rats and pigs (Anke et al., 1974; Nielsen et al., 1975b; Schnegg & Kirchgessner, 1975a).

(c) Histological parameters

Nielsen & Sauberlich (1970) described a nickel-deficiency syndrome in chickens, characterized by changes in the pigmentation of the shank skin, thicker bones, swollen joints, and a light-coloured liver. However, these findings are not consistent with those observed by other authors (Sunderman et al., 1972; Nielsen, 1974). Sunderman et al., (1972) and Nielsen & Ollerich (1974) observed ultrastructural lesions in the hepatocytes of chickens. In mini-pigs fed a nickel-deficient diet, Anke (1974) observed parakeratosis-like damage to the epithelium. Skin eruptions were also seen in nickel-deficient goats; the hair of the animals was brittle, there were fissures of the mouth and legs (Anke et al., 1976, 1980b). Offspring of nickel-deficient rats had an anaemic appearance (Schnegg & Kirchgessner, 1975a, 1980a; Nielsen et al., 1979a,b).

(d) Rumen activity

Nickel seems to be essential for ruminants, because urease activity in the rumen depends on nickel. Spears & Hatfield (1977) demonstrated disturbances in metabolic parameters in lambs maintained on a low-nickel diet, including reduced oxygen consumption in liver homogenate preparations, increased activity of alanine transaminase, decreased levels of serum proteins, and enhanced urinary nitrogen excretion. In a follow-up study, Spears et al., (1978) found that these animals had significantly lower microbial urease activity. It was possible to increase urease activity in the rumen contents by means of nickel supplementation.

(e) Disturbance of iron metabolism

Schnegg & Kirchgessner (1975b; 1976a,b) showed that nickel deficiency in rats led to a reduced iron content in organs, reduced haemoglobin and haematocrit values, and anaemia. Iron supplementation did not cure this anaemia (Nielsen et al., 1979a; Nielsen & Shuler, 1981), indicating a markedly impaired iron absorption. Nickel-deficient goats eliminated 33% more iron via the faeces than control animals (Anke et al., 1980). Spears et al., (1984) found that additional nickel could improve the iron status of neonatal pigs. The mechanism through which nickel might enhance iron absorption is still unclear. While nickel might act enzymatically to convert ferric to ferrous iron (a form more soluble for absorption), it might also promote the absorption of iron by enhancing its complexing with a molecule that can be absorbed (Nielsen, 1984).

(f) Nickel/calcium interaction

Anke (1974) found that nickel-deficient mini-pigs excreted more calcium renally than corresponding control animals. The skeletons of nickel-deficient animals contained less calcium than those of animals on a nickel-rich diet. Kirchgessner & Schnegg (1980a) confirmed this effect of nickel deficiency in 30-day-old rats and showed that more magnesium, instead of calcium, was incorporated into bones.

(g) Nickel/zinc interaction

Analysis showed that different organs and body fluids of nickel-deficient goats and pigs suffering from parakeratosis-like changes of the skin and hair, were not only poor in calcium, but also in zinc. There were single cases of dwarfism in goats (Anke, 1974; Anke et al., 1980, 1981). In rats, nickel deficiency also resulted in a significantly decreased zinc concentration in organs, demonstrated by the reduced size of the organs (Nielsen & Shuler, 1979; Kirchgessner & Schnegg, 1980a).

(h) Enzyme activities

The effects of nickel deficiency on enzyme activity have been studied in rats by Schnegg & Kirchgessner (1975b, 1977a,b,c,d) and Kirchgessner & Schnegg (1979, 1980b). They found that, as a rule, the activity of a number of dehydrogenases and transaminases decreased by 40–75% (malate dehydrogenase (MDH), isocitrate dehydrogenase (ICDH), lactic dehydrogenase (LDH), glucose-6-phosphate dehydrogenase (G6PDH), glutamate dehydrogenase (GLDH), glutamic oxalate transaminase (GOT), glutamic pyruvic transaminase (GPT)) with LDH and G6PDH being influenced by secondary iron deficiency. Kirchgessner & Schnegg (1979, 1980b) also measured a significant 50% reduction in the activity of α-amylases in the liver and pancreas. The results of other studies suggest that nickel may serve as a co-factor for the activation of calcineurin, a calmodulin-dependent phosphoprotein phosphatase (King et al., 1985).

(i) Substrate and metabolite concentrations

Nickel deficiency mainly affects carbohydrate metabolism, and this has been demonstrated in nickel-deficient rats by Schnegg & Kirchgessner (1977c). The glucose and glycogen contents of the liver in nickel-deficient rats was reduced by 90% and the triglycerides decreased by 40% compared with those in control animals. Similar values were found in the serum of the rats, and also in that of goats. Anke et al., (1980b) reported a reduced triglyceride level in the serum of ruminants, but the cholesterol level was unchanged. The significantly increased α-lipoprotein and reduced β-lipoprotein concentrations were probably connected with a normal cholesterol level, and a disturbance of triglyceride metabolism, because the α-fraction was rich in cholesterol and the β-fraction, rich in triglyceride.

The influence of nickel deficiency on the plasma cholesterol concentration and on the fat content of the liver has been studied (Nielsen, 1971; Sunderman et al., 1972; Nielsen et al., 1974, 1975a,b; Schnegg & Kirchgessner, 1977c). However, the results were inconsistent, because the cholesterol level was not affected in nickel-deficient animals. Nielsen (1971) found a decrease in the fat content of the liver in chickens, but Anke et al., (1977) did not find this in mini-pigs.

8.1.2 Acute exposures

8.1.2.1 Nickel carbonyl

Acute lethal concentrations of nickel carbonyl for laboratory animals are shown in Table 25. The lethal doses range from an LC_{50} of 0.1 mg nickel carbonyl/litre air for a 20-min inhalation exposure of the rat, to an LC_{50} of 2.5 mg/litre air for the dog following inhalation exposure of 30 min. The LD_{50}s via other routes range from 13 to 65 mg/kg, the intraperitoneal route being the most toxic.

Animals acutely exposed to nickel carbonyl vapour show pulmonary effects and lesions similar to those observed in human cases of industrial poisoning. The lung is the primary target organ for nickel carbonyl in animals, and pulmonary effects are rapid at high exposure levels, oedema occurring within 1 h of exposure. Subsequently, proliferation and hyperplasia of the bronchial epithelium and alveolar lining cells develop. Several days after exposure, severe intra-alveolar oedema with focal haemorrhage and alveolar cell degeneration occur. In animals that survive the acute effects, regression of cytological changes with fibroblastic proliferation within the alveolar interstitium occurs.

Pathological lesions in other organs after acute exposure of animals to nickel carbonyl are less severe than those in the lung. However, focal haemorrhage, congestion, oedema, hydropic degeneration, mild inflammation, and vacuolization have been reported in the brain, liver, kidney, adrenals, spleen, and pancreas. In hepatic parenchymal cells, dilatation of rough endoplasmic reticulum is the most prominent and consistent ultrastructural abnormality. Nucleolar alterations also develop in hepatocytes, 2–24 h after exposure to nickel carbonyl.

Pathological lesions of tubules and glomeruli have been seen in rats exposed to nickel carbonyl (Kincaid et al., 1953; Sunderman et al., 1961; Hackett & Sunderman, 1967).

Table 25. Acute toxicity studies of nickel carbonyl in experimental animals [a]

Species	Route	Lethal dose	Observations in surviving animals	Observation period after exposure	References
Rabbit Cat Dog	inhalation	LC_{80} = 1.4 mg/litre 50 min LC_{80} = 3.0 mg/litre 75 min LC_{80} = 2.7 mg/litre 75 min	*Lungs*: intra-alveolar haemorrhage, oedema, and exudate and alveolar cell degeneration *Adrenals*: haemorrhages *Brain*: perivascular leukocytosis and neuronal degeneration	1-5 days (rabbit)	Armit (1908)
Rat	inhalation	LC_{80} = 0.9 mg/litre 30 min	*Lungs*: at 2-12 h, capillary congestion and interstitial oedema, at 1-3 days, massive intra-alveolar oedema, at 4-10 days, pulmonary consolidation and interstitial fibrosis	2 h-several months	Barnes & Denz (1951)
Mouse Rat Cat	inhalation	LC_{50} = 0.067 mg/litre 30 min LC_{50} = 0.24 mg/litre 30 min LC_{50} = 0.19 mg/litre 30 min	*Lungs*: at 1 h, pulmonary congestion and oedema, at 12 h-6 days, interstitial pneumonities with focal atelectasis and necrosis, and peribronchial congestion;	0.2 h-6 days (rat)	Kincaid et al. (1953)

Table 25 (continued)

Species	Route	Dose	Effects	Time	Reference
Mouse Mouse	inhalation	$LC_{100} = 0.2$ mg/litre 120 min $LC_{100} = 0.01$ mg/litre 120 min	*Liver, spleen, kidneys, pancreas*: parenchymal cellular degeneration with focal necrosis		Kincaid et al. (1953) Sanotskii (1955)
Rat Rat	inhalation	$LC_{100} = 0.3$ mg/litre 20 min $LC_{50} = 0.1$ mg/litre 20 min			Ghiringhelli (1957)
Mouse Rat	inhalation	$LC_{80} = 0.048$ mg/litre 30 min $LC_{65} = 0.50$ mg/litre 30 min			West & Sunderman (1958)
Dog	inhalation	$LC_{90} = 2.5$ mg/litre 30 min			Sunderman et al. (1961)
Rat Dog	inhalation		*Lungs*: at 1-2 days, intra-alveolar oedema and swelling of alveolar lining cells, at 3-5 days inflammation, atelectasis, and interstitial fibroblastic proliferation; *kidneys and adrenals*: hyperaemia and haemorrhage	1-6 days 1-7 days	Sunderman et al. (1961)

Table 25 (continued)

Species	Route	Lethal dose	Observations in surviving animals	Observation period after exposure	References
Rat	inhalation	$LC_{50} = 0.51$ mg/litre; 30 min			Sunderman (1964)
Rat Rat Rat	intravenous subcutaneous intraperitoneal	$LD_{50} = 22$ mg/kg $LD_{50} = 21$ mg/kg $LD_{50} = 13$ mg/kg	*Lungs*: at 1-4 h, perivascular oedema, at 2-5 days severe pneumonitis with intra-alveolar oedema, haemorrhage, subpleural consolidation, hypertrophy and hyperplasia of alveolar lining cells, and focal adenomatous proliferation, at 8 days, interstitial fibroblastic proliferation, *Liver, kidneys, adrenals*: congestion, vacuolization, and oedema	1 h-21 days	Hackett & Sunderman (1967)
Rat	intravenous	65 mg/kg	*Lung*: ultrastructural alterations, including oedema of endothelial cells at 6 h and massive hypertrophy of membranous and granular pneumocytes at 2-6 days	0.5 h-8 days	Hackett & Sunderman (1968)

Table 25 (*continued*)

Mouse	inhalation	LC$_{100}$ = 0.1 mg/litre; 120 min	*Liver*: ultrastructural alterations of hepatocytes including nucleolar distortions at 2–24 h, dilatation of rough endoplasmic reticulum at 1–4 days, and cytoplasmic inclusion bodies at 4–6 days	0.5 h–6 days	Sanina (1968)

[a] From: NAS (1975)

8.1.2.2 Other nickel compounds

LD$_{50}$ data for some other nickel compounds are listed in Table 26.

Diarrhoea, respiratory distress, and lethargy were noted in rats and mice dying 2–3 hours after receiving nickel acetate or nickelocene by the oral or intraperitoneal route (Haro et al., 1968).

Benson et al. (1986) investigated the effects of single intratracheal doses of nickel subsulfide (3.2, 32, or 320 µg/kg body weight), nickel oxide (3, 30, or 300 µg/kg body weight), nickel sulfate (10.5, 105.2, or 1052 µg/kg body weight), and nickel chloride (9.5, 95.2, or 952 µg/kg body weight) on rats; 24 h after dosing, no effects were observed. However, at 7 days, multifocal alveolitis with some type II hyperplasia was observed in animals treated with nickel chloride, nickel sulfate, or nickel subsulfide. Lung lavage fluid contained increased numbers of neutrophils and macrophages in the medium- and high-dose groups of nickel chloride and nickel sulfate and in the highest nickel subsulfide dose group. While increased levels of enzymes, total protein, and sialic acid occurred in rats exposed to nickel chloride, nickel sulfate, or nickel subsulfide, no such changes were seen in rats receiving nickel oxide.

Effects on kidney function in rabbits were studied by Foulkes & Blanck (1984), who reported a reduction in the maximum tubular transport rate for aspartase following ip injection of 20 µmol nickel chloride/kg body weight. Intraperitoneal injection of 3 or 6 mg nickel/kg body weight in male Wistar rats induced a decrease in Bowman's space, dilated tubules, loss of brush border, flattened epithelia, and some regenerative activity (Sanford et al., 1988).

Intraperitoneal administration of NiCl$_2$ to mice and rats caused a rapid decrease of body temperature (Gordon,1989; Gordon & Stead, 1986; Gordon et al., 1989). The nickel chloride treatment resulted in hypothermia that lasted for more than 1 h, with a reduction in colonic temperature of 3–4 °C at 20 °C ambient temperature. The Ni^{2+}-induced hypothermia was accentuated at lower ambient temperature (10 °C) and ameliorated at higher ambient temperature (30 °C).

Hopfer & Sunderman (1988) monitored core body temperature and physical activity by radiotelemetry from a thermistor probe

Table 26. Acute toxicity of nickel compounds in experimental animals

Species (sex)		Substance	Route of administration	LD$_{50}$ (mg/kg) and confidence limits	References
Rat	(male)	nickel acetate	oral	360 (410-316)	Haro et al. (1968)
	(female)			350 (403-304)	
Mouse	(male)			410 (500-336)	
	(female)			420 (515-336)	
Rat			intraperitoneal	23 (28-19)	Haro et al. (1968)
Mouse				32 (37-28)	
Rat	(female)	nickel chloride	intraperitoneal	29	Horak et al. (1976)
	(pregnant female)		intramuscular	71	Sunderman et al. (1978a)
				98	
			intraperitoneal	38 (34-41)	Mas et al. (1985)
Rat	(male)	nickelocene	oral	490 (510-471)	Haro et al. (1968)
	(female)			500 (525-474)	
Rat			intraperitoneal	50 (59-42)	
Mouse			oral	600 (660-545)	
			intraperitoneal	86 (102-72)	

implanted in the peritoneal cavity. After injection of nickel chloride (250 μmol/kg body weight), core body temperature diminished to a minimum at 1.5 h and returned to baseline at 4 h; core body temperature at 1.5 h after dosing averaged 3.0 ±0.5 °C below the simultaneous value in control rats. During the 8–80 h following dosing, the mean body temperature of NiCl$_2$-treated rats did not differ from that of the controls, but the amplitude of the diurnal cycle of body temperature was dampened and the acrophase of the temperature cycle was delayed from 10.32 pm to 03.00 am. These parameters returned towards the control values during the 80–152 h period following dosing.

Acute thymic involution occurred in male Fischer 344 rats following a single subcutaneous injection of nickel chloride (500 μmol/kg body weight) (Knight et al., 1987). In nickel-treated rats, the mean thymic weight, which was significantly decreased at 24 h, continued to diminish at 48 h and reached 24% of the control value at 72 h. The concentration of lipoperoxides in the thymus increased 2-fold at 48 h and 7-fold at 72 h. Histological examination showed marked degenerative changes. Gitlitz et al., (1975) noted aminoaciduria and proteinuria in rats given single injections of nickel chloride (2 mg/kg body weight), the response being dose-dependent. Proteinuria was seen initially with aminoaciduria at higher doses. These effects, transitory in duration, were associated with morphological changes in the glomeruli.

8.1.2.3 Possible mechanisms of acute nickel toxicity

The mechanisms of nickel toxicity are not well understood, but studies by Knight et al. (1987), Sunderman et al. (1987), and Sunderman (1987) suggest that acute Ni^{2+} toxicity in rats is associated with lipid peroxidation in target organs. The chemical reactions whereby Ni^{2+} induces lipid peroxidation *in vivo* have not yet been explained, however, Sunderman (1987) proposed the following four possible mechanisms:

i. An indirect mechanism owing to Ni^{2+} displacement of iron and copper from intracellular binding sites;

ii. An indirect mechanism, by which Ni^{2+} inhibits cellular defences against peroxidative damage, mediated by catalase, superoxide dismutase, glutathione peroxidase, aldehyde dehydrogenase, or other enzymes that protect against free-radical injury or that metabolize products of lipid peroxidation;

iii. Generation of oxygen-free radicals by the redox couple:

Ni^{2+}/Ni^{3+}

$Ni^{2+} + H_2 \rightarrow Ni^{3+} + OH^- + OH\cdot$ (Fenton reaction)

$Ni^{3+} + O^-_2 \rightarrow Ni^{2+} + O_2$

$H_2O_2 + O^-_2 \rightarrow OH^- + OH\cdot + O_2$ (Haber-Weiss reaction)

iv. Ni^{2+} may accelerate the degradation of lipid hydroperoxides to form lipid-oxygen radicals, propagating autocatalytic peroxidation of polyenoic fatty acids.

Experiments performed by Inoue & Kawanishi (1989) using electron spin resonance (ESR) (spin traps 5,5-dimethylpyroline-N-oxide and α-(4-pyridyl 1-oxide)-N-tert-butylnitrone) indicate that hydroxyl radical adducts are produced *in vitro* by the decomposition of hydrogen peroxide in the presence of the nickel (II) oligopeptide Gly Gly His. These investigators suggested that Gly Gly His plus hydrogen peroxide produce superoxide in addition to the hydroxyl radical. The experimental findings support the conclusion that the nickel-dependent formation of an activated oxygen species is a primary molecular event in acute nickel toxicity and carcinogenicity.

8.1.3 Short- and long-term exposures

8.1.3.1 Effects on the respiratory tract

(*a*) In vivo *studies*

Data on the chronic respiratory effects of nickel carbonyl are summarized in section 8.1.2.1.

Long-term inhalation studies have been performed on guinea-pigs, rats, and mice (Hueper, 1958), rats (Bingham et al., 1972), and hamsters (Wehner et al., 1975, 1981). Exposure extended over more than one and a half years in all studies. The compounds tested were metallic nickel powder, nickel subsulfide, nickel oxide, and nickel-enriched fly ash. Hueper (1958) exposed animals to metallic nickel dust at 15 mg/m^3, and noted nasal sinus inflammation and ulcers in rats, and signs of lung irritation in guinea-pigs and rats. A common finding was an accumulation of adenomatoid cell formations. In mice, there were signs of lung irritation, but not to the same extent as in guinea-pigs and rats.

Bingham et al., (1972) used aerosols of soluble nickel chloride at 109 μg/m^3 and nickel oxide at 120 μg/m^3 and observed hyperplasia of bronchiolar and bronchial epithelium with peribronchial lymphocyte infiltrates.

Ottolenghi et al. (1974) found a number of lung changes, such as abscesses as well as metaplastic changes, when rats were exposed for 78 weeks to nickel subsulfide by inhalation. Wehner et al. (1975) exposed rats to a concentration of 53 mg nickel oxide/m^3 (type of oxide not specified) for the life span and found that particulate material accumulated on the alveolar septa. Emphysema was observed early in the exposure period. With longer exposure, the cellular response increased and pneumoconiosis developed gradually. However, there was no reduction in the life span. Wehner et al. (1981) exposed hamsters through inhalation to nickel-enriched fly ash or fly ash at concentrations of 17 or 70 μg/m^3 for up to 20 months. Lung weights and volumes were significantly increased in the 70 μg/m^3 fly ash exposure group. The severity of anthracosis, interstitial reaction, and bronchiolization was dose-dependent.

Friberg (1950) exposed rabbits for 6 months to nickel-graphite dust (nickel hydrate, nickel content approx. 50%) at a concentration of 100 mg/m^3 (3 h/day, 5 days/week) and found emphysematous and inflammatory changes in the nasal mucous membranes and the trachea, bronchitis, and sometimes slight fibrosis in the lung.

Port et al. (1975) reported that intratracheal injection of a suspension of nickel oxide (5 mg, particle size < 5 μm, type of nickel oxide not specified) into Syrian hamsters, treated 48 h previously with influenza A/PR/8 virus, resulted in significantly increased mortality compared with controls. Surviving animals at this and lower doses showed mild to severe acute interstitial infiltration by polymorphonuclear cells and macrophages, several weeks later. Additional pathological changes included bronchial epithelial hyperplasia, focal proliferative pleuritis, and adenomatosis.

Short-term inhalation studies (12 days) with nickel sulfate (3.5–60 mg/m^3) and nickel subsulfide (0.6–10 mg/m^3) were performed on rats and mice (Benson et al., 1987, 1988). Nickel sulfate caused lesions in the lung, nose, bronchial, and mediastinal lymph nodes in the surviving animals at 3.5 mg/m^3. In the nickel subsulfide-exposed animals, similar changes occurred in the respiratory tract with extensive lesions in the lung, including necrotizing pneumonia. Emphysema developed in rats exposed to 5 or 10 mg/m^3 and fibrosis was seen in mice exposed to 5 mg nickel subsulfide/m^3. Degeneration of the respiratory epithelium and atrophy of the olfactory epithelium occurred in all nickel subsulfide dose groups except in mice exposed to 0.6 mg/m^3. Clinical signs included laboured respiration, emaciation, dehydration, and reduced body weight gain in rats and mice exposed to nickel sulfate or nickel subsulfide. A 12-day inhalation exposure of rats and mice to nickel oxide (1.2–30 mg/m^3) also caused lung lesions in the higher dose groups. Lung lesions included hyperplasia of alveolar macrophages, focal suppurative inflammation, focal interstitial cellular infiltrate and particles in alveoli and alveolar macrophages. In mice, lung lesions were less severe (Dunnick et al., 1988). On the basis of these 12-day studies, relative toxicity ranking was $NiSO_4 \cdot 6H_2O > Ni_3S_2 > NiO$ (Dunnick et al., 1988).

The effects of nickel on the cellular respiratory system defence mechanisms were studied by exposing guinea-pigs, rats, or rabbits to various concentrations (0.05–2 mg/m^3, for 1–8 months) of nickel dust, nickel oxide, or nickel chloride (Waters et al., 1975; Graham et al., 1975a; Aranyi et al., 1979; Johansson et al., 1980, 1981, 1983a,b; Castranova et al., 1980; Casarett-Bruce et al., 1981; Lundborg & Camner, 1982, 1984; Wiernik et al., 1983; Murthy et al., 1983; Takenaka et al., 1985; Glaser et al., 1986). It was concluded that the overall effects had some features in common with the pulmonary alveolar proteinosis described in human beings. There was no difference in the effect pattern between exposure to insoluble and soluble nickel compounds. Effects, such as changes in the morphology and function of alveolar macrophages and type II alveolar epithelial cells, and in the composition of lung lavage, were found, with the severity of effects depending on the concentration and duration of exposure.

The phagocytic activity of alveolar macrophages was increased in rabbits following 4 weeks inhalation of metallic nickel dust (0.5–2.0 mg/m^3; Camner et al., 1978; Yarstrand et al., 1978) and in rats following 1–4 months inhalation of nickel oxide (produced by pyrolysis of nickel acetate at 550 °C) (Spiegelberg et al., 1984). No change in the phagocytic activity of alveolar macrophages was observed in rats following exposure to 0.13 mg metallic nickel dust/m^3 for 4–8 months (Johansson et al., 1983a), and in rabbits following exposure to nickel chloride aerosols (0.3 mg/m^3) for 1 month (Wiernik et al., 1983).

Alveolar macrophages varied in size and, ultrastructurally, had an active cell surface with numerous slender microvilli and long protrusions. Laminated structures were regularly found in the alveolar macrophages as well as in the lung fluid (Camner et al., 1978; Johansson et al., 1980, 1983a; Wiernik et al., 1983). Spiegelberg et al., (1984) found an increase in size and number of polynucleated cells in the lungs of exposed rats. Morphometric studies on the lungs of exposed rabbits showed an increase (up to 3-fold) in the volume density of type II cells. Cells contained many lamellar bodies and vesicles; the endoplasmic reticulum had a slightly dilated appearance (Johansson & Camner, 1980; Johansson et al., 1981, 1983b). There were changes in the pulmonary lipid content and composition of the lung fluid of rabbits inhaling metallic nickel dust at levels ranging from 0.13 to 1.7 mg/m^3 (Casarett-Bruce et al., 1981; Curstedt et al., 1983, 1984). There was a twofold increase in the concentration of total phospholipids, mainly due to an elevated level of phosphatidylcholines, especially disaturated species, and phosphatidylinositols. Lundborg & Camner (1982, 1984) exposed rabbits to nickel dust (0.1 mg/m^3, for 4 and 6 months) and nickel chloride (0.2–0.6 mg/m^3 for 4–6 weeks) and found markedly reduced lysosomal enzyme activity compared to control animals. When rats were exposed to aerosols of nickel oxide (type of oxide not specified) at 120 μg/m^3 or nickel chloride at 109 μg/m^3, Murthy et al., (1983) found that various hydrolytic enzymes were reduced in alveolar macrophages, but, in contrast, the enzyme activities were significantly increased in lung lavage.

Parenteral administration of nickel chloride can also cause toxic effects in alveolar macrophages, as demonstrated by Sunderman et al., (1989c) in rats that had received single subcutaneous injections of 8–65 mg (62–500 μmol) nickel chloride/kg. Alveolar macrophages showed morphological and biochemical signs of activation, functional impairment, and lipid peroxidation. In alveolar macrophages from ^{63}NiCl$_2$-treated rats, ^{63}Ni was primarily located in the cytoplasm bound to high- and low-relative molecular-mass constituents.

At the ciliated epithelium level, nickel significantly depressed normal ciliary activity in a hamster tracheal ring assay following *in vivo* exposure to 100 μg nickel/m^3 as nickel chloride for 2 h (Adalis et al., 1978; Olsen & Jonsen, 1979b).

Fisher et al. (1984) evaluated the effects of particle size and dose regimen on the toxicity of intratracheally instilled nickel subsulfide in mice, and showed the importance of physical form in the evaluation of pulmonary toxicity. The median lethal dose for a single exposure to larger particles (MMAD 13.3 μm) was 12 times that of fine particles (MMAD 1.8 μm), i.e., 50 versus 4 mg/kg. Repeated exposures (once/week for 4 weeks) resulted in a 2-fold greater median lethal dose of coarse particles compared with fine particles, i.e., 2 versus 1 mg/kg. Thus, repeated exposure to fine particles resulted in a cumulative lethality equivalent to the single exposure, while coarse particle lethality was enhanced with repeated exposure. Alveolar macrophages from mice exposed to fine nickel subsulfide showed depressed cellular function, 14 days after a single administration.

Reichrtova et al., (1986a,b) performed 2 inhalation studies on rats to compare the effects of different types of exposure on alveolar macrophages. In both studies, the nickel content of the aerosol was only 0.36% (NiO), thus, the effects are unlikely to be nickel-specific responses. Increases in the alveolar macrophage count and lysosomal enzyme activities (acid phosphatase and β-glucuronidase) were found in Wistar rats, exposed to an aerosol of solid waste from a nickel refinery dump, under chamber conditions, for 6 months (aerosol concentration 0.1 g/m^3, 4 h/day, 5 days/week). However, the activity of alveolar macrophage plasma membrane enzymes decreased. Cells contained in lung lavage had a pleomorphic appearance. When rabbits were environmentally exposed at a biomonitoring station, situated in the direction of the prevailing wind from a nickel refinery dump, for 6 months, with an average dust fallout of 5.5 g/m^2 in 30 days, the number of alveolar macrophages was significantly increased (Reichrtova et al., 1986b). A significant increase in lysosomal enzyme activity also occurred, but these changes were not as pronounced as the effects obtained in the chamber studies on rats. In contrast to the increased alveolar macrophage activity, a significant reduction occurred in antibody-mediated rosette formation by alveolar macrophages in the environmentally exposed rabbits.

(b) In vitro *cytotoxicity studies*

Alveolar macrophages exposed to nickel (1.1 mmol/litre) *in vitro* showed depressed phagocytic activity (Graham et al., 1975a; Aranyi et al., 1979). Waters et al. (1975) correlated changes in cell viability with changes in the morphology and the specific activity of a lysosomal enzyme (acid phosphatase). At 4.0 mmol/litre, cell viability was reduced to approximately 50% of that of controls. Castranova et al., (1980) demonstrated that nickel affected oxygen metabolism in the rat alveolar macrophages; at rest, and during phagocytosis, oxygen consumption and glucose metabolism were depressed.

The *in vitro* cytotoxicity of nickel chloride on a human pulmonary epithelium cell line (A549) was reported by Dubreuil et al. (1984). Nickel chloride in amounts ranging from 0.1 to 1.0 mmol/litre produced decreases in cell growth rate and in the levels of cell adenosine triphosphate (ATP), and loss of viability, at the highest concentration. No changes were seen in transmission electron microscopy preparations.

In vitro toxicity studies on bovine alveolar macrophages indicated that nickel subsulfide was 10 times more toxic than solubilized nickel subsulfide or soluble nickel chloride (Fisher et al., 1984).

Nickel subsulfide, nickel sulfate, nickel chloride and nickel oxide were tested for their relative toxicity in beagle dog and rat alveolar macrophages *in vitro*. Toxicity ranking was $Ni_3S_2 > NiCl_2 \approx NiSO_4 > NiO$ (Benson et al., 1986).

8.1.4 Relationship of nickel toxicity and mixed metal exposure

As both nickel production and some end uses of nickel involve mixed exposure of workers to nickel and copper, nickel and chromium, nickel, chromium, and manganese, and so on, the problem of the combined toxic effects of these metals is of great practical importance. However, the relevant information is rather scarce.

Johansson et al. (1988) showed that the cytotoxic effects of a trivalent chromium salt on alveolar macrophages of the rabbit impaired their dealing with pulmonary surfactant, the latter being hyperproduced as a result of nickel's action on the type II alveolar epithelial cells. However, analysis of these experimental data showed that, as far as the cytotoxicity of these metals for alveolar macrophages was concerned, their joint effect was antagonistic.

Davydova et al. (1981) demonstrated that subadditivity or even clear antagonism was the main type of combined acute toxicity for rats of nickel and chromium, nickel and manganese, and even a triple exposure of rats to nickel and cobalt. On the contrary, the long-term exposure of rats to nickel and cobalt in the drinking-water demonstrated the additive long-term toxicity of these two metals (Nadeenko et al., 1988).

8.1.5 Endocrine effects

Bertrand & Macheboeuf (1926a,b) reported that parenteral administration of nickel chloride or nickel sulfate to rabbits or dogs antagonized the hyperglycaemic action of insulin. Later investigators observed that after parenteral (iv or ip) injection to rabbits, rats, or chickens, or oral administration to rabbits, there was a rapid increase in plasma glucose concentrations, which returned to normal within 4 h (Kadota & Kurita, 1955; Gordynia, 1969; Clary & Vignati, 1973; Freeman & Langslow, 1973; Horak & Sunderman, 1975a,b). In the pancreatic islets of Langerhans, Kadota & Kurita (1955) noted marked damage of α-cells, and, to a lesser degree, degranulation and vacuolization of β-cells. Ashraf & Sybers (1974) noted lysis of pancreatic exocrine cells in rats fed nickel acetate (0.1%). In adrenalectomized or hypophysectomized rats, the hyperglycaemic effect was greatly depressed, but was not completely prevented. Concurrent administration of insulin antagonized the hyperglycaemic effect (Horak & Sunderman, 1975a,b). LaBella et al., (1973a,b) showed that nickel also affects the hypothalamus of animals. Nickel salts specifically inhibited the release of prolactin *in vivo* (in the rat) and *in vitro* from bovine pituitary. Subcutaneous injection of 10 or 20 mg nickel chloride/kg in rats initially produced a drop in serum prolactin over the short term, but resulted in a sustained elevation of the hormone after 1 day, which lasted up to 4 days. The elevation was due to reduced levels of the prolactin-inhibiting factor (Clemons & Garcia, 1981). Carlson (1984) demonstrated that nickel antagonized the stimulation of both prolactin and growth hormone by barium; thus, the basis of antagonism may be the competitive inhibition of calcium uptake. Dormer et al., (1973) showed that the nickel ion is a potent inhibitor of secretion *in vitro* in the parotid gland, (amylase), the islets of Langerhans (insulin), and the pituitary gland (growth hormone). Inhibition of growth hormone secretion at nickel concentrations comparable to those observed by LaBella et al., (1973b) to enhance release, may reflect differences in tissue preparation prior to assay. Dormer et al., (1973) suggested that nickel may block exocytosis by interfering with either secretory granule migration or membrane fusion and microvilli formation.

The effects of nickel on thyroid function were reported by Lestrovoi et al. (1974). Nickel chloride, given orally (0.5–5.0 mg/kg per day, for 2–4 weeks) or by inhalation (0.05–0.5 mg/m^3) to rats, significantly decreased iodine uptake by the thyroid, the effect being more pronounced with inhaled nickel.

8.1.6 Cardiovascular effects

The serum nickel level increased in patients with acute myocardial infarction, stroke, and burns (D'Alonzo & Pell, 1963; McNeely et al., 1971; Leach et al., 1985).

Rubanyi & Kovach (1980) found that micromolar concentrations of nickel chloride (0.1–1 μmol/litre) increased cardiac contractility in the isolated rat heart. At higher doses, there was depressed myocardial contractile performance. Ultrastructural damage was found (Rubanyi et al., 1980).

Nickel chloride at a concentration of 1 μmol/litre in the perfusate produced tonic contraction in the isolated canine coronary artery (Rubanyi et al., 1982b).

Nickel is released from the ischaemic dog myocardium (Rubanyi et al., 1981); exogenous nickel, in doses comparable to the amount released endogenously from the heart, induced coronary vasoconstriction in rat and dog hearts. In further studies, the possible involvement of Na/K ATPase inhibition (Rubanyi et al., 1982c) and/or stimulation of adrenergic receptors in the coronary vessels (Rubanyi et al., 1982d) were discussed as mechanisms of nickel-induced coronary vasoconstriction. Rubanyi & Inovay (1982) studied the effects of nickel ions on spontaneous, electrically-, and norepinephrine-stimulated isometric contractions in the isolated portal vein of the rat. They found that low concentrations of nickel (1–10 μmol/litre) inhibited spontaneous isometric force development and decreased basal tone, but significantly increased the frequency of contractions. Inhibition of the effect of selective stimulation of adrenergic nerves was significantly more pronounced than the depression of contractions evoked by exogenous norepinephrine.

In the *in situ* heart (open-chest anaesthetized dogs), a decrease in coronary vascular flow with a low intravenous dose of nickel (0.02 mg nickel chloride/kg body weight) was reported, while, at higher dose levels (0.2, 2, or 20 mg nickel chloride/kg body weight), further reduction in coronary blood flow, depression of heart rate, and a decrease in left ventricular contractility were observed. Coronary vasoconstriction may be regarded as a local action of nickel on coronary vessels (Ligeti et al., 1980).

In another *in situ* study on dogs (intravenous bolus injection of 0.02 mg nickel/kg, or intracoronary infusion of 0.04 mg nickel chloride/min per kg body weight), Rubanyi et al. (1984) found that vasoconstriction was induced when coronary arteries were dilated by low-flow ischaemia, arterial hypoxaemia, and adenosine infusion. Nickel inhibited post-occlusion reactive hyperaemia and vasorelaxation in response to arterial hypoxaemia or intracoronary infusion of adenosine. It was postulated that vasoactivity might be related to the existence of positive feedback loops triggered by alterations in the level of nickel.

Endogenous nickel release from damaged tissues and its implications for ischaemic heart disease have been examined with respect to the pathology of acute carbon monoxide poisoning and acute burn injury. Significant endogenous nickel ion accumulation was noted in the heart muscle of rats and rabbits, when the CO-Hb level was above 30% (Balogh et al., 1983).

8.1.7 Effects on the immune system

The effects of nickel on alveolar macrophages have been described in section 8.1.3.1.

Koller (1980) noted that nickel exposure of animals could reduce host resistance to both viral and bacterial infections, and suppress the phagocytic capacity of macrophages.

Other cellular and humoral immune responses following nickel treatment were studied by Smialowicz et al. (1984, 1985). Single or multiple intramuscular injections of nickel chloride in mice caused a significant reduction in a variety of T-lymphocytes and natural killer cell-mediated immune functions. Suppression of the lymphoproliferative responses to the T-cell mitogens, phytohaemagglutinin and concanavalin A, and a reduction in the number of theta-positive T-lymphocytes were observed in the spleens of nickel chloride-injected mice (18.3 and 36.6 mg/kg body weight). Reductions in the primary antibody response to T-lymphocyte-dependent antigen sheep red blood cells, but not T-lymphocyte-independent antigen polyvinyl-pyrrolidone, were observed following a single injection of 18.3 mg nickel chloride/kg (Smialowicz et al., 1984).

Smialowicz et al., (1985) demonstrated that suppression of natural killer cell activity could be detected by *in vitro* and *in vivo* assays and that reduction of natural killer cell activity was not associated with either a significant reduction in spleen cellularity or the production of suppressor cells. A further demonstration of the effect of nickel chloride on natural killer cell activity was the enhancement of the development of lung tumour colonies in mice injected with B16-F10 melanoma cells, following a single injection of 18.3 mg nickel chloride/kg. Unlike nickel chloride, manganese chloride was found to enhance natural killer cell activity, when injected into mice (Smialowicz, 1985). No alteration in natural killer cell activity was observed in mice injected with magnesium, calcium, or zinc (Smialowicz et al., 1987). Manganese chloride was considered to have an antagonistic effect on nickel chloride-suppression of natural killer cell activity.

The effects of nickel compounds on natural killer cells are of particular interest, because of the suspected function of these cells in nonspecific defence against certain types of infections and tumours.

The studies performed by Smialowicz et al. (1984, 1985, 1986, 1987) confirmed the findings of other investigators on the immunosuppressive effects of nickel salts on circulatory antibody titres to T_1 phage in rats (Figoni & Treagan, 1975), on antibody response to sheep erythrocytes (Graham et al., 1975b), on interferon production *in vitro* (Treagan & Furst, 1970) and *in vivo* in mice (Gainer, 1977), and on susceptibility to induced pulmonary infection in mice following inhalation of nickel chloride (Adkins et al., 1979).

In cynomolgus monkeys that had been previously immunized and repeatedly challenged with sheep red blood cells, instillation of 10.6 mg nickel subsulfide/kg lung in one immunized and one control lobe of each animal increased target cell killing by conjugate-forming natural killer cells and decreased macrophage phagocytic activity (Haley et al., 1987).

The results of *in vitro* studies have shown that nickel can replace magnesium, which is essential for the proper functioning of the complement system, in both the classical and the alternative pathway. Nickel has been shown to result in a more efficient formation of the C3b,Bb enzyme, which theoretically may lead to disturbance of a well balanced complement cascade. The significance of this finding *in vivo* is not known (Fishelson et al., 1983).

8.1.8 Skin and eye irritation and contact hypersensitivity

8.1.8.1 Skin and eye irritation

Repeated skin application of 40–100 mg nickel/kg (as nickel sulfate), daily for 30 days, produced skin atrophy, acanthosis, and hyperkeratinization in rats (Mathur et al., 1977). No data are available on the eye irritancy of soluble nickel salts or on the skin and eye irritancy of insoluble nickel compounds.

8.1.8.2 Contact hypersensitivity

Experimental sensitization to nickel in guinea-pigs has been reported (Walthard, 1926; Stewart & Cromia, 1934; Vinson & Choman, 1960; Jansen et al., 1964; Gross et al., 1968; Magnusson & Kligman, 1970). Nilzén & Wikström (1955) reported a method for sensitizing laboratory animals to nickel by repeated topical applications of aqueous nickel sulfate solutions containing sodium lauryl sulfate. However, Samitz & Pomerantz (1958) were unable to demonstrate sensitization with this technique and attributed the effect to local irritation, rather than true allergenic reaction. Samitz et al., (1975) were unable to induce sensitization in guinea-pigs using nickel compounds from the complexation of the nickel ion with amino acids or guinea-pig skin extracts. Furthermore, sensitization of experimental animals was not found by Hunziker (1960) or Bühler (1965).

The sensitivity of giunea-pigs was increased by repeated intradermal injections and skin painting with nickel sulfate solutions during the sensitization phase. The responses were significantly greater than in control animals (Wahlberg, 1976).

Turk & Parker (1977) reported sensitization to nickel manifested as allergic-type granuloma formation. This required the use of Freund's complete adjuvant followed by weekly intradermal injections of 25 µg of the salt after 2 weeks. Delayed hypersensitivity reactions developed in 2 out of 5 animals at 5 weeks when a split-adjuvant method was used. Suppression of the delayed hypersensitivity occurred when intratracheal intubation of nickel sulfate was also performed on these animals (Parker & Turk, 1978). Möller (1984) reported that mice could easily be sensitized to a potent antigen, such as picryl chloride, but response to nickel could only be achieved by repeated epicutaneous application of a strong (20%) nickel salt solution for 3 weeks.

The study of the allergic properties of nickel in experimental animals is a problem, because of difficulties involved in reproducing the phenomenon of allergodermatosis and because of lack of a uniform approach to reproducing the experimental contact allergic dermatitis model. Duyeva (1983) recommended a single intracutaneous administration of 100–200 μg nickel chloride in the ear of the guinea-pig. Sensitization developed as early as days 4–10, reaching a peak on day 20.

8.1.9 Reproduction, embryotoxicity, and teratogenicity

8.1.9.1 Effects on the male reproductive system

Data on the effects of nickel on the male reproductive system are limited. Hoey (1966) examined the effects of nickel sulfate on the testis and epididymis of Fischer rats and reported histological changes in the testis and adnexae. Following a single intracutaneous injection of 0.04 μmol nickel sulfate/kg body weight, histological examination, 18 h after exposure, revealed shrinkage of the tubules and complete degeneration of the spermatozoa. Infertility was observed in rats after 120 days of daily ingestion of 25 mg nickel sulfate/kg (Waltschewa et al., 1972).

Mathur et al., (1977) studied the dermal exposure of male rats to nickel sulfate at daily levels of 40, 60, or 100 mg nickel/kg body weight for 15 and 30 days. Tubular damage and spermatozoal degeneration were observed in the testis following exposure to 60 mg nickel/kg for 30 days. These changes were more severe with exposure at 100 mg/kg for 30 days. There were no effects on the testis following exposure to 40 mg/kg for 30 days or at any dose level after 15 days exposure.

In vitro embryo cultures were used to study the effects of nickel nitrate on male germ cells (Jacquet & Mayence, 1982). BALB/C mice were injected intraperitoneally with 40 or 56 mg nickel nitrate/kg body weight and were then allowed to mate with superovulated females, at weekly intervals, for 5 weeks following treatment. The embryos were isolated and those at the 2-cell stage were cultured for 3 days. These embryos were then classified according to their ability to develop to the blastocyst stage. A dose of 40 mg/kg body weight did not affect the fertilizing capacity of the spermatozoa or the ability of the fertilized egg to cleave, but a dose of 56 mg nickel nitrate/kg body weight yielded a significant proportion of uncleaved unfertilized eggs. Cleaved eggs from this treatment group were able to develop into blastocysts, suggesting that nickel nitrate treatment reduced the fertilizing capacity of the spermatozoa, presumably because of a toxic effect of nickel on spermatogenesis. Deknudt & Leonard (1982) conducted a dominant lethal mutation test for nickel chloride and nickel sulfate in BALB/C mice. Both compounds produced a reduction in implantations in the matings performed 2, 3, or 4 weeks after treatment, but the post-implantation loss was not increased. It was considered that this was attributable to a toxic effect of nickel treatment on male germ cells.

8.1.9.2 Effects on the female reproductive system

Insertion of nickel wire into one uterine horn of rats on day 3 of pregnancy produced a decrease in the number of implants and an increase in the number of resorption sites, compared with the untreated contralateral horn (Chang et al., 1970).

Nickel has been reported to localize within the pituitary and the hypothalamus of rats and to inhibit prolactin secretion (La Bella et al., 1973 a,b). It may, therefore, modify interactions between the hypothalamus and the pituitary, needed to maintain pregnancy.

No effects on fertility were observed in 3 generations, when breeding rats were exposed to 5 mg nickel/litre as a soluble salt (not specified) in the drinking-water (Schroeder & Mitchener, 1971).

8.1.10 Embryotoxicity and teratogenicity

Nickel has been shown to cross the placental barrier and enter the fetus (section 6.1.5).

Sunderman et al. (1978a) studied the embryo- and fetotoxicity of nickel chloride and nickel subsulfide in Fischer 344 rats. Single intramuscular injection of nickel chloride (12 or 16 mg nickel/kg body weight) on day 8 of gestation significantly reduced the mean number of live pups per dam and resulted in reduced body weight in fetuses on day 20 of gestation and in pups, 4–8 weeks after birth. When nickel chloride was administered in repeated intramuscular doses of 1.5 or 2.0 mg/kg body weight on days 6–10 of gestation, the higher dose caused significant intrauterine mortality, but did not cause any reduction in the mean body weight of live pups.

Intramuscular injection of nickel subsulfide (80 mg nickel/kg body weight) reduced the mean number of live pups per dam. Exposure to nickel chloride or nickel subsulfide did not produce any skeletal or visceral anomalies.

Lu et al. (1979) observed a dose-related increase in fetal deaths and a higher incidence of malformations in pregnant CD-1 mice following intraperitoneal injection of single doses of nickel chloride (1.2, 2.3, 3.5, 4.6, 5.7, or 6.9 mg nickel/kg) between days 7 and 11 of gestation. Fetal death and some general malformation (not described in detail) were reported in hamsters after intravenous injection of nickel acetate at 30 mg/kg body weight on day 8 of pregnancy (Ferm, 1972). Nadeenko et al., (1979) demonstrated a dose-dependent embryotoxic effect of ^{27}Ni given to female rats for 7 months (before and during pregnancy) in concentrations ranging from 5×10^{-1} to 5×10^{-4} mg/kg body weight.

The effects of nickel chloride on early embryogenesis were studied by Storeng & Jonsen (1980) *in vitro*. Mouse embryos at the 2-, 4-, and 8-cell stages were cultured in media containing nickel chloride hexahydrate at a concentration of 10–1000 μmol/litre. A concentration of nickel chloride hexahydrate of 10 μmol/litre adversely affected the development of 2-cell stage embryos whereas a concentration of 300 μmol/litre was needed to affect 8-cell stage embryos. No effect was observed at 100 μmol/litre. In a subsequent study (Storeng & Jonsen, 1981), a single intraperitoneal injection of nickel chloride hexahydrate (20 mg/kg) was given to mice on days 1–6 of gestation. On day 19, implantation frequency in females treated with nickel chloride hexahydrate on day 1 was significantly lower compared with the controls. Litter size was significantly reduced in females treated on days 1, 3, and 5 of gestation. The incidence of abnormalities, such as haematomas and exencephaly, in the fetuses of treated females was higher (statistical significance not reported) than in the controls.

Sunderman et al., (1983) used intrarenal injection of nickel subsulfide to assess the effects on the progeny of rats. Administration of 30 mg nickel subsulfide/kg by intrarenal injection, 1 week prior to breeding, produced intense erythrocytosis in the dams, but not in the pups. These findings indicate that the release of maternal erythropoietin by the maternal kidneys, caused by nickel subsulfide, did not stimulate erythropoiesis in the pups. Nickel subsulfide was associated with a significant decrease in mean body weights of pups, 2 and 4 weeks postpartum.

In a series of studies, Sunderman et al., (1978b,c, 1979a, 1980, 1983) demonstrated that nickel carbonyl, administered by inhalation or injection before, or a few days after, implantation produced various types of malformations in hamsters and rats. Intravenous injection of nickel carbonyl (11 mg nickel/kg body weight) into Fischer rats on day 7 of gestation caused fetal mortality, reduced body weight in live pups, and a 16% incidence of fetal malformations, including anophthalmia, microphthalmia, cystic lungs, and hydronephrosis (Sunderman et al., 1983). There was no maternal toxicity. Similar effects were observed in rats following inhalation of nickel carbonyl at a concentration of 0.16 g/m^3 on day 7 or 8 of gestation and a concentration of 0.3 g/m^3 on day 7 of gestation (Sunderman et al., 1979a).

In a dominant lethal mutation test on male rats, administration of nickel carbonyl, by the inhalation route (50 mg nickel/m^3 for 15 min), 2–6 weeks prior to breeding, did not affect fertilization rates or reproductive yield; administration of nickel carbonyl by intravenous injection during the same period (22 mg nickel/kg) caused reduced numbers of live pups in litters sired during the fifth week (Sunderman et al., 1983). In hamsters, inhalation exposure to 60 mg nickel carbonyl/m^3, for 15 min on day 4 or 5 of gestation, led to decreased fetal viability and increased numbers of fetuses with malformations including cystic lungs, exencephaly, and haemorrhages in serous cavities (Sunderman et al., 1980).

Gilani & Marano (1980) injected nickel chloride into chicken eggs (0.02 and 0.7 mg per egg) on days 0, 1, 2, or 3 of incubation. Examination on day 8 revealed a number of malformations, such as exencephaly, everted viscera, short and twisted neck or limbs, microphthalmia, haemorrhage, and reduced body size. The incidence of malformations was highest in embryos treated on day 2.

When rats were exposed long-term to nickel chloride or nickel sulfate in the diet or drinking-water, an increased frequency of runts and greater prenatal and neonatal mortality were observed (Schroeder & Mitchener, 1971; Ambrose et al., 1976; Nadeenko et al., 1979). Berman & Rehnberg (1983) observed spontaneous abortions, loss of fetal mass in survivors, and loss of maternal mass in mice.

8.2 Mutagenicity and related end-points

Genotoxicity data on nickel and nickel compounds are summarized in Table 27.

The data collectively show that nickel compounds are generally inactive in bacterial assays, but active in systems using eukaryotic organisms, and that positive responses were observed regardless of the nickel compounds used; particularly when these were compared at equitoxic concentrations (Hansen & Stern, 1983; Swierenga & McLean, 1985). The ability of nickel compounds to inhibit DNA synthesis and excision repair should also be noted (Table 27).

8.2.1 Mutagenesis in bacteria and mammalian cells

Nickel compounds were generally inactive in bacterial mutation assays. In some cases, where standard procedures were modified by, for example, using fluctuation assays (LaVelle & Witmer 1981; Pikalek & Necasek, 1983) or co-mutagenesis tests with other carcinogenic agents (Dubins & LaVelle, 1986; Ogawa et al., 1987), positive results were obtained.

The mutagenic effects of nickel chloride at the hypoxanthine-guanine phosphoriboxyl transferase (HGPRT) locus were studied in cultured V79 Chinese hamster cells and Chinese hamster ovary (CHO) cells (Hsie et al., 1979). Hsie et al. (1979) claimed positive results without reporting the detailed data. Miyaki et al. (1979) obtained a negative response. However, Hartwig & Beyersmann (1989) obtained a positive response with the same cell line, but the response occurred only in a serum-free medium.

Amacher & Paillet (1980) studied nickel chloride at the thymidine kinase (TK) locus in L51784Y mouse lymphoma cells and found a dose-dependent increase in the number of mutants of up to 4 times that in the controls. The possibility that this response was due to chromosome damage, also detected in this assay, cannot be excluded.

Table 27. Summary of genotoxic effects of nickel compounds

Species/Strain	Test system	Compound	Concentration[a]	Result	Reference
PROKARYOTIC SYSTEMS					
Prophage	Prophage induction	$Ni(CH_3CO_3)_2$	160–640 μmol/litre	+ (weak)	Rossman et al. (1984)
T4 Bacteriophage	Bacteriophage–forward mutation	$NiSO_4$	300 mg/litre	–	Corbett et al. (1970)
Salmonella					
TA1535	*Salmonella typhimurium* reverse mutation (fluctuation test)	$NiCl_2$	0.01–0.1 g/litre	+	LaVelle & Witmer (1981)
TA1535	*S. typhimurium* reverse mutation			–	Haworth et al. (1983)
TA1525	*S. typhimurium* reverse mutation	$NiCl_2$ $NiSO_4$		–	Arlauskas et al. (1985)
TA1537	*S. typhimurium* reverse mutation			–	Haworth et al. (1983)
TA1537	*S. typhimurium* reverse mutation	$NiCl_2$ $NiSO_4$		–	Arlauskas et al. (1985)
TA1535	*S. typhimurium* reverse mutation	Ni salts	10–1000 μg/plate	+ (weak)	Saichenko & Sharapora (1987)

Table 27 (continued)

TA98	S. typhimurium reverse mutation		-	Haworth et al. (1983)
TA98	S. typhimurium reverse mutation	NiCl$_2$ NiSO$_4$	-	Arlauskas et al. (1985)
TA100	S. typhimurium reverse mutation		-	Haworth et al. (1983)
TA100	S. typhimurium reverse mutation	NiCl$_2$ NiSO$_4$	- -	Arlauskas et al. (1985)
Several strains	S. typhimurium reverse mutation	Ni(CH$_3$CO$_2$)$_2$ NiCl$_2$ Ni(NO$_3$)$_2$	- - -	DeFlora et al. (1984)
Frameshift tester strains	Co-mutagenesis with 9-aminoacridine	NiCl$_2$	+	Ogawa et al. (1987)
	Co-mutagenesis with alkylating agents (base pair substitution)	Ni(II)	+	Dubins & Lavelle (1986)
Escherichia coli WP2	E. coli, reverse mutation	NiCl$_2$	-	Green et al. (1976) 0, 5, 10, 25 mg/litre
WP2 WP2 uvra CM571	E. coli (fluctuation) E. coli (fluctuation) E. coli (fluctuation)	NiCl$_2$ NiCl$_2$ NiCl$_2$	50 mmol/litre - 50 mmol/litre - 50 mmol/litre -	Nishioka (1975)

Table 27 (continued)

Species/Strain	Test system	Compound	Concentration[a]	Result	Reference
PROKARYOTIC SYSTEMS (continued)					
Escherichia coli (continued)					
WP67	Differential toxicity assay	$NiCl_2$	0, 200, 500, 1000 mg/litre	+ dose response	Tweats et al. (1981)
CM871	Differential toxicity assay	$NiCl_2$	0,200,500, 1000 mg/ml	+ dose response	Tweats et al. (1981)
WP2 (repair deficient)	Differential toxicity assay	$NiCl_2$	0,200,500, 1000 mg/ml	+ dose response	
	Differential toxicity assay	$NiCl_2$	0, 200, 500, 1000 mg/ml	+ dose response	
WP2	*E.coli*, reverse mutation	$NiCl_2$ $NiSO_4$		- -	Arlauskas et al. (1985)
Bacillus subtilis					
H17(rec$^+$) M45(rec$^-$)	*B. subtilis*, rec. strains differential toxicity	$NiCl_2$	50 mmol/litre	-	Nishioka (1975)
H17(rec$^+$) M45(rec$^-$)	*B. subtilis*, rec. strains differential toxicity	$NiCl_2$ NiO Ni_2O_3	5-50 mmol/litre 5-50 mmol/litre 5-50 mmol/litre	- - -	Kanematsu et al. (1980)

Table 27 (continued)

Corynebacterium SP887	Reverse mutation fluctuation test	$NiCl_2$	0.03–10 mg/litre 8 doses	+ dose response at ≥ 5.0 mg/litre	Pikalet & Necasek (1983)
Salmonella G46(his)	Host-mediated assay in mouse (NMRI strain)	$NiCl_2$	50 mg/kg	-	Buselmaier el al. (1972)
Serratia marcescens a21 (leu⁻)	Host-mediated assay in mouse (NMRI strain)	$NiCl_2$	50 mg/kg	-	Buselmaier et al. (1972)
Paramecium	Paramecium species mutation	Ni_3S_2		+ at 0.5 µg/ml	Smith-Sonneborn et al. (1983)
		NiS particles (1.8 µm)		+ at 0.5 µg/ml	
	Paramecium species chromosome aberration	NiS particles (1.8 µm)		+ at 0.57 µg/ml	Smith-Sonneborn et al. (1983)
Yeast (Saccharomyces)					
D7(diploid strain)	S. cerevisiae, gene conversion	$NiCl_2$	3 or 10 mmol/litre 24 h	+	Fukunaga et al. (1982)
D7(diploid strain)	S. cerevisiae, gene conversion	$NiSO_4$	5,10,20,40 mmol/litre	?	Singh (1984)

Table 27 (continued)

Species/Strain	Test system	Compound	Concentration[a]	Result	Reference
Yeast 19 haploid strains	S. cerevisiae, growth inhibition	$NiCl_2$	26 mmol/litre	+	Egilsson et al. (1979)
Vicia faba	Chromosome aberrations	NiO		+	Glaess (1956a)
	Mitotic effects	NiO		+	Glaess (1956b)
Vicia faba	Mitotic effects	$NiCl_2$		+	Komczynski et al. (1963)
		$Ni(NO_3)_2$		+	
		$NiSO_4$		+	
Pisum	Chromosome aberrations	$Ni(NO_3)_2$		+	Van Rosen (1954)
Allium cepa	Chromosome aberrations	NiO		+	Levan (1945)

INSECT SYSTEM

Drosophila

D. melanogaster white males BASC females	Sex-linked recessive lethal mutations	$NiSO_4$	200,300,400 mg/litre in 5% sucrose i.p.	+	Rodriguez-Arnaiz & Ramos (1986)

Table 27 (continued)

D. melanogaster $X^{C2}Y$ B/SC^8Y males $Y^2 W^a$ females	Sex chromosome loss assay	NiSO$_4$	200,300,400 mg/litre in 5% sucrose i.p.	+ (weak)	Rodriguez-Arnaiz & Ramos (1986)
D. melanogaster eggs from C(1)DX females X SC Z W$^+$f males	Somatic eye colour test system	Ni(NO$_3$)$_2$ NiCl$_2$	0.14 mmol/litre 0.21 mmol/litre	-- --	Rasmuson (1985)
D. melanogaster	mutation	Ni(NO$_3$)$_2$ NiCl$_2$	3.4-6.9 mmol/litre 4.2 mmol/litre	+? --	Vogel (1984)

MAMMALIAN CELLS: DNA DAMAGE

Rat Sprague-Dawley *in vivo*	DNA strand breaks, X links, alkaline elution, *in vivo* exposure	NiCO$_3$	5-40 mg/kg	+	Ciccarelli et al. (1981) Ciccarelli & Wetterhahn (1985)
Primary hepatocyte	DNA strand breaks, X links			(+)	Sina et al. (1983)

205

Table 27 (continued)

Species/Strain	Test system	Compound	Concentration[a]	Result	Reference
Hamster					
CHO	DNA strand breaks, alkaline sucrose gradients	$NiCl_2$ Ni_3S_2 (cryst.) NiS (cryst.) NiS (amorphous)	100-500 μmol/litre 10 mg/litre 1-10 mg/litre 10 mg/litre	+ + + -	Robison & Costa (1982)
CHO	DNA strand breaks X links			+	Costa et al. (1982)
CHO	DNA strand breaks X links	Ni_3S_2 $NiCl_2$	10 mg/litre, 24 h. 0.25-1.0 mmol/litre	+ +	Patierno & Costa (1985)
Human					
Normal human fibroblasts XP cells	DNA strand breaks alkaline elution	$NiSO_4$	10-2000 mg/litre	-	Fornace (1982)
Peripheral lymphocytes	DNA strand breaks FADU technique	$NiCl_2$	0.05 mmol/litre	?	McLean et al. (1982)

Table 27 (continued)

MAMMALIAN CELLS: DNA BINDING

Hamster

CHO	Binding to DNA in vitro (radioactive precursor)	^{63}NiS ^{63}NiCl$_2$	10 mg/litre 10 mg/litre	+ + (10x less)	Harnett et al. (1982)
CHO	Binding to DNA in vivo	^{63}NiCl$_2$ ^{63}Ni(CO)$_4$		+ to liver & + kidney DNA	Hui & Sunderman (1980)
CHO	Binding to RNA, protein	^{63}NiS ^{63}NiCl$_2$	10 mg/litre 10 mg/litre	+ + (10x less)	Harnett et al. (1982)
L132 pulmonary cells	Binding, X-ray microprobe analysis	Ni$_3$S$_2$		Ni bound to phosphate groups of DNA, RNA of cell membranes	Chirali et al. (1982)

Human

HeLa	Binding, colorimetric assay (dimethyl glyoxime)			Ni localized in centrioles	Kovacs & Darvas (1982)

Mouse

FM3A cells (mouse mammary carcinoma)	Inhibition of DNA synthesis	Ni(CH$_3$CO$_2$)$_2$	0.6, 0.8, 1.0 mmol/litre	+	Umeda & Nishimura (1979) Nishimura & Umeda (1979)

Table 27 (continued)

Species/Strain	Test system	Compound	Concentration[a]	Result	Reference
CBA strain	Inhibition of DNA synthesis in vivo	$NiSO_4$	15-30% LD_{50}	- kidney epithelium + hepatic epithelium	Amlacher & Rudolf (1981)
Rat Fischer rat embryo cells	Inhibition of DNA synthesis; ^{3}H TdR uptake	$NiCl_2$		+	Basrur & Gilman (1967)
Rat liver epithelial cell line T51B	Inhibition of DNA synthesis; ^{3}H TdR uptake	Ni_3S_2 $NiCl_2$	5, 10, 10 mg/litre up to 150 mg/litre	+ equal + potency at equitoxic doses	Swierenga & McLean (1985)
Hamster CHO	S-phase inhibition, flow cytometry	$NiCl_2$ Ni metal Ni_3S_2 (cryst.) Ni_3Se_2 (cryst). NiO	40-120 µmol/litre 10-20 mg/litre 1-10 mg/litre 1-5 mg/litre 5 mg/litre	+ + + + +	Costa et al. (1980b, 1982)
Human Hela	Inhibition of DNA synthesis; ^{3}HTdR uptake	$NiCl_2$		-	Painter & Howard (1982)

Table 27 (continued)

Bronchial epithelial cells	Inhibition of DNA synthesis; ^{3}HTdR uptake	NiSO$_4$		+	Lechner et al. (1984)

MAMMALIAN CELLS: DNA REPAIR

Rat

Fischer rat primary hepatocytes	UDS	NiCl$_2$		Inhibition of MMS-induced UDS	Swierenga et al. (1987)

Hamster

CHO	Repair induction Cesium chloride gradients	NiCl$_2$	100–1000 μmol/litre	+	Robison & Costa (1982)
SHE		Ni$_3$S$_2$ (cryst.)	10 mg/litre	+	
		NiS (cryst.)	1–10 mg/litre	+	
		NiS (amorphous)	5–10 mg/litre	-	

MAMMALIAN CELLS: GENE MUTATION

Mouse

L5178Y mouse lymphoma cells	Thymidine kinase locus	NiCl$_2$	0.16–0.53 mmol/litre (5 concs.)	+ dose response (toxic)	Amacher & Paillet (1980)

Table 27 (continued)

Species/Strain	Test system	Compound	Concentration[a]	Result	Reference
L5178Y mouse lymphoma cells	Thymidine kinase locus	NiSO$_4$	300-800 mg/litre	+ (weak) at toxic levels	McGregor et al. (1988)
Rat					
T51B rat liver epithelial cells	HPRT locus	Ni$_3$S$_2$	5-20 mg/litre	+	Swierenga & McLean
NRK, normal rat kidney cells, with integrated viral gene	Frameshift and base pair substitution	NiCl$_2$	20-40 µg/litre	+	Biggart & Murphy (1988)
6m2 murine sarcoma virus-infected cells	Gene expression	NiCl$_2$	20-160 µmol/litre	+ dose dependent	
Hamster					
V79	HPRT locus	NiCl$_2$	0.4, 0.8 mmol/litre	-	Miyaki et al. (1979)
V79	HPRT locus	NiCl$_2$		+	Hartwig & Beyersmann (1989)
V79, HGPRT with integrated E.coli gpt gene	HPRT locus	Ni(II)		+	Christie & Tummolo (1988)

Table 27 (continued)

Cell	Assay	Compound	Concentration	Result	Reference
CHO	HPRT locus	$NiCl_2$		?	Hsie et al. (1979)
CHO	HPRT locus	Ni_3S_2		+ (weak)	Costa et al. (1980)
		NiS		+ (weak)	
SHE	Ouabain resistance	$NiSO_4$	0.019 mmol/litre	− + as comutagen with BaP	Rivedal & Sanner (1980)

MAMMALIAN CELLS: SISTER CHROMATID EXCHANGE (SCE)

Cell	Assay	Compound	Concentration	Result	Reference
Mouse					
FM3A cells mouse mammary carcinoma	SCE	$NiCl_2$	6×10^{-4}–10^{-3} mol/litre	+	Nishimura & Umeda (1979)
FM3A cells mouse mammary carcinoma		$Ni(CH_3CO_2)_2$	6×10^{-4}–10^{-3} mol/litre	+ ⎫ recovery period included	
		$NiK_2(CN)_4$	1–1.6×10^{-3} mol/litre	+	
		NiS	4–8×10^{-4} mol/litre	+	
		$NiSO_4$	2.3×10^{-6} to 2.3×10^{-3} mol/litre	+ ⎭	
P338D$_1$ mouse macrophage line	SCE	$NiSO_4$	10^{-4}	+	Andersen (1983)

Table 27 (continued)

Species/Strain	Test system	Compound	Concentration[a]	Result	Reference
Hamster					
Don cells	SCE	NiSO$_4$ NiCl$_2$	0.19 mmol/litre 0.13 mmol/litre	+ at LC$_{50}$ + at LC$_{50}$	Ohno et al (1982)
CHO	SCE	NiSO$_4$	0.95, 2.85 µmol/litre	+ at 0.75 mg/litre	Deng & Qu (1981)
CHO	SCE	NiS(cryst.) NiCl$_2$	5-20 mg/litre 1-1000 µmol/litre	+ +	Sen & Costa (1985)
SHE	SCE	NiSO$_4$	3.8, 9.5, 19 µmol/litre	+ dose response	Larramendy et al. (1981)
Human					
Peripheral blood lymphocytes, in vitro	SCE	NiCl$_2$	0.001-1.0 mmol/litre	+ dose response at 0.1 mmol/litre	Newman et al. (1982)
Peripheral blood lymphocytes, in vitro	SCE	NiSO$_4$	0.0023-2.3 mmol/litre	+ dose response at 0.02, 0.2 mmol/litre	Wulf (1980)

Table 27 (continued)

Peripheral blood lymphocytes, in vitro	SCE	NiSO$_4$	0.95, 2.85 µmol/litre	+ dose response	Deng & Qu (1981)
Peripheral blood lymphocytes, in vitro	SCE	NiSO$_4$	9.5, 19.0 µmol/litre	+ dose response	Larramendy et al. (1981)
Peripheral blood lymphocytes, in vitro	SCE	Ni$_3$S$_2$	10^{-4} mol/litre	+ (weak)	Andersen (1983)
Peripheral blood lymphocyte:	SCE	Ni$_3$S$_2$	0.001–100 mg/litre	?	Saxholm et al. (1981)
Peripheral blood lymphocytes	SCE	nickel	2–20 mg/litre	+ (weak)	Djachenko (1989)

MAMMALIAN CELLS: CHROMOSOME ABERRATIONS (CA)

Mouse

In vivo	CA, in Ni-induced rhabdomyosarcomas	NiS (cryst)		+ aneuploidy, marker chromosome	Christie et al. (1988)
BalbC in vivo	Micronucleus	Ni(NO$_3$)$_2$ NiCl$_2$	56 mg/kg 25 mg/kg	– –	Deknudt & Leonard (1982)

Table 27 (*continued*)

Species/Strain	Test system	Compound	Concentration[a]	Result	Reference
FM3A cells, mammary carcinoma	CA	Ni(CH$_3$CO$_2$)$_2$	0.2, 0.3, 0.6, 1 mmol/litre; 24, 48h recovery	+ gaps	Umeda & Nishimura (1979)
		NiCl$_2$	0.2, 0.3, 0.6, 1 mmol/litre; 24, 48h recovery	? gaps	
		NiS	0.2, 0.3, 0.6, 1 mmol/litre; 24, 48h recovery	+ gaps, breaks, exchanges	
FM3A cells, mammary carcinoma	CA	Ni(CH$_3$CO$_2$)$_2$	0.6, 0.8, 1.0 mmol/litre; 6, 24, 48h + recovery	+ gaps, breaks, exchanges	Nishimura & Umeda (1979)
		NiCl$_2$	0.6, 0.8, 1.0 mmol/litre; 6, 24, 48h recovery	+ gaps, breaks, exchanges	
		NiS	0.6, 0.8, 1.0 mmol/litre; 6, 24, 48h recovery	+ gaps, breaks, exchanges	
FM3A cells, mammary carcinoma	CA	Ni(CH$_3$CO$_2$)$_2$ NiCl$_2$	10^{-4} mol/litre (2-3 days)	+ gaps, breaks, exchanges	Morita et al. (1985)

Table 27 (continued)

Rat				
In vivo	CA, rat bone marrow CA, spermatogonial cells	NiSO$_4$	3, 6 mg/kg for 7 and 14 days	-	Mathur et al. (1978)
Hamster				
Chinese hamster *in vivo*	CA	NiCl$_2$	1/5, 1/10, 1/25 LD$_{50}$	+ dose response 1/10, 1/5	Chorvatovicova (1983)
Chinese hamster *in vitro*	CA	NiCl$_2$	1-1000 µmol/litre, 2 h + 24 h recovery	dose response gaps, breaks, exchanges	Sen & Costa (1985)
		NiS (cryst.)	5-20 mg/litre	+ at 20 mg/litre, gaps, breaks, exchanges	
Syrian hamster *in vitro*	CA	NiSO$_4$	0.019 mmol/litre 24 hr	+ gaps, breaks, exchanges, minutes, dicentrics	Larramendy et al. (1981)

Table 27 (continued)

Species/Strain	Test system	Compound	Concentration[a]	Result	Reference
Human					
Peripheral blood lymphocytes *in vitro*	CA	NiSO$_4$	0.19 mmol/litre	+ breaks, rings, minutes	Larramendy et al. (1981)
Peripheral blood lymphocytes *in vitro*	CA	Ni powder NiO		-	Paton & Allison (1972)
Peripheral blood lymphocytes *in vitro*	CA	Ni(CH$_3$CO$_2$)$_2$	10-100 mg/litre	-	Voroshilin et al. (1977)
Peripheral blood lymphocytes *in vitro*	CA	NiCl$_2$	2-20 mg/litre	+	Djachenko (1989)
MAMMALIAN CELLS: CELL TRANSFORMATION					
Mouse					
C3H10T	Transformed foci	Ni$_3$S$_2$	0.001, 0.1 mg/litre	+	Saxholm et al. (1981)
Mouse embryo fibroblasts	Inhibition of interferon production	NiS(cryst.) NiS (amorph.)	1-20 mg/litre 1-20 mg/litre	+ -	Sonnenfeld et al. (1983)

Table 27 (continued)

Rat

HRT cells (hereditary renal tumour)	Initiation & promotion test	NiSO$_4$	40 µmol/litre	+	Eker & Sanner (1983)
Rat embryo cells infected with Rauscher leukaemia virus	Transformed foci	NiSO$_4$	0.19, 0.38 mmol/litre	+	Traul et al. (1981)
NRK (normal rat kidney) cells, infected with Maloney murine sarcoma virus	Transformed foci	NiSO$_4$	38–152 µmol/litre	+ max at 10 mg/litre	Wilson & Khoobyarian (1982)
T51B rat liver epithelial cells	Calcium independence, growth control and morphology, differentiation induction	αNi$_3$S$_2$	2.5 mg/litre, long-term exposure	+	Swierenga et al. (1989)
Rat tracheal primary epithelial cells	Transformed foci	αNi$_3$S$_2$	5 mg/litre, 24 h	+	Feren & Reith (1988)

Table 27 (continued)

Species/Strain	Test system	Compound	Concentration[a]	Result	Reference
Hamster					
SHE	Transformed foci	NiSO$_4$	10, 19, 38 μmol/litre	+ dose response	Di Paulo & Costa (1979)
		Ni$_3$S$_2$	1.0, 2.5, 5.0 mg/litre	+ dose response	
		NiS (amorphous)	up to 20 mg/litre	-	
SHE	Transformed foci	NiSO$_4$	3, 8, 19, 76 μmol/litre	+ at 19, 76 μmol/litre co-carcinogen with BaP	Rivedal & Sanner (1980)
SHE	Transformed foci	NiSO$_4$	19 μmol/litre + BaP	+ at 19, 76 μmol/litre co-carcinogen with BaP	Rivedal & Sanner (1981)
SHE	Transformed foci	NiSO$_4$	19 μmol/litre	+ with cigarette smoke extract	Rivedal et al. (1980)
SHE	Transformed foci	Ni$_3$S$_2$ (crystalline)	0.1, 1.0 mg/litre	+ at subtoxic doses, confirmed in nude mice	Costa et al. (1979)

Table 27 (continued)

Cell	Endpoint	Compound	Dose	Result	Notes	Reference
Hamster SHE	Transformed foci	NiS (amorphous)	0.1, 1.0 mg/litre			Costa et al. (1979)-
SHE	Transformed foci	Ni NiO, Ni$_2$O$_3$ NiS (cryst.) NiS (amorph.) Ni$_3$S$_2$ (cryst.) Ni$_3$S$_2$ (cryst.)	5, 10, 20 mg/litre	+ + + + + +	crystalline compounds more potent	Costa et al. (1981a)[b]
SHE	Transformed foci	Inco Black NiO (jet black, 5 μm particle size)	5, 10, 20 mg/litre	+		Sunderman et al. (1987a)
BHK-21 (baby hamster kidney)	Anchorage independence	αNi$_3$S$_2$ Ni$_2$O$_3$ NiO Ni(CH$_3$CO$_2$)$_2$ MIG nickel welding fume	5–20 mg/litre 5–20 mg/litre 37.5–150 mg/litre 100–400 mg/litre 100–400 mg/litre	+ + + + +	anchorage independence at equitoxic concentrations	Hansen & Stern (1983)

Table 27 (continued)

Species/Strain	Test system	Compound	Concentration[a]	Result	Reference
Human					
Bronchial epithelial cells	Growth control and morphology	$NiSO_4$	5-20 mg/litre	+	Lechner et al. (1984)
Foreskin fibroblasts	Anchorage independence	$Ni(CH_3CO_2)_2$ NiO $NiSO_4$ Ni_3S_2	10 μmol/litre 50 μmol/litre 10 μmol/litre 10 μmol/litre	+ anchorage + indepen- + dence at + LG_{50} concentrations	Biedermann & Landolph (1986)
MAMMALIAN CELLS: OTHER TESTS					
Balb/c mouse in vivo	Dominant lethal	$NiCl_2$ $Ni(NO_3)_2$	25 mg/kg 56 mg/kg	- preimplantation loss observed	DeKnudt & Leonard (1982)
Balb/c mouse in vivo	Dominant lethal	$Ni(NO_3)_2$	8.1, 11.3 mg/kg	- preimplantation loss observed	Jacquet & Mayence (1982)
Rat	Dominant lethal	$NiCl_2$	0.005-50 mg/kg for 24 weeks	- preimplantation loss observed	Saichenko et al. (1985)

Table 27 (continued)

Rat embryo cells in vitro	Spindle disturbance	$NiCl_2$	1 mg Ni/litre culture medium	+	Swierenga & Basrur (1968)
Human lymphocytes	Spindle disturbance	$NiSO_4$	10^{-3} mol/litre	+	Andersen (1985)
NIH3T3 cells	Inhibition of intercellular communication (radioisotope transfer)	$NiSO_4$	0.5-20 mmol/litre	+ dose response	Miki et al. (1987)
Human polymorphonuclear leukocytes	Induction of free radicals (chemiluminescent technique)	Ni_3S_2 NiO		+ more potent +	Evans et al. (1988)

[a] No figures means no concentration data given.
[b] Transformation experiments with this system have been repeated numerous times for mechanistic studies. Findings include:
- crystalline but not amorphous nickel compounds are actively phagocytosed *in vitro* (Costa & Mollenhauer, 1980b);
- reduction of amorphous compounds with $LiA1H_4$ enhances phagocytosis (Costa & Mollenhauer 1980b);
- phagocytosis is Ca-dependent (Heck & Costa, 1982);
- phagocytosed particles are contained in cytoplasmic vesicles where they slowly dissolve, releasing nickel ions (Evans et al. 1982);
- nickel compounds selectively damage heterochromatin (Sen & Costa, 1985);
- NiS(cryst.) and $NiCl_2$ preferentially transform male SHE cells (Costa et al., 1981a);
- observed damage includes deletion of long arm of X chromosome (Conway & Costa, 1989).

Swierenga & McLean (1985) used epithelial cells from rat liver (T51B) to study the genotoxic effects of nickel chloride and also of nickel subsulfide, either as an aqueous suspension of washed particles or as an aqueous solution. All produced increased numbers of mutations at the hypoxanthine-guanine phosphoriboxyl transferase (HGPRT) locus, over a range of concentrations.

In all the gene mutation studies using mammalian cells, any response following exposure to nickel compounds was associated with considerable cell toxicity.

8.2.2 Chromosomal aberration and sister chromatid exchange (SCE)

Nishimura & Umeda (1979) compared the effects of nickel chloride, nickel acetate, potassium cyanonickelate and nickel sulfide on the induction of chromosomal aberrations in cultured FM3A mouse mammary carcinoma cells. The 4 compounds elicited similar inhibitory effects on the synthesis of protein, RNA, and DNA. The chromosomal aberrations were manifested as breaks, exchanges, and fragmentations. Treatment of Chinese hamster ovary (CHO) cells with crystalline nickel sulfide and nickel chloride induced chromosomal aberrations including gaps, breaks, and exchanges. In both cases, the heterochromatic centromeric regions of the chromosomes were preferred; nickel sulfide also caused selective fragmentation at the heterochromatic long arms of the X-chromosome.

Chromosomal damage and increased SCE were induced in Syrian hamster embryo (SHE) cells and human lymphocyte cultures treated with nickel sulfate (Larramendy et al., 1981).

Wulf (1980) showed that nickel sulfate and nickel subsulfide could produce SCE in human lymphocytes *in vitro*. A significant increase in SCE in human lymphocytes exposed to nickel subsulfide was reported by Saxholm et al. (1981). Newman et al. (1982) observed increased SCE in human lymphocytes exposed to nickel chloride. Ohno et al. (1982) investigated the induction of SCE by nickel sulfate and nickel chloride in Don Chinese hamster cells and found significantly increased frequencies of SCE.

8.2.3 Mammalian cell transformation

These assays may not necessarily have a "genotoxic" endpoint as they also predict the carcinogenic potential of compounds that act by non-genotoxic mechanisms.

DiPaolo & Casto (1979) studied the capacity of nickel sulfate, nickel subsulfide, and amorphous nickel monosulfide to induce morphological transformation in Syrian hamster embryo (SHE) cells *in vitro*. Amorphous nickel monosulfide did not produce transformation at plate concentrations as high as 20 μg/ml medium. Nickel sulfate, tested at plate concentrations of 2.5, 5, and 10 μg/ml medium, caused transformation in a dose-dependent manner. Nickel subsulfide produced a higher percentage of transformations, in a dose-dependent manner, than nickel sulfate, when tested at plate concentrations of 1.0, 2.5, and 5μg/ml medium. In a study by Costa et al. (1979), nickel subsulfide caused morphological transformation, in a dose-dependent manner, of Syrian hamster fetal cells, at plate concentrations of 0.1, 1.0, 5.0, and 10.0 μg/ml medium, but, under the same experimental conditions, nickel monosulfide, at plate concentrations of 0.1, 1.0, and 5.0 μg/ml medium, did not.

Costa et al. (1981) demonstrated that the incidence of morphological transformations of SHE and CHO cells, exposed to various nickel compounds, was directly correlated with phagocytosis. Subcutaneous implantation of transformed SHE cells led to the development of fibrosarcomas in nude mice (Costa et al., 1979). Nickel sulfate caused cell transformation in cultured Syrian hamster embryo cells (Pienta et al., 1977).

Sunderman et al. (1987a) compared nickel oxides with different physical and chemical properties in the SHE cell transformation assay, and *in vivo* responses (Table 28). Three out of the 4 compounds that were active in the transformation assay were also positive *in vivo* when injected intramuscularly.

Hansen & Stern (1984) compared the abilities of welding fume, nickel powder, nickel acetate, black nickel oxide, black nickel oxide catalyst (a commercial catalyst for organic reactions, which is a mixture of nickel (II) and nickel (IV) oxides ($NiO_{1.4} \cdot 3H_2O$)), and nickel subsulfide, in the *in vitro* transformation of Syrian baby hamster kidney BHK-21 cells. Although a wide range of transformation

Table 28. Comparison of activity of some nickel oxides with different physiochemical characterisitics in *in vitro* transformation assays and *in vivo* carcinogenesis

Compound		Concentration	Result [a]	No. of tumours observed (20 mg Ni/rat im) [b]
Nickel oxides	Calcination temperature			
INCO Black (jet black)	650 °C	5, 10, 20 mg/litre	+	6/15
Grey black	735 °C	Particle size 5 μm	+	0/15
Green	1045 °C		−	0/15
Nickel copper oxides	Ni-Cu ratio			
Maroon	2.5:1	5, 10, 20 mg/litre	+	13/15
Red-brown	5.2:1	Particle size 5 μm	+	15/15
αNiS (positive control)			+	15/15

[a] Sunderman et al. (1987a).
[b] Sunderman et al. (1988a).

potency was found, the compounds produced the same number of transformed colonies at the same degree of toxicity (50% survival). The authors concluded that this indicated that if toxicity is a direct measure of net available nickel, then nickel or the nickel ion is the ultimate transforming agent. In a subsequent BHK-21 mammalian cell assay, Stern et al. (1985) determined the 50% toxicity of the soluble and insoluble fraction of nickel welding fume and found that only the insoluble fraction showed a transformation potential.

Synergistic effects of nickel compounds and benzo(a)pyrene on morphological transformation of SHE cells have been reported. Costa & Mollenhauer (1980c) found that pretreatment of the cells with benzo(a)pyrene enhanced the cellular uptake of nickel subsulfide. Treatment with nickel sulfate and benzo(a)pyrene resulted in a transformation frequency of 10.7% compared with 0.5% and 0.6%, respectively, for the individual substances (Rivedal & Sanner 1980).

8.3 Other test systems

Smith-Sonneborn et al. (1986) used the ciliated protozoan *Paramecium* to quantify the effects of pure nickel powder, iron-nickel powder, and nickel subsulfide. Genotoxicity was indicated by significant increases in the fraction of non-viable offspring (presumed index of lethal mutation) found after autogamy in parents from the nickel-treated groups compared with the controls. Only nickel subsulfide consistently induced a significant increase in offspring lethality.

Rodriguez-Arnaiz & Ramos (1986) studied the mutagenic potential of nickel sulfate in the *Drosophila* sex-linked recessive lethal (SLRL) assay *in vivo*. Nickel sulfate induced SLRL in a dose-dependent way, whereas sex chromosome loss was only detectable in significant numbers at the highest concentration.

8.4 Carcinogenicity

8.4.1 Inhalation

Studies on inhalation exposure are summarized in Table 29. Hueper (1958) and Hueper & Payne (1962) studied the effects of exposure

Table 29. Experimental animal studies on the carcinogencity studies of nickel compounds administered by inhalation or tracheal instillation

Nickel compound	Animal (group size)	Dosage schedule	Lung tumours detected	Comments	References
Inhalation studies					
Nickel powder	C57B1 female mice (20)	15 mg/m^3, 6 h/day, 4-5 days/week, for 60 weeks	no lung tumours, 2 lymphosarcomas	all animals died by week 60, no control group	Hueper (1958)
Nickel powder	Wistar rats (108) Bethesda black rats (60)	15 mg/m^3, 6 h/day, 4-5 days/week, for 60 weeks	numerous multicentric adenomatoid proliferations in 15 animals	histology on 50 animals only, no control group	Hueper (1958)
Nickel powder	Guinea-pigs (42)	15 mg/m^3, 6 h/day, 4-5 days/week, for 60 weeks	adenomatous alveolar lesions in almost all animals	all animals died by 21 months	Hueper (1958)
Nickel powder + powdered limestone + sulfur dioxide	Bethesda black rats (120) Hamsters (100)	level not specified	no lung tumours		Hueper & Payne (1962)
Nickel powder	Wistar rats (77)	3.1 mg/m^3, 6 h/day, 5 days/week, for 21 months (?)	carcinoid lung tumour in 2 exposed and in 1 control rat		Kim et al. (1969)

Table 29 (continued)

Ni(CO₄)	Wistar rats (64)	30 mg/m³, 30 min/day, 3 days/week, for 52 weeks	1 lung carcinoma	no lung tumours in a control group of 41 animals	Sunderman et al. (1959)
Ni(CO₄)	Wistar rats (32)	60 mg/m³, 30 min/day, 3 days/week, for 52 weeks	1 lung carcinoma	9 animals survived for 2 years	Sunderman et al. (1959)
Ni(CO₄)	Wistar rats (80)	single exposure to 250 mg/m³ (30 min ?)	2 lung carcinomas		Sunderman et al. (1959)
Ni(CO₄)	Wistar rats (64)	30 mg/m³, 30 min/day, 3 days/week, for lifetime	1 lung adeno-carcinoma		Sunderman & Donelly (1965)
Ni(CO₄)	Wistar rats (285)	single exposure to 600 mg/m³ for 30 min	1 lung adeno-carcinoma	214/285 animals died within 3 weeks	Sunderman & Donelly (1965)
Nickel subsulfide	F344 rats (226)	0.97 mg/m³, 6 h/day, 5 days/week, for 78 weeks, followed by observation for 30 weeks	14 malignant and 15 benign lung tumours in the exposed, 1 malignant and 1 benign in the control animals $P < 0.01$	survival: 5% in nickel-exposed rats; 31% in controls	Ottolenghi et al. (1974)

227

Table 29 (*continued*)

Nickel compound	Animal (group size)	Dosage schedule	Lung tumours detected	Comments	References
Black nickel oxide	Syrian golden hamsters (102)	53.2 mg/m^3, 7 h/day, 5 days/week, for 2 years	no significant increase in incidence of respiratory tumours	massive pneumoconiosis in exposed animals; part of animals also exposed to tobacco smoke	Wehner et al. (1975)
Green nickel oxide	Wistar rats (6)	0.6 mg/m^3, 6 h/day 5 days/week, for 4 weeks, followed by observation for 80 weeks	1 adenocarcinoma and 1 adenomatous pulmonary lesion	5 control animals	Horie et al. (1985)
Green nickel oxide	Wistar rats (8)	8 mg/m^3, 6 h/day, 5 days/week, for 4 weeks, followed by observaation for 80 weeks	1 adenomatous lesion in lungs		Horie et al. (1985)
Nickel-enriched coal fly ash (Ni acetate added to coal before burning)	Syrian golden hamster (102)	fly ash containing 6% Ni, 70 mg/m^3, 6 h/day, 5 days/ week, for 20 months	no lung tumours	20 of the animals removed from exposure for other studies	Wehner et al. (1981)

Table 29 (continued)

Nickel-enriched coal fly ash (Ni acetate added to coal before burning)	Syrian golden hamsters	fly ash containing 6% Ni, 17 mg/m^3, 6 h/day, 5 days/week for 20 months	no lung tumours	Wehner et al. (1981)
		fly ash containing 0.3% Ni, 70 mg/m^3, 6 h/day, 5 days/week for 20 months	no lung tumours	Wehner et al. (1981)
Nickel oxide	Wistar rats (60)	60 µg/m^3 continuous exposure for 18 months	no lung tumours	Glaser et al. (1986)
		200 µg/m^3 continuous exposure, doe, 18 months	no lung tumours	
Intratracheal instillation				
Nickel powder	Wistar rats (39)	0.3 mg Ni x 20, at weekly intervals	1 adenoma, 1 adenocarcinoma, 8 squamous cell carcinomas	no lung tumours in 40 control animals	Pott et al. (1987)
	Wistar rats (32)	0.9 mg Ni x 10, at weekly intervals	3 adenocarcinomas, 4 squamous cell carcinomas, 1 mixed tumour	Pott et al. (1987)

229

Table 29 (*continued*)

Nickel compound	Animal (group size)	Dosage schedule	Lung tumours detected	Comments	References
Nickel oxide (not specified)	Wistar rats (37)	5 mg × 10, at weekly intervals	4 adenocarcinomas, 4 squamous cell carcinomas, 2 mixed tumours		Pott et al. (1987)
	Wistar rats (38)	15 mg × 10, at weekly intervals	12 squamous cell carcinomas		
Black nickel oxide	Hamsters (50)	4 mg × 30, at weekly intervals	no lung tumours	3 hamsters in both groups survived for 12 months	Farrell & Davis (1974)
Black nickel oxide	Rats (26)	single administration 20-40 mg	1 squamous cell carcinoma of the lung	the nickel oxide dust contained 64.7% NiO, 0.13% NiS, 0.18% Ni; no pulmonary tumours in 47 controls	Sakmyn & Blokhin (1978)
Nickel subsulfide	B6C3F1 mice (20)	0.024, 0.056, 0.156, 0.412, or 1.1 mg/kg, four doses at weekly intervals, follow-up of 27 months	no neoplastic or non-neoplastic lesions		Fisher et al. (1986)

Table 29 (continued)

Ni_3S_2	Rats (47)	0.063 mg × 15, at weekly intervals	2 adenocarcinomas, 5 squamous cell carcinomas	no pulmonary tumours in control rats	Pott et al. (1987)
Ni_3S_2	Rats (45)	0.125 mg × 15, at weekly intervals	3 adenocarcinomas, 6 squamous cell carcinomas, 4 mixed tumours	no pulmonary tumours in control rats	Pott et al. (1987)
Ni_3S_2	Rats (40)	0.25 mg × 15, at weekly intervals	7 adenocarcinomas, 4 squamous cell carcinomas, 1 mixed tumour	no pulmonary tumours in control rats	Pott et al. (1987)
Powder Ni_3S_2 Pentlandite Cr-Ni(55)Alloy Cr(55)Alloy	Hamsters, (30–40 per sex)	0.8 mg powder, 0.1 mg αNi_3S_2, 3 mg pentlandite 3 or 9 mg Cr-Ni(55) alloy, 9 mg Cr(55) alloy; 12 times at 14-day intervals	1 lung tumour (nickel powder group) 1 lung tumour (pentlandite group)	no pulmonary tumours in positive control group	Muhle et al. (1988)

to airborne concentrations of elemental nickel. Hueper (1958) exposed 20 female C57 BL mice, 50 male and 50 female Wistar rats, 60 female Bethesda black rats, and 32 male and 10 female guinea-pigs to an atmosphere containing 99% pure nickel powder (particle size 4 μm or less) at 15 mg/m^3, for an exposure period of 6 h/day, 4–5 days/week, for up to 21 months. There were no control groups. There were no lung tumours, but 2 lymphosarcomas were seen in the mice. However, most animals died before 15 months. Fifteen out of 50 rats of both strains that were studied histologically showed adenomatoid formations in the lung, which were classified as benign neoplasms. At death, most of the guinea-pigs exhibited adenomatoid proliferations. One animal had an intra-alveolar carcinoma. The results of a later study (Hueper & Payne, 1962) on rats and hamsters did not reveal lung tumours following inhalation exposure to 99% pure nickel together with 56–98 mg sulfur dioxide/m^3 (20–35 ppm) and powdered limestone.

In studies by Kim et al. (1969) a group of 77 male Wistar rats was exposed to metallic nickel dust, equivalent to 3.1 mg/m^3 for 6 h a day, 5 days/week, over 21 months, followed by exposure to air alone; 98% of the dust particles were less than 2 μm in diameter. Subgroups were exposed to the dust for various periods followed by periods of recovery. Two exposed rats developed lung tumours of a carcinoid pattern and a similar tumour was found in one rat in the unexposed control group.

Ottolenghi et al. (1974) found a significantly higher incidence of pulmonary hyperplastic and neoplastic lesions in 226 Fischer 344 rats of both sexes exposed to nickel subsulfide (0.97 mg nickel/m^3; 70% of particles smaller than 1 μm) for 6 h/day, 5 days/week, over 78 weeks. The overall incidence of lung tumours (adenoma and adenocarcinomas) in treated animals was 14% compared with 1% in the controls. The nickel sulfide-exposed rats also had a higher incidence of respiratory tract inflammation.

The pathogenic effect of inhaled black nickel oxide was investigated in hamsters by Wehner et al. (1975). Random-bred ENG: ELA Syrian golden hamsters (102 males) were exposed to respirable aerosols of nickel oxide (count median diameter 0.3 μm) at a concentration of 53.2 mg/m^3 for 7 h/day, on 5 days/week, for up to 2 years. Half of the animals were also exposed (nose-only) to cigarette smoke for 10 min, 3 times a day. Histopathological examination revealed increasing cellular response, proliferative and inflammatory, in animals dying late in the study. Histopathologically, there were no marked differences between the nickel-oxide-plus cigarette smoke and the nickel-oxide-only exposed groups. It was concluded that there was neither a significant carcinogenic effect of nickel oxide nor a co-carcinogenic effect of cigarette smoke. However, though not statistically significant, 3 malignant musculoskeletal tumours were found among nickel oxide-exposed hamsters (Wehner et al., 1975).

Wehner et al. (1981) exposed 4 groups, each of 102 male Syrian golden hamsters (outbred LAK:LVG), through inhalation to nickel-enriched fly ash (NEFA) for 6 h/day, 5 days/week, for up to 20 months. The first group was exposed to 70 mg NEFA/m^3 aerosol (4 mg nickel/m^3), the second group to 17 mg NEFA/m^3 aerosol (1 mg nickel/m^3), the third group was exposed to 70 mg fly ash (FA)/m^3 aerosol containing 0.21 mg nickel/m^3, and the fourth group was exposed to filtered air (control group). Five animals from each group were killed after 4, 8, 12, and 16 months of exposure. Additional groups of 5 animals were withdrawn from exposure at the same time intervals and maintained for observation up to the twentieth month of the study, when all the animals were killed. Dust deposition, interstitial reaction, and bronchiolization in the lungs were higher in the high-NEFA and FA groups than in the low-NEFA group, indicating that dust quantity rather than actual nickel content may be the major factor in determining tissue response. While 2 malignant pulmonary tumours were found in 2 hamsters of the high-NEFA group, no statistically significant carcinogenesis was observed.

Horie et al. (1985) studied the carcinogenic effects of inhalation of 8.0 or 0.6 mg green nickel oxide/m^3 on male Wistar rats (5-8 rats/dose group). The exposure time was 6 h/day, 5 days/week, for 1 month. Animals were killed 20 months after exposure. There was one adenocarcinoma in a low-dose animal. In a study by Glaser et al. (1986), male Wistar rats were exposed continuously for 18 months to nickel oxide aerosols (60 or 200 µg nickel/m^3). The nickel oxide aerosols were generated by atomization of aqueous nickel acetate solutions and subsequent pyrolysis. No lung tumours were observed.

Sunderman et al. (1959) and Sunderman & Donelly (1965) reported carcinogenesis in rats following inhalation exposure to nickel carbonyl. In the first study (Sunderman et al., 1959), groups of 64 and 32 male Wistar rats were exposed to 30 and 60 mg nickel carbonyl/m^3, respectively, for 30 min, 3 days/week, for one year. A further group was exposed once to 250 mg nickel carbonyl/m^3. Four out of the 9 animals that survived 2 years developed neoplasms of the lung; 2 of these animals were in the single-exposure group, the other 2 rats were in the repeated-exposure groups. No pulmonary tumours were seen in 41 control animals, the death rate of which was similar to that in the animals exposed to nickel carbonyl. Sunderman & Donelly (1965), using 285 male Wistar rats, observed pulmonary adenocarcinoma with metastases in one of the 35 rats that survived 2 or more years after a single 30-min inhalation of 600 mg nickel carbonyl/m^3. In a further group of 64 male rats, exposed to repeated inhalation of nickel carbonyl (30 mg/m^3 for 30 min, 3 days/week, until death), one pulmonary adenocarcinoma with metastases developed among 8 rats that survived 2 or more years. The control animals did not show any tumours. Because of the rarity of spontaneous pulmonary malignancies in Wistar rats, it was suggested that the tumours observed in the two studies were due to inhalation of nickel carbonyl. Survivability of treated and control animals was poor in both studies. Statistical analysis could not be performed because of small sample size.

Kasprzak et al. (1973) did not find any lung tumours in 13 rats following intratracheal instillation of 5 mg nickel subsulfide. Thirty weekly intratracheal injections of 4 mg black nickel oxide did not produce lung tumours in 50 hamsters, but only 3 animals survived (Farrell & Davis, 1974). In a study by Saknyn & Blokhin (1978) 26 rats were exposed to 20–40 mg nickel oxide by intratracheal instillation; 1 lung carcinoma was found. In a series of studies, Pott et al. (1987) examined the potential carcinogenic effect of a number of dusts, including nickel oxide, nickel powder, and nickel subsulfide, at various concentrations, in rats. All 3 nickel compounds produced lung tumours, including adenocarcinomas and squamous cell carcinomas, nickel subsulfide exhibiting the strongest effect in relation to dose. Lung tumour incidence was 14.9% (7 out of 47 animals), 28.9% (12 out of 45 animals), and 30% (12 out of 40 animals) following 15 weekly intratracheal injections each containing 0.063 mg, 0.125 mg, and 0.25 mg nickel, respectively. Nickel powder was given intratracheally in 20 weekly doses of 0.3 mg nickel; an additional group was given 10 weekly doses of 0.9 mg nickel. Lung tumours occurred in 10 out of 39 animals (25.6%) and in 8 out of 32 animals (25.0%), respectively. Ten weekly instillations of nickel oxide, each corresponding to a nickel content of 5 or 15 mg, produced 27.0% (10/37 animals) or 31.6% (12/38 animals) lung tumours, respectively. No lung tumours occurred in the controls (40 animals). In a study with nickel subsulfide using the intratracheal route, no increase in tumour incidence was observed compared with the controls. Doses ranged from 0.024 to 1.1 mg/kg, administered once a week for 4 weeks (Fisher et al., 1984).

Muhle et al. (1988) administered nickel powder (cumulative dose 9.6 mg), nickel subsulfide (cumulative dose 1.2 mg), pentlandite (cumulative dose 36 mg), chromium-nickel-steel (cumulative dose 36 or 108 mg) and chromium-containing stainless steel (cumulative dose 108 mg) to Syrian golden hamsters (30–40 animals per sex and dose group) by intratracheal instillation, 12 times every 14 days. One lung tumour was found after treatment with nickel powder and one tumour after treatment with pentlandite.

8.4.2 Oral

Schroeder et al., (1964, 1974) and Schroeder & Mitchener (1975) studied the carcinogenic potential for mice and rats of nickel salts in the drinking-water. Levels of 5 mg nickel/litre in the drinking-water did not produce a significantly higher incidence of tumours compared with controls.

Ambrose et al. (1976) added nickel sulfate hexahydrate fines (22.3% nickel) to the diet of Wistar derived rats for 2 years, at concentrations of 0, 100, 1000, or 2500 mg nickel/kg. There was no carcinogenic response.

8.4.3 Other routes

The results of carcinogenesis studies of nickel compounds in rodents are summarized in Table 30.

Table 30. Experimental carcinogenesis studies of various nickel compounds

Nickel compound	Animals	Route[a]	Tumours	Comments	Reference
Ni_3S_2	Fischer rats	im -implants 500 mg -diffusion chambers 10 mg -powder 10 mg -chips 500 mg	 14/17 14/17 19/20 5/7	Form of implant is important	Gilman & Herchen (1963)
Ni_3S_2	Rat	1 or 3 mg Ni_3S_2/gelatin pellet implanted 4 weeks post-grafting in tracheas grafted under dorsal skin of isogenic recipients	1 mg: 9/60 tumour (6 carcinomas, 3 sarcomas) 3 mg: 45/60 tumours (1 carcinoma, 44 sarcomas)	No tumours in control transplanted tracheas	Yarita & Nettesheim (1978)
Ni_3S_2	Rats	Intratesticular, 0.6-10 mg	16/19 (fibrosarcomas, rhabdomyosarcomas, malignant histiocytomas)	no tumours in controls	Damjanov et al., (1978)
Ni_3S_2	Fischer rats	intra-renal	Renal cancers in 18/24 rats at 10 mg dosage and 5/18 rats at 5 mg dosage	no tumours in controls	Sunderman et al. (1979b)
Ni_3S_2	NMRI mice	im, sc	Local sarcomas in 19 of 32 mice at 10 mg dosage		Oskarsson et al. (1979)

Table 30 (continued)

Compound	Animal	Route	Findings	Reference
Ni_3S_2	Pregnant rats	im, on day 6	Local sarcomas in all dams; no excess tumours in progeny	Sunderman et al. (1981)
Ni_3S_2	Rabbits	im	Leiomyosarcomas with myosin light-chains typical of fetal smooth muscle, rhabdomyosarcomas	Hildebrand & Tetaert (1981); Hildebrand & Biserte (1979a,b)
Ni_3S_2	Fischer rats	intraocular	Retinoblastomas, gliomas, and melanomas	Albert et al. (1980, 1982)
Ni_3S_2	Fischer & hooded rats	im	Rhabdomyosarcomas, leiomyosarcomas, fibrosarcomas and lymphosarcomas	Yamashiro et al. (1980, 1983)
Ni_3S_2	Fischer rats	im	Ni_3S_2 carcinogenesis inhibited by simultaneous injection of manganese dust	Sunderman & McCully (1983)
Ni_3S_2, $Ni(OH)_2$, $NiSO_4$	Wistar rats	im	Sarcomas induced by Ni_3S_2 and crystalline $Ni(OH)_2$, but not by $NiSO_4$ or amorphous $Ni(OH)_2$	Kasprzak et al. (1983)
Various nickel compounds	Fischer rats	im	Sarcomas induced by nickel compounds	Sunderman (1984)

Table 30 (continued)

Nickel compound	Animals	Route[a]	Tumours	Comments	Reference
Ni$_3$S$_2$, NiO Ni$_3$S$_2$	Wistar & Fischer rats	intra-pleural implantation	Rhabdomyosarcomas, malignant tumours of pluripotential origin		Skaug et al. (1985) Lumb et al (1985)
Ni$_3$S$_2$	Fischer rats	im	Rhabdomyosarcomas, fibrosarcomas		Kasprzak & Poirier (1985)
Ni$_3$S$_2$	F344 rats	sc, im, if, ia	Malignant soft tissue tumours in 18/19 (sc), 19/20 (im), 9/20 (if) 16/19 (ia)	No synovial sarcomas	Shibata et al. (1985a)
Ni$_3$S$_2$	Male Fischer rats 10 groups 20 each	im (single dose)	Mg carbonate inhibited Ni$_3$S$_2$ carcinogenicity from 100 to 55%	A possible involvement of NK and phagocytic cells in this inhibition	Kasprzak et al. (1987)
Ni$_3$S$_2$	Fischer rats	im	Local sarcomas	Simultaneous admin. of ZnO or Zn acetate delayed the onset of sarcomas	Kasprzak et al. (1988)
Ni$_3$S$_2$	F344 NCT	im	Local sarcomas 100%		Kasprzak & Rodriguez (1988)

Table 30 (continued)

Ni_3S_2 + iron: Fe·/Ni = 0.5 Fe·/Ni = 1.0 Fe·/Ni = 2.0	Rat	im	Local sarcomas 60% 40% 10% Local sarcomas	Interaction role of inflammation, active oxygen species, enzyme system	Kasprzak & Rodriguez (1988)
Fe^{3+} = 0.5 Fe^{3+} = 1.0 Fe^{3+} = 2.0			95% 45% 35%		Kasprzak & Rodriguez (1988)
Ni_3S_2 + remote Fe, or $Fe^{3+}/Ni=2$			100%	Iron effect is purely local	Kasprzak & Rodriguez (1988)
Ni_3S_2	in vitro cell free			Deactivation of catalase and glutathione peroxidase	Kasprzak & Rodriguez (1988)
Ni_3S_2	Japanese common newts (Cynops) pyrrhogaster	intra-ocular	Malignant melanoma-like tumour in 7/8 at 400–1100 µg/newt		Mitsumasa (1987)
NiO	Wistar rats	im 20 mg	26 local tumours in 32 rats (80%)	No controls used	Gilman (1962)

Table 30 (continued)

Nickel compound	Animals	Route[a]	Tumours	Comments	Reference
NiO	Swiss mice	im, 5 mg	35% mice with tumours		Payne (1964)
NiO	C3H mice	im, 5 mg	23% mice with tumours		
NiO	NIH black rats	im implants, 7 mg	4 sarcomas in 35 rats after 18 months		
Nickel oxides and nickel-copper oxides	Rats	Instillation in pleural cavity	Rhabdomyosarcomas in 8/220 (4/20 in positive control)	Negative study	Eilertsen et al. (1988)
Nickel refinery flue dust	Hooded Wistar rats	im, 20 or 30 mg, one or both thighs	52/66 rats with sarcomas 8/20 rats with sarcomas		Gilman & Ruckerbauer (1962)
	Mice	im, 10 mg each thigh	23/40 mice with sarcomas		
Feinstein dust	Rats	ip 90–150 mg dust/ rat	6/39 rats with injection site sarcomas		Saknyn & Blokhin (1978)

Table 30 (continued)

Nickel powder	Hooded rats	im, 28.3 mg in 0.4 ml fowl serum	10/10 rats with local rhabdomyosarcomas	Heath and Daniel (1964)
Nickel powder	Osborne-Mendel rats	Intrafemoral, 21 mg	4/17 rats with tumours 1 squamous cell carcinoma, 3 osteosarcomas	Hueper (1952)
	Wistar rats	Intrafemoral implant, 50 mg	28% of treated rats with tumours of injected thighs, No tumours in control rats	
	Dutch rabbits	Intrafemoral implant, 54 mg/kg	1/6 rabbits with fibrosarcomas	
	C57BL mice	iv, weekly for 2 weeks	No tumours	
	Rabbits	iv, 6 times of a 1% Ni suspension in 2.5% gelatin at a rate of 0.5 ml/kg	No tumours	
	Wistar rats	iv, 6 times of 0.5% Ni suspension in saline at 0.5 ml/kg	7/25 rats with tumours	

Table 30 (continued)

Nickel compound	Animals	Route[a]	Tumours	Comments	Reference
Nickel powder	Fischer rats	im, 5 monthly injections of 5 mg Ni in 0.2 ml trioctanoin	38/50 rats with fibrosarcomas		Furst & Schlauder (1971)
	Hamsters	im, 5 monthly injections of 5 mg Ni in 0.2 ml trioctanoin	2/50 hamsters with fibrosarcomas		Furst & Schlauder (1971)
Nickel powder	Fischer rats	im, 3.6 mg/rat 14.4 mg/rat	0/10 rats with local tumours 2/10 rats with local tumours		Sunderman & Maenza (1976)
Nickel powder	Fischer 344 rats	Intrathoracic, 5 monthly doses of 5 mg/animal	Mesothelioma, pleural cavity in 2 out of 10 rats	Only 10 animals	Furst et al. (1973)
Nickel powder	Fischer 344 rats	ip, 16 injections twice monthly, 5 mg/animal	30-50% intraperitoneal tumours		Furst & Cassetta (1973)

Table 30 (continued)

Compound	Animal	Route/Dose	Result	Notes	Reference
Nickel powder	Fischer 344 rats	Intrathoracic 5 monthly dose, of 5 mg/animal	40% mesothelioma		Furst & Cassetta (1973)
Ni(CO)$_4$	Sprague-Dawley rats	iv, 6 × 50 µl/kg (9 mg Ni/kg)	19/121 rats with malignant tumours at various sites, 2/47 rats with pulmonary lymphomas < 0.05)		Lau et al. (1972)
NiCO$_3$	NIH black rats	Muscle implant, 7 mg	4/35 implant site sarcomas		Payne (1964)
Nickelocene	Fischer rats	im injections, 12 × 12 mg or 12 × 25 mg nickelocene	18/50 rats or 21/50 rats with fibrosarcomas	No tumours in controls	Furst & Schlauder (1971)
	Hamsters	im injections, 1 × 25 mg or 8 × 5 mg nickelocene	4/29 hamsters with fibrosarcomas at 1 × 25 mg	No tumours at 8 × 5 mg	
Ni(CH$_3$CO$_2$)$_2$	Mice, Strain A	ip, 24 injections, 3/week at 72, 180, 360 mg/kg	Lung adenomas and adenocarcinomas (significant for 360 mg/kg group)		Stoner et al. (1976)

Table 30 (continued)

Nickel compound	Animals	Route[a]	Tumours	Comments	Reference
Ni(CH$_3$CO$_2$)$_2$	Mice, Strain A	ip	average 1.5 lung adenomas per mouse		Poirier et al. (1984)
Ni(CH$_3$CO$_2$)$_2$	F344 rats	sc., 150 mg/kg	Injection site sarcomas		Teraki & Uchiumi (1988)
Nickel powder NiO, Ni$_3$S$_2$	Wistar rats	ip	46/48, 48/47 and 27/42 rats with sarcomas, mesotheliomas, or carcinomas		Pott et al. (1987)
Nickel powder NiO, Ni$_3$S$_2$ NiCl$_2$, Ni(CH$_3$CO$_2$)$_2$ NiSO$_4$, NiCO$_3$ nickel alloy	Wistar rats	ip	High incidence of sarcomas and mesotheliomas in Ni$_3$S$_2$-, nickel powder-, nickel alloy-, and NiO-groups		Pott et al. (1987)

[a] ia = intra-articular
  if = intra-retroperitoneal fat
  im = intramuscular
  ip = intraperitoneal
  iv = intravenous
  sc = subcutaneous.

Nickel subsulfide is the nickel compound that has been most studied. All routes of administration, except submaxillary gland injection into Fischer rats and local application to the cheek pouches of Syrian hamsters (Sunderman et al., 1978d), produced tumours. Sunderman (1983b) and Gilman & Yamashiro (1985) reviewed experimental data on factors influencing the carcinogenetic activity of nickel subsulfide and other nickel compounds, including species and strain, susceptibility of organs, route of administration, physical form of nickel compound, and dose-response characteristics. Dose-effect relationships could be demonstrated for nickel subsulfide in rats following intramuscular and intrarenal injections and in hamsters following intramuscular and intratesticular injections. Intraocular and intramuscular routes of administration yielded the highest tumour incidence. Local tumours after intraocular injection included malignant melanomas, retinoblastomas, gliomas, a phakocarcinoma, and unclassified tumours. Following intramuscular injection, local tumours were mostly rhabdomyosarcomas, fibrosarcomas, and undifferentiated sarcomas. Rats are more susceptible to the induction of sarcomas by nickel subsulfide than mice, hamsters, or rabbits, though Sunderman (1983b) considered that absolute species susceptibility could not be defined, because of differences in experimental conditions. For strain differences in rats, Gilman & Yamashiro (1985) ranked susceptibility to tumour induction by nickel subsulfide as: Hooded > Wistar > Fischer > Sprague Dawley. Yarita & Nettesheim (1978) implanted nickel subsulfide pellets into heterotopic tracheas grafted in Fischer 344 rats and reported tracheal tumours including sarcomas (2 fibrosarcomas, 1 leiomyosarcoma) and carcinomas (5 squamous cell carcinoma, 1 undifferentiated carcinoma). Malignant testicular neoplasms developed in Fischer 344 rats after intratesticular injection of 10 mg nickel subsulfide (Damjanov et al., 1978). Gilman & Herchen (1963) found corresponding numbers of rhabdomyosarcomas in rats after intramuscular implantation of Ni_2S_3 in implants of different forms, injection of powder, or exposure in a diffusion chamber. Shibata et al. (1989) injected nickel subsulfide (single injection, 0.5 mg) subcutaneously (sc), intramuscularly (im), in retroperitoneal fat, and intra-articularly (ia), in male Fischer rats, and observed the animals for 48 weeks. Malignant soft-tissue tumours developed in 18/19 (sc) 19/28 (im) 9/20 (retroperitoneally), and 16/19 (ia) animals. No synovial sarcomas were detected. Magnesium carbonate was observed to decrease sarcoma incidence due to Ni_3S_2 from 100 to 55% in a dose-related manner, when both compounds were injected into the thigh muscles of male Fischer 344 rats (Kasprzak et al., 1987). Kasprzak et al. (1988) found that, 40 weeks after injection, zinc, either as zinc oxide or zinc acetate, delayed the onset, but not the final incidence of local sarcomas when injected intramuscularly with Ni_3S_2. All of the rats given Ni_3S_2 developed tumours compared with only 60% of those that received Ni_3S_2 plus zinc. Kasprzak & Rodriguez (1988) found that iron counteracted the ability of nickel subsulfide to induce cancer and that it inhibited catalase and glutathione peroxidase, leading to a build up of hydrogen peroxide (H_2O_2), which may be related to carcinogenesis.

Iron can decompose H_2O_2 so reversing the action of nickel subsulfide. Mitsumasa (1987) administered Ni_3S_2 to lentectomized Japanese common newts (*Cynops pyrrhogaster*). A single injection of 40–100 μg/newt produced malignant melanoma-like tumours in the injected eyes of 7 or 8 newts.

Nickel oxide produced local tumours (mainly rhabdomyosarcomas, fibrosarcomas, and undifferentiated sarcomas) following intramuscular injections in rats and mice (Gilman, 1962; Sunderman, 1984) and a few tumours (not specified) following implantation in rats (Payne, 1964). It was also carcinogenic when administered by intrapleural injection in Wistar rats (Skaug et al., 1985). Eilertsen et al. (1988) tested the carcinogenicity in the rat of a variety of nickel oxides given by the intrapleural route but obtained negative results.

When Gilman & Ruckerbauer (1962) injected nickel refinery flue dust (20% nickel sulfate, 57% nickel subsulfide, 6.3% nickel oxide) into the thigh muscles of rats and mice, they found a high induction rate of sarcomas. Saknyn & Blokhin (1978) injected "feinstein" dust (an intermediate product of nickel refining containing nickel sulfide, nickel oxide, and metallic nickel) intraperitoneally at a dose of 90–150 mg/rat and found sarcomas at the injection sites in 6 out of 39 rats.

Intramuscular injection of nickel powder produced sarcomas in rats at the injection sites (Heath & Daniel, 1964; Furst & Schlauder, 1971; Sunderman & Maenza, 1976; Sunderman, 1984). In hamsters, intramuscular injection of nickel powder produced sarcomas at the site of injection, but the incidence was low (2 out of 50 animals) (Furst & Schlauder, 1971). Intravenous injection of nickel powder produced local sarcomas in rats, but not in mice and rabbits (Hueper, 1955). No controls were used. Hueper (1952, 1955) injected nickel powder into the femurs of rats and rabbits and observed sarcomas at the site of injection. Intrathoracic injections of nickel powder produced mesotheliomas in Fischer 344 rats (Furst et al., 1973; Furst & Cassetta, 1973). Following intraperitoneal injection of nickel powder, tumours were seen in Fischer 344 rats (Furst & Cassetta, 1973).

Besides its carcinogenic potential in inhalation studies, administration of nickel carbonyl to rats by repeated intravenous injection was also associated with an increased tumour rate (Lau et al., 1972). Malignant tumours included pulmonary lymphomas, undifferentiated sarcomas in the lung, pleura, liver, pancreas, uterus, and abdominal wall, fibrosarcomas in the neck, pinna, and orbit, carcinomas of liver, breast, and kidney, and one haemangioendothelioma.

Local sarcomas were induced in rats with repeated intramuscular injections of crystalline nickel hydroxide, but not with colloidal nickel hydroxide or nickel sulfate, under similar experimental conditions (Kasprzak et al., 1983).

Implantation of 7 mg nickel carbonate hydroxide/rat produced local sarcomas in 4 out of 35 rats (Payne, 1964).

Nickelocene, an organic nickel compound used as a laboratory reagent, induced fibrosarcomas in rats and hamsters, when injected intramuscularly (Furst & Schlauder, 1971).

It was reported by Stoner et al. (1976) that 24 intraperitoneal injections of nickel acetate to strain A mice (total dose 72–360 mg/kg body weight) induced lung tumours at the highest dose. Poirier et al. (1984) observed an increased incidence of lung adenomas in Strain A mice following intraperitoneal administration of nickel acetate, though these authors also observed an increase in tumour incidence in animals treated with calcium acetate or lead acetate. Teraki & Uchiumi (1988) reported that nickel acetate causes injection site tumours in rats.

Single intraperitoneal doses of nickel powder, nickel subsulfide, or an unspecified nickel oxide induced a high frequency (46 out of 48 animals, 27 out of 42 animals, and 46 out of 47 animals, respectively) of abdominal cavity tumours in Wistar rats (Pott et al., 1987). Pott et al., (1988) investigated nickel carcinogenicity by intra-peritoneal injection in rats and produced a variety of responses (see Table 30). They concluded that the model is good for screening nickel compounds and alloys for carcinogenicity.

In a study by Sunderman (1984), 18 nickel compounds were investigated under standardized experimental conditions to relate their physical, chemical, and biological properties to their carcinogenic activity. Marked differences were observed in the incidence of sarcomas in male Fischer rats within 2 years following single intramuscular injections of equivalent doses (14 mg nickel/rat). The results of these bioassays are summarized in Table 31. They provide an adequate basis for ranking the relative carcinogenicity of different nickel compounds in animals. Sarcoma incidences in these bioassays were significantly correlated with the mass-fractions of nickel in the respective compounds. They were not significantly related to the dissolution half-times of the nickel compounds in rat serum or renal cytosol (Kuehn & Sunderman, 1982), or to the susceptibilities of the compounds to phagocytosis by rat peritoneal macrophages *in vitro* (Kuehn et al., 1982). On the other hand, there was a strong correlation between the sarcoma incidences and the capacity of nickel compounds to induce erythrocytosis after intrarenal administration to rats, suggesting that nickel-stimulated production of erythropoietin and nickel carcinogenesis may be related in some way (Sunderman & Hopfer, 1983).

Six nickel oxides and 4 nickel-copper oxides were categorized according to the criteria shown in Table 32 (Sunderman et al., 1987a). These compounds were assayed as follows: (*a*) *in vitro* dissolution in water and body fluids; (*b*) *in vitro* phagocytosis tests in Chinese hamster ovary and C3H-10T1/2 cells; (*c*) morphological transformation and cytotoxicity tests in cultured Syrian hamster embryos (SHE) cells; (*d*) erythropoiesis stimulation assay by intrarenal administration to Fischer 344 rats; and (*e*) scoring the renal histopathological responses in rats killed 3 months after injection. The extent of agreement among the physical, chemical, and biological attributes of the test compounds was tested by Kendalls' non-

Table 31. Summary of survival data and sarcoma incidences in carcinogenesis tests of 18 nickel compounds[a]

Category	Test substance	Survivors at 2 years/total number of rats (% survival)	Rats with local sarcomas/total number of rats (% survival)	Median tumour latency (weeks)	Median survival period (weeks)	Rats with metastases/rats with sarcomas (% survival)
Controls	Glycerol vehicle	25/40 (63%)	0/40 (0%)	-	>100	-
	Penicillin vehicle	24/44 (55%)	0/44 (0%)	-	>100	-
	All controls	49/84 (58%)	0/84 (0%)	-	>100	-
Class A	Nickel subsulfide (αNi_3S_2)	0/9[d] (0%)	9/9[d] (100%)	30	39[c]	5/9 (56%)
	Nickel monosulfide (BNiS)	0/14[d] (0%)	14/14[d] (100%)	40	48[c]	10/14 (71%)
	Nickel ferrosulfide (Ni_4FeS_4)	0/15[d] (0%)	15/15[d] (100%)	16	32[c]	10/15 (67%)
Class B	Nickel oxide (NiO)	0/15[d] (0%)	14/15[d] (93%)	49	58[c]	4/14 (29%)
	Nickel subselenide (Ni_3Se_2)	0/23[d] (0%)	21/23[d] (91%)	28	38[c]	18/21 (86%)
	Nickel sulfarsenide (NiAsS)	0/16[d] (0%)	14/16[d] (88%)	40	57[c]	10/14 (71%)
	Nickel disulfide (NiS_2)	0/14[d] (0%)	12/14[d] (86%)	36	47[c]	6/12 (50%)
	Nickel subarsenide (Ni_5As_2)	0/20[d] (0%)	17/20[d] (85%)	22	44[c]	9/17 (53%)
Class C	Nickel dust	4/20[c] (20%)	13/20[d] (65%)	34	42[c]	6/13 (40%)
	Nickel antimonide (NiSb)	9/29[c] (31%)	17/29[d] (59%)	20	66[c]	10/17 (59%)
	Nickel telluride (NiTe)	12/26 (46%)	14/26[d] (54%)	17	80[c]	8/14 (57%)
	Nickel monoselenide (NiSe)	7/16 (44%)	8/16[d] (50%)	56	72[c]	3/8 (38%)
	Nickel subarsenide ($Ni_{11}As_8$)	5/16[b] (31%)	8/16[d] (50%)	33	88[c]	6/8 (75%)

Table 31 (continued)

Category	Test substance	Survivors at 2 years/total number of rats (% survival)	Rats with local sarcomas/total number of rats (% survival)	Median tumour latency (weeks)	Median survival period (weeks)	Rats with metastases/rats with sarcomas (% survival)
Class D	Amorphous nickel monosulfide (NiS)	5/25[c] (20%)	3/25[c] (12%)	41	71[c]	3/3 (100%)
	Nickel chromate (NiCrO$_4$)	10/16 (63%)	1/16 (6%)	72	>100	1/1 (100%)
Class E	Nickel monoarsenide (NiAs)	13/20 (65%)	0/20 (0%)	–	>100	–
	Nickel titanate (NiTiO$_3$)	11/20 (55%)	0/20 (0%)	–	>100	–
	Feronickel alloy (NiFe$_{1.6}$)	11/16 (75%)	0/20 (0%)	–	>100	–

[a] From: Sunderman (1984d).
[b] $P < 0.05$ versus corresponding vehicle controls.
[c] $P < 0.01$ versus corresponding vehicle controls.
[d] $P < 0.001$ versus corresponding vehicle controls.

Table 32 Physical and chemical characteristics of some test compounds: 6 Ni oxides and 4 Ni-Cu oxides[a]

Test compound and color	Calcination temp. (°C)	Surface area (m²/g)	Elemental composition (% by weight)						Constituents and speciation (% by weight)					Relative height of XRD peaks for Ni oxides[d]
			Ni	Cu	O[b]	S	Fe	Co	Ni(III)	SO$_4$	Ni$_3$S$_2$	CuO	NiO[c]	
A (jet black)	<650	41.5	77.7	<.001	22.3	<0.02	<0.01	<0.01	0.81	<0.06	ND[e]	ND	SS[f]	0.31
B (grey-black)	735	3.0	78.1	<.001	21.9	<0.02	<0.01	<0.01	0.05	<0.06	ND	ND	SS	0.74
C (grey)	800	1.4	78.2	<.001	21.8	<0.02	<0.01	<0.01	0.03	<0.06	ND	ND	SS	0.87
D (grey)	850	0.7	78.2	<.001	21.8	<0.02	<0.01	<0.01	<0.03	<0.06	ND	ND	SS	0.88
E (dark green)	918	0.2	78.4	<.001	21.6	<0.02	<0.01	<0.01	<0.03	<0.06	ND	ND	SS	0.90
F (green)	1045	<0.2	78.6	<.001	21.4	<0.02	<0.01	<0.01	<0.03	<0.06	ND	ND	SS	1.00
G (dark maroon)	850	<0.2	43.8	27.9	23.4	0.69	2.98	1.17	[g]	1.86	0.3	14	77	
H (maroon)	850	<0.2	51.7	20.6	23.6	1.02	2.02	1.09	[g]	2.67	0.5	3	89	
I (red-brown)	850	<0.2	62.6	12.6	22.3	0.30	1.20	1.04	[g]	0.06	1.0	ND	95	
J (brown)	850	<0.2	68.8	6.9	22.1	0.42	0.75	0.97	[g]	0.06	1.5	ND	96	

a From Sunderman et al. (1987a).
b Estimated by subtracting the percentages for other elemental constituents from 100%.
c NiO (bunsenite) can accommodate up to 25% CuO (tenorite) as a solid solution within its crystal lattice at 850C.
d Heights of XRD peaks of NiO (bunsenite) at 1.48, 2.09, and 2.41 Å, relative to peak heights obtained with compound F.
e ND = not detected by XRD.
f SS = sole species detected by XRD.
g Presence of sulfur in compounds G-J precluded assays for Ni(III).

parametric test for concordance of rankings. On the basis of a highly significant concordance of ranked results in the assays ($P < 0.001$), 6 biological attributes of the compounds were identified: (i) dissolution half- times in rat serum and renal cytosol; (ii) phagocytosis by C3H-10T1/2 cells; (iii) morphological transformation of SHE cells; (iv) erythropoiesis stimulation in rats; (v) induction of tubular hyperplasia in rat kidneys; and (vi) induction of arteriosclerosis in rat kidneys. A strong rank correlation between results of the cell transformation and erythropoiesis stimulation was observed. The presence of high surface area and demonstrable Ni (III) were the two physicochemical characteristics associated with the greatest biological effects of the nickel oxides.

Compounds A, B, and F (nickel oxides) and compounds H and I (nickel-copper oxides) were selected for carcinogenesis testing in male Fischer 344 rats (20 mg, intramuscular injection, right thigh) (Sunderman et al., 1988a). Injection site sarcomas (524 months) developed in 6 out of 15 rats that received compound A (INCO black NiO), versus 0 out of 15 rats that received compound B (NiO calcined at 735 °C), and 0 out of 15 rats that received compound F (NiO calcined at 1045 °C). Sarcomas developed in 13 out of 15 rats that received compound H (NiO/CuO, 2.5/1 ratio) and 15 out of 15 rats that received compound I (NiO/CuO, 5.2/1 ratio). Injection site sarcomas developed in 0 out of 15 vehicle control rats and 15 out of 15 positive controls that received intramuscular injection of Ni_3S_2.

8.4.4 Interactions with known carcinogens

Several studies have been performed to investigate the possible interactions of different nickel species with known carcinogens, either in classical two-stage carcinogenesis studies, or using simultaneous administration.

In a two-stage carcinogenesis assay, orally administered nickel chloride enhanced the renal carcinogenicity of N-ethyl-N-hydroxyethyl nitrosamine in rats, but not the hepatocarcinogenicity of N-nitrosodiethylamine, the gastric carcinogenicity of N-methyl-N-nitro-N-nitrosoguanidine, the pancreatic carcinogenicity of N-nitrosobis(2-oxopropyl)amine, or the skin carcinogenicity of 7,12-dimethylbenzanthracene (Hayashi et al., 1985).

Single intratracheal injections of nickel subsulfide did not increase the carcinogenicity of benzo(a)pyrene, also instilled intratracheally, in rats (Kasprzak et al., 1973).

Metallic nickel powder seemed to enhance the lung carcinogenicity of 20-methylcholanthrene (MC) when both were administered intratracheally to rats. The tumour incidences in different groups, 12 weeks after the instillation were: (i) 5 mg MC: 2/7; (ii) 5 mg MC + 10 mg Ni: 3/5; (iii) 1 mg MC: 1/8; (iv) 1 mg MC + 10 mg Ni 2/7; and (v) 10 mg Ni only: 0/7 (Mukubo, 1978). Nickel oxide, administered intratracheally, did not enhance the carcinogenic response to diethylnitrosamine in the lungs or nasal cavity of hamsters (Farrell & Davis, 1974).

In studies on combined intramuscular injection of nickel subsulfide and benzo(a)pyrene, latency times to local tumour development were shorter, and the proportion of malignant tumours higher, than when either carcinogen was injected alone (Maenza et al., 1971).

In a two-stage carcinogenicity study, Ou et al. (1981) gave a single injection of dinitrosopiperazine (9 mg) to rats, followed by administration of nickel sulfate, either in the drinking-water (3.7 mg/day), or topically in the nasopharynx as a gelatin solution (0.02 ml of 0.5% Ni SO$_4$). Two out of 12 rats in both groups receiving both nickel and the nitrosamine developed nasopharyngeal tumours (1 squamous cell carcinoma, 1 fibrosarcoma, 1 papilloma, 1 "early carcinoma"), while none was seen in the 24 animals treated with the nitrosamine alone, or in the two groups of 12 animals treated with nickel sulfate. The same authors (abstract only) reported an extension of the study, using topical nickel sulfate as a promoter. Five out of 22 rats given the combined treatment developed carcinomas of the nasopharynx, nasal cavity, or hard palate, while these tumours were not found in the other groups (Liu et al., 1983).

Ou et al. (1983) reported a study in which N-nitrosopiperazine was given to female rats on the 18th day of pregnancy, and the pups were given nickel sulfate orally. Five out of 21 pups with combined treatment, and 1 out of 4 with nitrosamine treatment only developed nasopharyngeal carcinoma, and two other pups developed other tumours.

8.4.5 Possible mechanisms of nickel carcinogenesis

Both *in vivo* and *in vitro* studies have shown varying carcinogenic responses, depending on the bioavailability of certain nickel compounds. Several factors influencing bioavailability have been suggested, such as the physical form of the compound (crystalline or amorphous), solubility in water or serum, and its ability to be phagocytosed by target cells (Costa et al., 1982) or pass through calcium channels (Sunderman, 1989a). *In vitro* studies have shown that various nickel compounds had similar transforming potencies at equitoxic concentrations (Hansen & Stern, 1984).

The transformation potency of a number of particulate nickel compounds has been correlated with phagocytosis (Costa et al., 1981b). The incidence of sarcomas and the capacity of nickel compounds to produce erythrocytosis are also related (Sunderman & Hopfer, 1983; Sunderman et al., 1987a). Erythrocytosis can be induced in rats by intrarenal injection of nickel subsulfide, probably due to increased production of the stimulating hormone erythropoietin (Jasmin & Ripolle, 1976; Morse et al., 1977; Hopfer et al., 1978). Intrarenal injection of nickel subsulfide caused renal cancers (Sunderman et al., 1984b). Induction of erythrocytosis and carcinogenesis in rats by nickel subsulfide were both inhibited by the administration of manganese dust (Sunderman et al., 1976b). *In vitro* studies have demonstrated that nickel compounds cause (*a*) DNA damage in a wide variety of organisms, (*b*) mutagenicity, possibly related to cytotoxicity, in mammalian cells, and (*c*) clastogenicity and morphological transformation in mammalian cells including human cells. From *in vitro* and *in vivo* studies, it is known that nickel interacts with DNA and proteins, causing lesions, such as DNA-protein (possibly heterochromatin) cross-links and DNA single strand breaks (Ciccarelli et al., 1981; Ciccarelli & Wetterhahn 1982; Robison et al., 1982; Patierno et al., 1985), resulting in DNA damage and other cytotoxic effects that can lead to altered gene expression in surviving cells. For example, Sen & Costa (1986) showed that nickel preferentially induced decondensation (presumably by DNA- protein cross-linking) and fragmentation of the long arm of the X-chromosome in CHO cells. Other consequences of cross-linking, such as the appearance of irreversible, but heritable, lesions of intermediate filaments (a family of cytoskeletal proteins linking the nucleus with the cell membrane), related to disturbance of normal growth control processes, have been observed early after nickel intoxication *in vitro* (Swierenga & McLean, 1985; Swierenga et al., 1987, 1989). Nickel-induced DNA damage/alterations that have been reported include: conformational transitions of the DNA from the normal right-handed B-helix to the left-handed Z helix (van de Sande et al., 1982; Boutayre et al., 1984), infidelity of DNA synthesis, as shown with cell-free systems using synthetic templates (Sirover & Loeb, 1977; Zakour et al, 1981), and inhibition of DNA synthesis in a wide variety of organisms extending to inhibition of DNA excision repair (Table 27). The latter may be an important mechanism in the ability of nickel to enhance the carcinogenic activity of other compounds (Maenza et al., 1971; Kurokawa et al., 1985). Other *in vitro* evidence indicates that nickel disrupts intercellular communication, a property of tumour promoters (Miki et al., 1987).

Studies by Nieboer et al. (1986) indicated that, in the presence of certain peptides, the Ni^{3+}/Ni^{2+} redox couple can participate in dioxygen radical reactions *in vitro*. As chemical promoters of cancer appear to have the capacity to stimulate phagocytic cells to produce dioxygen radicals, which may damage DNA, proteins and lipids, and as metal ions catalyze processes involving molecular oxygen, this observed effect of nickel may have implications for nickel carcinogenesis.

Sunderman & Barber (1988) have suggested a mechanism for the interaction of nickel with DNA. This model proposes that the Zn^{2+} binding sites of DNA binding proteins known as "finger loop domains", which have been identified on some proto-oncogens, are likely molecular targets for metal toxicity. As Ni^{2+} has a similar ionic radius to Zn^{2+}, substitution is possible, interfering with regulation of gene expression. The replacement of Zn^{2+} in finger loops might affect the conformation and stability of DNA-binding structures interfering with expression and induce site-specific free radical reactions, causing cleavage of DNA, formation of DNA-protein cross links, and disturbances of mitosis.

8.4.6 Factors influencing nickel carcinogenesis

Maenza et al. (1971) reported a synergistic action between nickel subsulfide and benzo(*a*)pyrene. After intramuscular injection, the latency period for sarcoma induction was significantly reduced. Kasprzak et al., (1973) found an increased incidence of premalignant changes in the lungs of rats receiving intratracheal injections with a mixture of benzo(*a*)pyrene and nickel subsulfide compared with groups receiving one compound at a time.

Manganese in metallic form has been shown to antagonize the tumourigenic effect of nickel subsulfide by intramuscular injection (Sunderman et al., 1976b) and intrarenal injection (Sunderman et al., 1979b). Kasprzak & Poirier (1985) found that simultaneous intramuscular injection of magnesium carbonate inhibited the development of muscle tumours induced by nickel subsulfide. Calcium carbonate, under the same experimental conditions, was ineffective. Poirier et al. (1984) reported that both magnesium acetate and calcium acetate inhibited lung adenoma formation in Strain A mice treated with intraperitoneal nickel acetate. Manganese, magnesium, calcium, copper, and zinc were shown by Kasprzak & Poirier (1985) and Kasprzak et al. (1986) to inhibit binding of nickel to DNA. They also found that amino acids were very effective inhibitors of nickel binding to DNA, but their eventual role in moderating nickel carcinogenicity has not been investigated.

When rats with muscular implantations of nickel subsulfide were injected intraperitoneally with sodium diethyldithiocarbamate, tumour incidence was significantly reduced compared with that in rats that had received the nickel subsulfide implant alone (Sunderman et al., 1984b).

9. EFFECTS ON HUMAN BEINGS

9.1 Systemic effects

9.1.1 Acute toxicity - poisoning incidents

9.1.1.1 Nickel carbonyl

In terms of human health, the most acutely toxic nickel compound is nickel carbonyl. The clinical manifestations of nickel carbonyl poisoning have been described in detail by Sunderman (1970) and Vuopala et al. (1970). A summary of 179 cases of nickel carbonyl poisoning in China since 1961 has been published by Shi (1986).

The acute toxic effects occur in 2 stages, immediate and delayed. The immediate symptomatology includes frontal headache, vertigo, nausea, vomiting, insomnia, and irritability, followed by an asymptomatic interval before the onset of delayed pulmonary symptoms. These include constrictive chest pains, dry coughing, dyspnoea, cyanosis, tachycardia, occasional gastrointestinal symptoms, sweating, visual disturbances, and weakness. The symptomatology resembles that of a viral pneumonia. In men who died, pulmonary haemorrhage and oedema or pneumonitis were observed accompanied by derangement of alveolar cells, degeneration of the bronchial epithelium, and the appearance of a fibrinous intra-alveolar exudate. The pathology of the pulmonary lesions was similar to that observed in animal studies. Other affected organs included the liver, kidneys, adrenal glands, and spleen, where parenchymal degeneration was observed. Cerebral oedema and punctate cerebral haemorrhages were noted in men dying after inhalation of nickel carbonyl. Shi (1986) reported leukocytosis in 25 out of 179 cases. Urinary nickel concentrations were determined in 27 cases and ranged from 0.003 to 0.66 mg/litre. Air concentrations were measured and were reported to have exceeded 50 mg nickel carbonyl/m^3 with exposure periods ranging from 30 min to more than 2 h. Recovery time was 7–40 days, depending on the severity of exposure. In some cases, symptoms persisted for 3–6 months. In lethal cases, death occurred between the third and thirteenth days following exposure.

9.1.1.2 Other nickel compounds

Information on poisoning with other nickel compounds is limited. Daldrup et al. (1983) reported a case of fatal nickel poisoning. A 2 1/2-year-old girl ingested about 15 g of nickel sulfate crystals. On admission to hospital, she was somnolent with wide and unresponsive pupils, a high pulse rate, and pulmonary rhonchi. Cardiac arrest occurred after 4 h. Autopsy findings revealed increased nickel levels (7.5 mg/kg in blood, 50 mg/litre in urine, 25 mg/kg in liver tissue). Nickel poisoning was reported in a group of 23 dialysed patients, when leaching from a nickel-plated stainless steel water heater tank contaminated the dialysate. Symptoms occurred during, and after, dialysis, at plasma-nickel concentrations of approximately 3 mg/litre. Symptoms included nausea (37 out of 37), vomiting (31 out of 37), weakness (29 out of 37), headache (22 out of 37), and palpitation (2 out of 37). Recovery occurred spontaneously, generally 3–13 h after cessation of dialysis (Webster et al., 1980).

Thirty-two workers in an electroplating plant accidentally drank water contaminated with nickel sulfate and nickel chloride (1.63 g nickel/litre). Twenty workers rapidly developed symptoms (e.g., nausea, vomiting, abdominal discomfort, diarrhoea, giddiness, lassitude, headache, cough, shortness of breath) that typically lasted a few hours, but persisted for 1–2 days in 7 cases. The nickel doses in workers with symptoms were estimated to range from 0.5 to 2.5 g. In 15 exposed workers, who were investigated on day 1 after exposure, serum-nickel concentrations ranged from 12.8 to 1340 μg/litre (average 286 μg/litre) with urinary nickel concentrations ranging from 0.23 to 37.1 mg/litre (average 5.8 mg/litre) compared with control values for nickel-plating workers of 2.0–6.5 μg/litre (average 4.0 μg/litre) for serum nickel and 22–70 μg/litre (average 50 μg/litre) for urinary nickel. Laboratory tests showed transiently elevated levels of blood reticulocytes (7 workers), urine-albumin (3 workers), and serum-bilirubin (2 workers) (Sunderman et al., 1988b).

9.1.2 Short- and long-term exposure

9.1.2.1 Respiratory effects

A chemical engineer, who had been exposed for a long period to low levels of nickel carbonyl, developed asthma and Löffler's syndrome. In addition to pulmonary infiltrations and eosinophilia, which are markers in Löffler's syndrome, the patient had an eczematous dermatitis of the hands (Sunderman & Sunderman, 1961b).

There have been systematic investigations of the health state of the respiratory tract of workers in the nickel industry. Tatarskaya (1960) examined 486 workers from nickel-refining plants, exposed mainly to nickel sulfate during electrorefining operations. Exposure levels were not given. Rhinitis was observed in 10–16%, chronic rhinitis in 5.3%, nasal septal erosions in 13%, perforations in 6.1%, and ulceration in 1.4% of the workers. The frequencies of hypo-osmia and anosmia were 30.6 and 32.9%, respectively. Kucharin (1970) studied 302 workers who had been exposed to nickel sulfate, accompanied by sulfuric acid vapour, for at least 10 years, at concentrations of 0.02–4.5 mg/m^3. Clinical and radiological examination revealed chronic sinusitis in 83% of the workers. Among these workers, severe injuries were apparent, such as septal erosion and perforation in 41.4 and 5.6%, respectively.

Hypo- or anosmia was noted in 46% of the workers with chronic sinusitis. In a nickel refinery where hydrometallurgical methods were used, 37 out of 151 workers (24.5%) exhibited pathological changes of the nasopharynx. Septal erosions were noted in 14 workers (9.3%). Exposure concentrations of soluble nickel salts, during the years 1966 and 1970, ranged from 0.035 to 1.65 mg/m^3 (Sushenko & Rafikova, 1972).

Other studies on the nasopharyngeal effects of inhalation of nickel salt aerosols, conducted in order to detect precancerous lesions, are described in the section 9.5.

Data are available on the development of pulmonary changes with fibrosis in workers inhaling nickel dust or fumes. Zislin et al. (1969) studied respiratory function in 13 workers, who had been exposed to nickel dust for periods of 12.9–21.7 years. Exposure levels were not given. There was decreased pulmonary residual capacity, increased respiratory frequency, and radiography showed a diffuse fibrosis, considered to be nickel pneumoconiosis.

Industrial dusts, collected in the workrooms of nickel refineries, induced the development of pneumoconiosis of an interstitial type, together with bronchitis, in experimental animals (Saknyn et al., 1978; Grebnev & Yelnichnykh, 1983). The fibrogenicity of these dusts did not depend on their silica content, and the dust that was almost pure nickel suboxide (Ni_2O_3) was fibrogenic, when instilled intratracheally. Pneumoconiosis of an interstitial and micronodular type had been detected in a number of workers in the same industry (Zislin et al., 1969).

Peto et al. (1984) reported mortality experience in Welsh nickel refinery workers. Men first employed before 1925, who had worked for 5 or more years in the process areas where lung and nasal cancer risks were highest, also suffered a significantly elevated mortality from non-malignant respiratory disease (20 deaths, 11.1 expected); less heavily exposed workers suffered no such excess (43 deaths, 51.2 expected).

Jones & Warner (1972) studied respiratory symptoms, chest X-rays, and lung ventilation capacities in 12 steelworkers, who had been employed as de-seamers of steel ingots for periods of up to 16 years. Total fume concentration at the workplace ranged from 1.3 to 294.1 mg/m^3, either from iron oxide, chromium oxide, and nickel oxide in proportions of 6:1:1 (stainless steel fume) or 98.8% iron oxide (special steel fume). Airborne nickel oxide concentrations of 0.15–34 mg/m^3 were calculated from these values. Two of the workers had measurable loss of pulmonary function. In 5 men, definite signs of pneumoconiosis were detected radiographically.

Effects on the respiratory system have also been observed in welders. Between 1955 and 1979, Zober (1981a,b) found a total of 47 cases of histologically verified fibrotic changes in the lungs of electric arc welders. Though the findings were very heterogeneous, indicating a diversity of causes, welding fumes were found to be a concomitant cause of fibrotic changes in 20 cases. Zober (1982) studied 40 welders (22, non-smoking and 18, smoking), who were using metal inert gas and electric arc welding techniques on chromium- and nickel-containing materials. Concentrations of airborne nickel did not exceed 0.5 mg/m^3, except in one working process, where there was a concentration of 1.2 mg nickel/m^3. Respiratory-tract effects, including acute bronchitis and bronchitic pulmonary rhonchi, were increased, mainly in welders who smoked tobacco.

Asthmatic lung disease in nickel-plating workers exposed to soluble nickel has been reported (Tolot et al., 1956; McConnell et al., 1973; Block & Yeung, 1982; Malo et al., 1982; Dolovich et al., 1984). Cirla et al. (1985) examined 12 workers with respiratory problems in a nickel-plating industry. In 6 workers, allergic asthma could be provoked by the inhalation of nickel sulfate aerosols. Novey et al. (1983) reported that a metal-plating (chromium and nickel) worker developed acute asthma in response to inhalational challenge with nickel sulfate and chromium sulfate. Radioimmuno-assays incorporating the challenge materials revealed specific IgE antibodies to the provocative agents.

Five stainless steel welders, suffering from respiratory distress, developed asthmatic symptoms during provocation tests with stainless steel fumes (Keskinen et al., 1980).

Inhalation testing using 0.01–0.001% nickel chloride solution has been carried out to detect hypersensitivity to nickel and its compounds in patients occupationally exposed to nickel and suffering from atopic bronchial asthma. In inhalation testing, it is necessary to take pneumotachometric readings before, and after, the test, at various times over a period of 24 h (Izmerov, 1983).

9.1.2.2 Renal effects

Sanford et al. (1988) assessed different parameters for kidney dysfunction in urine samples from electrolytic nickel refinery workers (renal clearance, protein, specific gravity, and qualitative "dipstick" tests). In male workers who accidentally ingested drinking-water contaminated with nickel sulfate and chloride, Sunderman et al., (1988b) found elevated urine-albumin concentrations (68, 40, and 27 mg/g creatinine), which returned to normal on the fifth day after exposure. The findings suggested a mild transient nephrotoxicity (section 9.1.1).

9.1.2.3 Cardiovascular effects

Patients with acute myocardial infarction and unstable angina pectoris develop high serum-nickel levels, which may be followed by coronary vasoconstriction (Leach et al., 1985). The mechanism and source of nickel release are not yet known, but Sunderman (1983a) recommended that the amount of nickel in intravenous solutions should not exceed 5 µg/kg per day. Peto et al. (1984) observed elevated cerebrovascular mortality (16 deaths, 8.5 expected) among nickel refinery workers who had worked for 5 years in the process area, where lung and nasal cancer rates were highest. Mortality from all other cardiovascular causes was similar to local population rates.

9.1.2.4 Other effects

In a study by Shi et al. (1986), nickel carbonyl refinery workers were compared with a control group. Exposure ranged from 0.007 to 0.52 mg $Ni(CO_4)/m^3$. A decrease in the MAO (serum monoamine oxidase) activity, and EEG abnormalities were observed in the most severely exposed. A number of non-specific symptoms in persons exposed for long periods to low concentrations (unspecified) were reported, but no details were given.

Immune reactions elicited in the sera of individuals exposed to nickel and cobalt were assessed according to changes in the concentrations of the serum immunoglobulins, IgG, IgA, and IgM, and the serum proteins, $alpha_2$ macroglobulin (A_2M), transferin (TRF), $alpha_1$-antitrypsin (A_1AT), ceruloplasmin (CPL), and lysozyme (LYS)(Bencko et al., 1983). Examinations were carried out on workers occupationally exposed to nickel (38 individuals) or cobalt (35 individuals) and in groups of non-occupationally exposed children living in areas with different levels of air pollution from a nearby source of nickel and cobalt emissions.

Non-exposed controls were represented by a group of 42 male adults, matched by age, and by a group of 48 children from a non-polluted area. Significantly increased average values were obtained for IgG, IgA, and IgM, in the group of workers exposed to nickel, for IgA, in workers exposed to cobalt, and for A_1AT, A_2M, CPL and LYS, in both groups of occupationally-exposed adults

($P > 0.001$ and $P > 0.005$, respectively). Among non-occupationally exposed children, the most highly exposed group had significantly elevated average values for A_2M and A_1AT, which were higher than those recorded in the groups of "less exposed" and control children ($P > 0.02$ and $P > 0.05$, respectively) (Bencko et al., 1983).

9.2 Skin and eye irritation and contact hypersensitivity

9.2.1 Skin and eye irritancy

Primary skin irritation reactions to nickel salts in solution were observed when human volunteers were patch tested. When aqueous solutions of nickel chloride were applied to the back, the threshold concentrations for irritancy were 1% with occlusion and 10% without occlusion (Vandenberg & Epstein, 1963).

A 5% aqueous solution of nickel sulfate was also irritant in some individuals, when applied to the back, under occlusion (Kalimo & Lammintausta, 1984). On the unoccluded human forearm, the threshold concentrations of aqueous nickel sulfate producing irritation were 20% on unbroken skin and 0.13% on scarified skin, the nickel solutions being applied once daily for 3 days (Frosch & Kligman, 1976).

9.2.2 Contact hypersensitivity

Nickel and its water-soluble salts are potent skin sensitizers. Nickel ions pass the skin barrier and bind to carrier proteins to form the allergen. In the pathogenesis of allergic contact dermatitis (a type IV reaction), two steps can be recognized. First, sensitization, after which, subsequent exposure of fully sensitized skin, even to minute quantities of nickel or its water-soluble salts, will elicit or provoke an eczematous response which begins within 12 h, is fully developed at 48–72 h, and then subsides. After the elicitation phase, the sensitivity will remain, usually for years, and frequently for a lifetime. Sensitization is only known to occur after dermal exposure to nickel, whereas provocation can take place after exposure of the skin and after ingestion of nickel. The clinical manifestations of nickel-associated allergic contact dermatitis comprise both primary and secondary eruptions. The primary eruptions usually appear at the site of sensitization, and the secondary eruptions will frequently be more or less generalized, often symmetrical, maculopapular vesicules of the pompholyx type eczema, which affects the flexures of the elbows, the sides of the neck, the axillae, the eyelids, and the genital area. A vesicular hand eczema can also develop. Nickel contact dermatitis has been documented in the general population and in nickel workers, especially in the nickel-plating industry (NAS, 1975). Nickel contact hypersensitivity was originally considered to be a problem in occupational medicine, but more recent epidemiological and clinical data indicate that it could also be a problem in the general population, because of the increasing number of nickel-containing commodities in everyday use. The prevalence of nickel sensitivity in the general population can be evaluated by patch-testing with 5% nickel sulfate in petrolatum or by an interview technique. Peltonen (1979) and Prystowsky et al. (1979) studied volunteers from different subpopulations in Finland and the USA. The prevalence of nickel sensitivity was about 10% for women and about 1% for men. In women, Prystowsky et al. (1979) found a high correlation of sensitivity to nickel with the practice of ear piercing. Peltonen (1979) reported that hand eczema is very common in patients with nickel sensitivity. Christensen & Möller (1975a) reported that pompholyx was the predominant type of hand eczema. Nickel allergy was investigated in a stratified sample of the Danish female population using an interview technique (Menné et al., 1982). Of those responding, 14.5% reported a history of nickel allergy. The prevalence rate was highest in the younger age groups and declined after the age of 50 years. Although the use of the interview technique had certain limitations, the data were in good agreement with data obtained through other test methods (e.g., patch testing).

Edman & Möller (1982) reported a study on a patient population of 8933, who had been patch-tested over a 12-year period. Nickel sensitization increased during this period in both male and female patients, females producing a higher rate of positive reactions than males.

Oral provocation of hand eczema in nickel-allergic patients with vesicular hand eczema has been studied (Nielsen, in press). Oral provocation, using nickel sulfate in lactose capsules, was performed by Christensen & Möller (1975), Kaaber et al. (1978), Cronin et al.

(1980), Jordan & King (1979), Burrows et al. (1981), and Gawkrodger et al. (1986). The last three studies were double-blind, weeks with provocation and weeks with placebo following each other randomly. Evaluation of the eczema was performed in the 3 days following exposure and it was assumed that no delayed response occurred. The doses ranged from 0.5 to 5.6 mg nickel; lower doses were used in the double-blind studies. An exacerbation of the eczema was observed with doses of 2.5 mg nickel or more. The response to nickel sulfate in lactose differed from nickel contained in the diet. A provocation study was performed with a diet naturally containing about 500 µg nickel. Two conclusions were drawn: (i) a diet naturally high in nickel content (chocolate cake, soya bean stew, and oatmeal) resulted in an exacerbation of vesicular hand eczema, when ingesting the diet daily for 4 days, (ii) exacerbation of hand eczema was observed on day 7 after finishing the 4-day provocation period. This delayed reaction may be the reason why reactions to nickel were not reported in previous double-blind studies in which weeks with nickel exposure and weeks with placebo randomly followed each other.

The response to oral intake of nickel salts may not be uniform and hyposensitization may be possible. In 2 controlled studies, doses of 5.0 mg nickel sulfate given once a week, for 6 weeks, significantly lowered the degree of contact allergy, as measured by patch test reactions before, and after, nickel administration (Sjövall et al., 1987). Further adaptation to oral doses of 2.24 mg nickel sulfate was observed after the volunteer subjects were given gradually increasing daily doses over a 3-month period (Santucci et al., 1988). These reported responses to the oral intake of inorganic nickel do not necessarily contradict one another, but the mechanisms are not known. However, they indicate that hand eczema can benefit from a diet low in nickel.

Allergic reaction as well as urticarial and eczematous dermatitis in nickel-sensitive persons can be produced by implanted prostheses and other surgical devices made from nickel-containing alloys (NAS, 1975; Fisher, 1977). Deutman et al. (1977) found nickel sensitivity in 11 out of 85 patients following hip prosthesis implantation. Three of the patients had had previous implants. Oakley et al. (1987) described 4 patients with a history of metal intolerance, who developed dermatitis following surgical wound closure with disposable skin clips (nickel content about 10%). Two positive reactions were reported by Waterman & Schrik (1985) in 85 patch-tested patients, who had undergone hip arthroplasty.

Metal alloys, used in dental surgery, contain various amounts of nickel. Data on the relationship between nickel allergy and dental alloys are limited. In a retrospective study, Moffa et al. (1983) found that the incidence of nickel allergy in patients with intraoral exposure to nickel alloys was not significantly higher than that in patients with no intraoral exposure.

Treatment of nickel-sensitive patients who were suffering from hand eczema of the pompholyx type, with the nickel-chelating agent diethyldithiocarbamate was successful in 10 out of 11 patients (Christensen & Kristensen, 1982). However, hepatotoxicity was observed in some patients, and this form of treatment is no longer used.

9.3 Reproduction, embryotoxicity, and teratogenicity

No data are available on the reproductive and developmental effects of nickel in human beings.

9.4 Genetic effects in exposed workers

Elevated levels of chromosome aberrations, mainly gaps, were reported by Waksvik & Boysen (1982) in 2 groups of nickel workers, but SCE was not observed; one group was involved in nickel crushing/roasting/smelting processes and exposed mainly to nickel oxide and nickel sulfide, and the other was involved in electrolysis and exposed mainly to nickel chloride and nickel sulfate. In a further study on 9 ex-nickel workers, who had been exposed for an average duration of 25 years, but retired for at least 8 years, Waksvik et al. (1984) showed the persistence of a low level of gaps, as well as chromatid breaks, when plasma nickel levels were still 2 μg/litre. Deng et al. (1983, 1988) showed elevated levels of both SCEs and chromosome aberrations (gaps, breaks, fragments) in electroplating workers. However, these effects were not observed in a study by Decheng et al. (1987) on workers exposed to nickel carbonyl. Low exposure and good workplace conditions were cited by the latter authorities.

9.5 Carcinogenicity

9.5.1 Epidemiological studies

9.5.1.1 Nickel refining industry

(a) Clydach nickel refinery, Clydach (Wales)

The existence of an increased risk of nasal cancer in workers from the Clydach nickel refinery was recognized in 1928. Processes at Clydach involved: the crushing and grinding of nickel copper matte, received from Canada, which consisted of the phases Ni_3S_2, CuS, and metallic nickel and copper, in a 3:4:1 ratio (Boldt & Queneau, 1967); the calcining of the crushed matte to produce nickel and copper oxides (the copper oxide is dissolved in the metal oxide and described as $(Ni,Cu)O$ (Boldt & Queneau, 1967)); the extraction of the copper by sulfuric acid; reduction to nickel oxide; and refining with nickel carbonyl. The first epidemiological study on Clydach plant workers was carried out by Hill in 1939 (quoted by Morgan, 1958). Hill found that the observed to expected ratios (O/E) of deaths were 16:1 for lung cancer and 22:1 for nasal cancer in the period 1929–38. Morgan (1958) and Doll (1958) extended the observation period from 1902 to 1957 and identified 131 deaths from lung cancer and 61 deaths from nasal cancer in 9340 workers. Excess mortality occurred only in workers first employed before 1924. The highest risk of lung cancer appeared to have occurred in the calcination and copper sulfate departments. Deaths from nasal cancer occurred most frequently in the calcination department. There were no nasal cancers among workers entering employment after 1924. The decline in the numbers of deaths from nasal and lung cancer was attributed to improvements in work practices and refinery processes in the early 1920s. These improvements included the wearing of gauze masks, improvements in ventilation, and the use of arsenic-free sulfuric acid in refining. Follow-up studies were carried out by Doll et al. (1970) on a cohort of 819 workers, who had been employed for at least 5 years at Clydach between 1902 and 1944, and who were still employed in 1934. Thirty-nine cases of nasal cancer and 113 cases of lung cancer were identified. The reported O/E ratio for lung cancer was 10.5 for workers starting between 1910 and 1915, 1.8 for workers employed between 1925 and 1929, and 1.1 for those employed between 1930 and 1944. In this study, no nasal cancer was reported in workers starting employment after 1925. Doll et al., (1977) extended the follow-up period to 1971 and confirmed that, since 1925, the risk of respiratory cancer had decreased sharply; they attributed this to the improvements in work practices and refinery processes.

Cuckle et al., (1980) studied the causes of death in a small cohort of 297 men who were mainly exposed to soluble nickel salts between 1937 and 1960. They did not find any nasal sinus cancers. Thirteen cases of lung cancer were found, a significant excess above regional rates (expected number 7.54), but not significant when compared with the rates for England and Wales. The small size of the group limited this study.

Estimated concentration data for different nickel species in different areas of Clydach nickel refinery are shown in Table 33.

Table 33. Estimated concentration (mg/m^3) before 1930, of major forms of nickel in high-risk areas of the Clydach nickel refinery, South Wales [a]

	Metal	Oxide	Ni_3S_2	Soluble
Calcining	5.3	18.8	6.8	0.8
Furnaces	5.6	6.4	2.6	0.4
Copper plant	-	13.1	0.4	1.1
Hydrometallurgy	0.5	0.9	0.1	1.4

[a] From, Peto (1988).

Peto et al., (1984) extended the follow-up to 1981. The cohort included 968 men. One hundred and fifty-nine lung cancer deaths and 56 nasal cancer deaths were identified. The duration of employment in 4 areas showed a statistically significant association with lung or nasal cancer or both. These were the calcining furnace area, the calcining/crushing area, the copper sulfate area, and the furnace area. Peto et al. (1984) also related age, time of exposure, and exposure levels, to lung and nasal sinus cancer incidence in pre-1925 workers. Exposure level was defined on the basis of duration of work in high-risk areas. Both lung and nasal cancer showed increasing risk with increasing duration of work in high-exposure areas. Nasal cancer showed a strong positive relationship with age at first exposure, whereas lung cancer did not. The latency period was 20–50 years for nasal cancer and somewhat shorter for lung cancer. The risk of nasal cancer was highest for workers with first exposure between 1910 and 1915 and declined thereafter, whereas the risk for lung cancer was highest for workers between 1920 and 1925.

(b) *Falconbridge refinery, Kristiansand (Norway)*

Loken (1950) reported 3 cases of lung cancer in workers from the Falconbridge refinery. This plant began operations in 1910, processing matte from Ontario via roasting, smelting, and electrolysis. Although workers in the roasting/smelting departments were, and are, exposed to sulfides, oxides, and nickel-copper oxides in dust and fumes, the processes have changed in recent years and exposure levels have been reduced dramatically. The electrolytic workers are exposed to aerosols containing, *inter alia*, nickel sulfate and nickel chloride (Hogetveit & Barton, 1976). These workers are, as cited by Hogetveit & Barton (1976), "exposed largely to soluble nickel aerosols". That the soluble form predominates in the exposure of electrolysis workers is also supported by the results in Table 34. This table shows that, even when the air values are lower in the electrolysis department than in roasting/smelting department, the plasma and urine values are higher.

The first epidemiological study was carried out on a cohort of 1916 workers who were employed for at least 3 years before 1961 (Pedersen et al., 1973). Exposure was defined according to department or work in which there had been the longest employment. The authors used 4 work categories: roasting and smelting, electrolysis, other processes, other work groups. All occupational groups were associated with an excess risk of respiratory cancer. Risk was greatest in the roasting and smelting department and the electrolysis department. Workers in the electrolysis department showed the highest risk of lung cancer ($O/E = 26/3.6$) and an excess risk of nasal cavity cancer ($O/E = 6/0.2$). The risk of nasal cancer was greatest ($O/E = 5/0.1$) in the roasting and smelting department,

Table 34. Summary of biological tests and personal sampler measurements [a]

Departments	Number of employees	Mean plasma ± SD (μg Ni/litre)	Mean urine ± SD (μg Ni/m^3)	Mean concentration in air ± SD (μg Ni/m^3)
No. 01-19 (Roasting-smelting)	24	7.2 ± 2.8	65.0 ± 57.6	0.86 ± 1.20
No. 22-44 (Electrolytic)	90	11.9 ± 8.0	129.2 ± 105.6	0.23 ± 0.42
Other process departments	13	6.4 ± 1.9	44.6 ± 26.7	0.42 ± 0.49

CORRELATION BETWEEN RESULTS

Departments	Number of employees	Plasma vs. air Corr. factor R	Plasma vs. air Corr. factor R	Plasma vs. urine Corr. factor R
No. 01-19	24	-0.55 -0.11	-0.71 +0.14	3.72 +0.76
No. 22-44	90	1.98 0.21	2.99 0.31	7.30 0.77
Other	13	2.41 0.67	1.70 0.47	2.27 0.63

[a] Modified from Hogetveit et al. (1978).

but there was also an increased risk of lung cancer in this department (O/E = 12/2.5).

Magnus et al. (1982) updated the study of Pedersen et al. (1973) and confirmed the results. They also analysed data on smoking habits and found evidence for an additive effect of smoking and nickel exposure.

The cohort was further followed up by Andersen (1988). During the follow-up period 1953–87, 163 new cases of respiratory cancer were observed against 42 expected. There was a decreasing trend in ratio between observed/expected from 6.9 in 1926–35 to 2.4 among those first employed in 1956–65. No risk for any other cancer was recorded, in fact, the observed number is lower than the expected number in all periods (207/248 all periods combined). No cases of nasal cancer have been observed in workers employed after 1955, but the expected number is only 0.20.

Following technological improvements in the plant, Andersen (1988) formed a new cohort of 1237 employees with a total employment of one year or more, the first entry in 1968 or later, and exposure estimated according to nickel species and exposure concentration. For entry in the period 1968–72, 7 new cases of lung cancer were found, versus 1 expected. Actual measurements reported in 1979 (Torjussen & Andersen, 1979) were $0.1–0.5$ mg/m^3 for electrolysis workers and $0.1–1.0$ mg/m^3 for the roasting/smelting category. Estimated concentrations for the period 1968–77 were about $0.5–2.0$ mg/m^3 and about $2.0–5.0$ mg/m^3, respectively. In the electrolysis department, the exposure was mainly to soluble nickel, even if there might have been up to 60% NiS and NiO present in some working areas. No exposure to soluble compounds was considered to take place in the roasting/smelting department, but the possible presence of NiSO$_4$, as an intermediate, was not appreciated.

The risk of lung cancer for a 1953–84 cohort of 3250 workers was also estimated according to department and to cumulative exposure to different nickel species (nickel subsulfide, nickel oxide, soluble nickel). Although an increased risk was found for both electrolysis and roasting, smelting, and calcining, the highest risk and steepest dose-response relationship (based on exposure time) was found for electrolysis, and the highest risk was found for the highest cumulative exposure to soluble nickel. It must be remembered, however, that most workers were exposed to more than one form of nickel.

Torjussen et al. (1979a,b) examined nasal biopsy specimens from 318 active and 15 retired workers with at least 8 years of employment. Epithelial dysplasia was found in 14 (14%) of the 97 roasting/smelting workers, in 16 (11%) of the 144 electrolysis workers, in 8 (10%) of the 77 non-process workers and in 7 (47%) of the retired workers. One out of 57 controls, a carpenter, had dysplasia. Two co-workers from the roasting/smelting department had nasal carcinoma. Based on the methods employed by Torjussen et al., (1979a,b), histopathological reinvestigations of nasal biopsy specimens from workers, previously included in the Torjussen study, were carried out by Boysen et al. (1980, 1982, 1984) and similar results were found. The histopathological changes in the nasal epithelium, including dysplasia, persisted in active workers, despite reduction of airborne nickel concentrations in the working environment. The frequency of dysplasia was also still elevated in retired workers. The real significance of nasal epithelial dysplasia is not known, but it could be considered as a precancerous lesion because of its prevalence in groups with increased risk of nasal cancer. The role of nasal biopsies in identifying and monitoring occupational nickel exposure is unclear.

(c) *International Nickel Company (INCO), Ontario*

Sulfide nickel ore is mined and refined at several locations using different processes. The Ontario Health Department studied a cohort of 2355 refinery workers followed from 1930 to 1957 (Sutherland, 1959, reviewed by Mastromatteo, 1967). There were 7 deaths from sinus cancer (0.19 expected) and 19 from lung cancer (8.45 expected). Nearly all excess deaths from sinus cancer and lung cancer occurred in workers exposed to furnace dusts in calcining and sintering. Mastromatteo (1967) extended Sutherland's study to the period between 1930 and 1960. The risk of nasal cancer was still elevated (16 sinus cancer deaths observed, 0.166 expected) as well as the risk of lung cancer (37 lung cancer deaths observed, 12.71 expected). According to Mastromatteo (1967), the investigations of Sutherland led to major process changes, among them the termination of the sinter plant operations, in 1962. The Sutherland study was updated by Rigaut(1986) and the US EPA, (1986a) to cover the period between 1930 and 1970. By this time, sinus and lung cancer deaths amounted to 24 (SMR = 5106) and 76 (SMR = 1861), respectively.

On the basis of the unpublished work by Sutherland (1959), Chovil et al. (1981) identified a cohort of 495 men, who had been employed at the Copper Cliff sinter plant for a period between 1948 and 1962, and were alive in 1963. However, only 75% of the cohort were successfully followed through 1977 and 1978. Six cases of sinus cancer were identified (no expected numbers given), and 54 cases of lung cancer of which 37 were fatal, compared with 4.25 expected deaths. Despite some methodological weaknesses, an excess risk of lung and sinus cancer is suggested.

A large cohort study was carried out by Roberts et al. (1984). The cohort consisted of 54724 INCO workers employed in Port Colborne refinery and in the extraction and refining of copper and nickel in two Sudbury sinter plants (Coniston smelter and Copper Cliff plant). The cohort was followed for mortality between 1950 and 1976. In Sudbury workers not exposed to sintering operations, the SMRs for nasal and lung cancer were 144 (O/E = 3/2.08) for nasal cancer and 108 (O/E = 222/204.98), respectively, compared with SMRs of 2174 (O/E = 2/0.09) for nasal cancer and 463 (O/E = 42/9.08) for lung cancer in workers exposed to sintering operations. In workers from the Port Colborne leaching, calcining, and sintering departments, SMRs for nasal and lung cancer were 9412 (O/E = 16/0.17) and 298 (O/E = 49/16.44), respectively, compared with SMRs of 78 (O/E = 13/16.65) for lung cancer in men who had never worked in this department. A single case of nasal cancer was observed compared to 0.16 expected, but, on further investigation, it was found that this man had had 20 years exposure to roasting/calcining operations elsewhere. Nasal and lung cancer mortality was found to increase with increasing duration of exposure. Lung cancer was already seen in excess 5–10 years after first exposure in Sudbury, but, in Port Colborne, it did not become apparent until 20 years after the first exposure. Nasal cancer began to appear after 15 years. Apparently the nasal cancer risk was higher at Port Colborne than at Sudbury, while lung cancer was more frequent at Sudbury.

Airborne nickel levels were much higher in the sintering department than in other areas of these refineries. Warner (1985) reported that a 40-h sample, taken on the operating floor of the Copper Cliff Sinter Plant in 1960, contained 46.4 mg total dust/m^3. The major nickel compounds in the dust were Ni_3S_2, NiO, and $NiSO_4$. Samples taken between 1948 and 1952 from dusty air escaping from roof monitors contained over 100 mg total dust/m^3, but it is not known if these were representative of exposures in working areas. Other processes are believed to have been very much less dusty. Mean levels in most areas were generally well below 1 mg nickel/m^3 in later years, though substantially higher levels occasionally occurred, particularly in matte grinding and in the handling of metallic nickel powders (Warner, 1984).

The composition of a sample of dust collected in 1959 from the calcining and sintering rooms of the Port Colborne, Ontario, refinery is shown in Table 35 (Gilman & Yamashiro, 1985). The major constituents were nickel subsulfide (57%), nickel sulfate (20%), nickel oxide (6%), and copper oxide (3%).

Table 35. Composition of dust collected in 1959 in the calcining and sintering rooms of the Port Colborne (INCO) nickel refinery Ontario [a]

Compound	Percentage
CuO	3.4
$NiSO_4.6H_2O$	20.0
Ni_3S_2	57.0
NiO	6.3
CoO	1.0
Fe_2O_3	1.8
SiO_2	1.2
Miscellaneous	2.0
Moisture	7.3

[a] From, Gilman & Yamashiro (1985).

Kaldor et al. (1986) analysed the same data using a case-control approach. In addition to the four areas identified by Peto et al. (1984), employment in nickel sulfate production was associated with increased lung and nasal cancer risk. Lung and nasal cancer mortality and exposure levels for various nickel compounds for men employed in major production areas were reported by Peto (1988). There were 5 lung cancer deaths (1.5 expected) and 4 nasal cancer deaths among men, employed for 5 or more years in hydrometallurgy (nickel sulfate production), who had spent less than a year in other high-risk areas (copper sulfate, calcining, furnaces). The principal exposure in hydrometallurgy was to soluble nickel (about 1.4 mg/m^3) with lower levels of nickel oxide (0.9 mg/m^3), nickel subsulfide (0.1 mg/m^3), and metallic nickel (0.5 mg/m^3). These estimates were provided by the technical staff of the refinery, but no measurements were made to verify them.

(d) Falconbridge, Ltd., Falconbridge (Ontario)

A mortality study was carried out on a cohort of 11 594 workers, employed for at least 6 months by the Falconbridge nickel mines, Sudbury, Ontario, between 1950 and 1976 (Shannon et al., 1984). In this plant, sintering techniques included a low temperature step. Nasal cancer was not observed. A slight excess of lung cancer (SMR = 123, O/E = 46/37.5) in both mines and service occupations was not statistically significant. However, laryngeal cancer deaths were significantly increased in miners (SMR = 261, O/E = 5/1.92). A slight, though not significant, increase in deaths from prostate cancer was noted in smelter workers.

(e) Sherritt Gordon Mines Ltd., Fort Saskatchewan, Alberta

Sherritt Gordon Mines, Ltd., established hydrometallurgical nickel refining in 1954. Nickel exposures were mainly in the form of oxide ore concentrate, nickel metal, and some soluble compounds. Egedahl & Rice (1984) conducted a study of cancer incidence and mortality with a cohort of 720 workers employed in hydrometallurgical refining and did not find any excess of respiratory cancer. Two cases of renal cancer occurred in workers with 11 and 16 years exposure to nickel concentrate, soluble nickel substances, and nickel metal, prior to developing their illness. Both were cigarette smokers.

(f) Nickel refinery, Huntington (West Virginia)

Copper-nickel matte was refined in Huntington during 1922–47 and a mortality study was performed on 1855 men, employed before

1947 who had worked for at least one year at the plant (Enterline & Marsh, 1982). Concentrations of airborne nickel in the area where nickel was crushed, ground, and handled, were estimated to range from 20 to 350 mg/m^3 and from 5 to 15 mg/m^3 in the calcining area. The SMR for nasal cancer was 2443.5 with 2 observed and 0.08 expected cases. Two nasal cancers were observed among nickel-exposed non-refinery workers. There was no significant increase in deaths from lung cancer among refinery workers (SMR = 118.5 with 8 observed and 6.75 expected cases) and it was suggested that the nickel exposures at the plant may have been considerably lower than in other plants. However, for refinery and non-refinery workers combined, and for all respiratory cancer cases combined, there was a dose-relationship with accumulated nickel exposure.

(g) Oxide ore refinery, New Caledonia

Nickel oxide ore has been mined on the South Pacific Island of New Caledonia since 1866 and smelted there since 1885. Lessard et al. (1978) identified 92 cases of lung cancer that had occurred in this area between 1970 and 1974. The observed rate was 3–7 times higher than rates in other Central or South Pacific areas, but similar to those seen in industrialized countries. A significant increase in lung cancer rates was observed with decreasing distance from the refinery. The authors found that, in New Caledonia, there was a 3-fold risk of lung cancer among workers compared with non-workers, independent of age and tobacco smoking, and considered that these results supported the etiological role of nickel in lung cancer. Goldberg et al. (1985a,b) carried out a retrospective cohort study and a case-control study and reported no excess of respiratory cancer in workers of the Society Le Nickel (SLN) compared to the general population of New Caledonia; however, these studies have weaknesses in statistical power and definition of the control population. Goldberg et al. (1988) concluded that, in a 10-year period, there was no excess of either lung or sinonasal cancer in the nickel workers, when compared with the island population on an age-matched basis. Both studies were considered to have methodological problems, but the Goldberg et al. (1988) study had more comprehensive health data.

(h) Other nickel refineries

In the Soviet Union, Saknyn & Shabynina (1970, 1973) reported an increase in lung and gastric cancer in workers in a nickel refinery plant in the Urals. The refinery was engaged in the preparation of nickel oxide ore, roasting, and smelting. They stated that the lung cancer death rate from 1955 to 1967 exceeded that of the nearby urban population by 180% and that the death rate from sarcomas and stomach cancer was also increased. Excess lung cancer was seen in workers in roasting/reducing and smelting, but also in electrolysis, where exposure was principally to nickel sulfate and nickel chloride, and, in the cobalt shop, where exposure was to various soluble salts of nickel and cobalt.

Tatarskaya (1965) reported 2 cases of nasal cancer among workers engaged in the electrolytic refining of nickel.

In workers in a Czech plant, where nickel oxide ore had been refined since 1963, 8 lung tumours were diagnosed between 1971 and 1980 (Olejar et al., 1982). All workers had smoked more than 25 cigarettes per day. These results must be treated with caution as the latency period from 1963 was very short.

Rockstroh (1959) reported 45 cases of lung cancer in workers in a nickel refinery in Aue in the German Democratic Republic between 1928 and 1956. Exposure was very complex: nickel, arsenic, cobalt, copper, bismuth, and benzo(*a*)pyrene, and workers were often involved in different processes, during their employment. One case of lung cancer was found among office personnel, but the size of the comparison group was not indicated.

9.5.1.2 Nickel alloy manufacturing

In a nickel alloy manufacturing plant in Hereford, England, the alloys are manufactured from materials consisting of metallic nickel and other metals (iron, copper, cobalt, chromium, molybdenum). Workers are exposed to metallic nickel and oxidic nickel. The average airborne nickel concentrations ranged from 0.84 mg/m^3 in the melting, fettling, and pickling area to 0.04 mg/m^3 in the process, stock handling, distribution, and warehouse. Cox et al. (1981) studied a cohort of 1925 men with at least 5 years of employment between 1953 and 1978. He did not find any cases of nasal sinus

EHC 108: Nickel

cancer. The lung cancer death rate was not increased compared with the numbers expected from national and local mortality rates. However, the study period was too short to exclude the risk of respiratory cancer completely.

Redmond (1984) reported a cohort mortality study on 28 261 workers employed at 12 high nickel alloy production plants between 1956 and 1960, including a follow-up to 1977. The predominant nickel species, to which the workers were exposed, appeared to have been nickel dust and nickel oxide. Average levels of airborne nickel ranged from 0.064 mg/m^3 in the cold working area to 1.5 mg/m^3 in the powder metallurgy area (Redmond et al., 1983, internal report). Overall, no statistically significant increased risk was observed for cancers of the lung, nasal sinuses, larynx, or kidney.

Another study dealt with the standardized proportionate mortality ratios of 851 foundrymen from 26 nickel/chromium alloy foundries compared with 141 unexposed foundrymen during the period 1968–79 (Cornell & Landis, 1984). There were no cases of nasal cancer. The percentages of respiratory cancer in the exposed group were not significantly different from the percentages of respiratory cancer in the unexposed group. When compared to national mortality data, the observed numbers of lung cancer and all cancer deaths in exposed men were not significantly different from the expected values.

Cornell (1984) performed a proportional mortality analysis on 4487 workers in 12 US plants engaged in the production of stainless steel and low nickel alloys. All deaths were included for, at least, a 5-year period, up to 1977. No cases of nasal cancer were observed and there was no excess of lung cancer. However, details of the study, such as the definition of exposure and latency period, were not provided.

9.5.1.3 Nickel plating industry

Silverstein et al. (1981) conducted a case-control study on deaths among workers at a Scandinavian die-casting and electroplating plant. The major operations in the plant were zinc alloy die-casting, buffing, polishing of zinc and steel parts, and electroplating with copper, nickel, and chrome. There were 238 deaths among workers with at least 10 years of employment. The proportionate mortality ratio (PMR) (38 deaths, PMR = 209) for lung cancer was significantly increased. However, because of heterogeneous exposure, the increased risk of lung cancer mortality could not be exclusively associated with nickel exposure.

The authors subsequently initiated a case-control study on the 28 white male and 10 white female lung cancer deaths. Cases and controls were compared for length of employment in individual departments. Odds ratios were estimated for work in 14 different departments. For males in three departments, the odds ratio increased with duration of work in the departments. This trend was most significant for the department where die-casting and plating were the major activities.

A study was carried out by Burges (1980) on a group of 508 nickel platers in an engineering firm in England. Nickel plating was done in a separate shop from chrome plating. The cohort was divided into groups: nickel bath workers with less than one year of exposure, more than one year of exposure, and other exposed workers. In the short-exposure group (less than one year), no excess mortality was found, whereas, in the longer-exposure group (more than one year), a significant excess of deaths was found for stomach cancer (O/E = 4/0.6, SMR = 623) and for non-malignant respiratory disease (O/E = 8/2.8, SMR = 286). Only one lung cancer death was observed in the longer-exposure group compared with 1.7 expected cases. The results of the study suggest a higher incidence of gastric cancer in nickel platers. Sorahan et al., (1987) reported a study on mortality in nickel and chromium platers in the United Kingdom. While an excess of certain cancers was associated with chromium bath workers, it was stated that nickel exposure was not a confounding factor.

9.5.1.4 Welding

Stern (1983) and Langard & Stern (1984) concluded that the excess risk of lung cancer in welders is approximately 30% greater than that in the non-welding population. It was suggested that the greater risk of lung cancer found among stainless steel welders, in some studies, could be related to high concentrations of nickel and hexavalent chromium in stainless steel welding fumes.

The role of nickel compounds as a cause of respiratory cancer is difficult to assess, because chromates and other carcinogens may occur in the work-place air during the welding of stainless steel and other nickel-containing alloys (IARC, 1990).

9.5.1.5 Nickel powder

In the Oak Ridge Gaseous Diffusion Plant, Tennessee, metallic nickel is used in the production of barrier material for the isotopic enrichment of uranium by gaseous diffusion. The median nickel concentration in air between 1948 and 1963 ranged between 0.1 and 1.0 mg/m^3 (Godbold & Tompkins, 1979). In a cohort of 814 workers, Godbold & Tompkins (1979) did not find any increased risk of respiratory cancer during the period 1948–72. In a follow-up study by Cragle et al. (1984) covering the years from 1972 to 1977, the lung cancer rate did not increase. The SMR for men employed for more than 5 years was 91, based on 8 deaths.

9.5.1.6 Nickel-cadmium battery manufacturing

Sorahan & Waterhouse (1983) examined 3025 nickel-cadmium battery workers for a possible excess of cancer mortality. Overall, the SMRs showed a statistically significant excess of respiratory cancer deaths, but it was not possible to attribute the excess to nickel or to cadmium, since exposure to both occurred at high levels. In studies by Kjellstrom et al. (1979) and Andersen et al. (1983), the mortality was examined in men working at a nickel-cadmium battery-manufacturing plant with mixed exposure to both metals and nickel hydroxide. Increased respiratory, renal, and prostate cancer mortality were noted, though there was no increase in overall cancer mortality. There were two cases of nasopharyngeal cancer. One started employment before 1948, the other between 1948 and 1961, when exposure to nickel powder was very high, exceeding 100 mg nickel/m^3 in the work area. The low risk ratio for lung cancer rendered the studies inconclusive for the effects of nickel compounds.

9.5.1.7 Case-control studies

The case-referent design has been used to associate the incidence of certain cancer types with occupations in nickel-producing or nickel-using industries.

In a British study (Acheson et al., 1981), an excess risk for nasal cancer was found in furnace and foundry workers, partly, because of several cases among Clydach refinery workers.

Bernacki et al. (1978b) carried out a small case-control study at an aircraft-engine factory where high nickel alloys were used. Nickel exposure was not found to be associated with lung cancer cases. Roush et al. (1980), using data from the Connecticut Tumor Registry and from City Directories, determined the occupational exposures of individuals reported to have cancer of the nasal sinuses in the years 1935–75. The study did not reveal any increase in nasal cancer deaths in occupations classified as having nickel exposure. A case-control study of 204 laryngeal cancer patients and 204 neighbourhood controls did not show any relation to occupational exposure to nickel, while tobacco, alcohol, and asbestos were major risk factors (Burch et al., 1981). However, Olsen & Sabroe (1984) found a statistically significant excess risk of laryngeal cancer related to potential nickel exposure from alloys, battery chemicals, and chemicals used in the production of plastics. A case-referent study covering the entire Montreal population showed a statistically significant odds ratio of 3.1 for lung cancer and nickel exposure and some indication of a dose-response relationship (Gerin et al., 1984), though individuals exposed to nickel were always exposed to chromium as well.

9.5.2 Carcinogenicity of metal alloys in orthopaedic prostheses

It was concluded by Sunderman (1989b) that case reports of sarcomas at sites of metal implants in human beings and domestic animals have implicated carcinogenesis as a potential, albeit rare, complication of metal orthopaedic prostheses (Table 36). Since physicians and veterinarians are not obliged to report adverse reactions to prostheses, the actual incidences of sarcomas in association with metal implants cannot be reliably estimated, and the

Table 36. Patients with tumours at sites of metal implants[a]

Patient age [years] (site)	Time lapse to tumour (years)	Implant type	Tumour type	Reference
12 (humerus)	30	Steel plate (FeCrNi) and 4 screws (FeCr), much corrosion	Ewing's sarcoma	McDougall (1956)
37 (tibia)	3	Osteosynthesis plate and 2 screws, alloy unknown	Osteochondrosarcoma	Delgado (1958)
53 (femur)	3	Steel plate and nail (6 months) then Moore prosthesis and trochanteric wires	Giant cell tumour	Castleman & McNeely (1956)
58 (tibia)	26	Steel (Type 316) plate and screws (Types 304 and 316)	Haemangiosarcoma	Dube & Fisher (1972)
56 (hip)	2.5	Mckee metal-to-metal arthroplasty, alloy not stated	Fibrosarcoma	Arden & Bywaters (1978)
4 (hip)	7	Sherman (CoCrMo) plate and 6 screws (implanted 1 year)	Ewing's sarcoma	Tayton (1980)
31 (tibia)	17	Sherman (CoCrMo) plate and screws	Histiocytic lymphoma	McDonald (1981)

Table 36 (continued)

75 (hip)	2	Charnley-Muller arthroplasty and trochanteric wires, alloy not stated	Malignant fibrous histiocytoma	Bago-Granell et al. (1984)
63 (hip)	2.4	McKee-Farrar arthroplasty (CoCrMo)	Malignant fibrous histiocytoma	Swann (1984)
75 (hip)	5	Ring arthroplasty (CoCrMo)	Osteogenic sarcoma	Penman & Ring (1984)
76 (knee)	4.5	CoCrMo arthroplasty	Epithelioid sarcoma	Weber (1986)
14 (hip)	29	Sherman screw (CoCrMo)	Malignant fibrous histiocytoma	Hughes et al. (1987)
40 (hip)	13	2 Screws (unspecified alloy) for 12 years; then hip prosthesis (CoCrMo stem and Al_2O_3 head)	Undifferentiated sarcoma	Ryu et al. (1987)

a Modified from: Sanderman (1989b).

possibility that the apparent associations are coincidental cannot be excluded. Carcinogenesis bioassays in rodents have shown that pure nickel or cobalt powders induce sarcomas at injection sites, and transformation assays in mammalian cells have yielded positive results with nickel powder, suggesting that surgical implantation of alloys containing these metals may pose carcinogenic risks.

10. EVALUATION OF HUMAN HEALTH RISKS AND EFFECTS ON THE ENVIRONMENT

10.1 Exposure

Nickel is an ubiquitous element and has been detected in different media in all parts of the biosphere.

Nickel is introduced into the environment from both natural and man-made sources and is circulated throughout all environmental compartments by means of chemical and physical processes, as well as through the biological transport mechanisms of living organisms.

Acid rain may leach nickel as well as other metals from plants and soil.

Atmospheric nickel is considered to exist mainly in the form of particulate aerosols.

Nickel is introduced into the hydrosphere by removal from the atmosphere, by surface run-off, by discharge of industrial and municipal wastes, and also following the natural erosion of soils and rocks.

A major source of nickel in the environment is the combustion of fossil fuels, particularly coal.

Uncontrolled emissions and waste disposal may have adverse effects on the environment.

Both the chemical and physical forms of nickel and its salts strongly influence bioavailability and toxicity.

Nickel from soil and water is absorbed and metabolized by plants and microorganisms and these small quantities of nickel are widely present in all foods and water.

Some foods, such as pulses and cocoa products, contain relatively high amounts of nickel, but these quantities have not been correlated with adverse health effects.

10.2 Human health effects

Nickel is normally present in human tissues and, under conditions of high exposure, these levels may increase significantly.

The general population is exposed to nickel via the diet and objects containing nickel, especially jewellery and coins.

Occupational exposure to nickel is important.

Inhalation is an important route of exposure to nickel and its salts in relation to health risks. The gastrointestinal route is of lesser importance.

Nickel absorption from the gastrointestinal tract is poor, though nickel in the drinking-water is absorbed to a greater extent in an empty stomach. This may be a risk for sensitized persons.

Smoking tobacco may contribute to nickel intake, but there is no agreement on the chemical nature of nickel in tobacco smoke and its health significance.

Target organs are the respiratory system, especially the nasal cavities and sinuses, and the immune system.

The percutaneous absorption of nickel is minor, but it is important in sensitization.

Nickel and its salts are potent skin sensitizers and possible respiratory sensitizers in human beings. Nickel dermatitis is a common result of nickel exposure, especially in women.

Primary skin and eye irritation reactions to high concentrations of soluble nickel salts have also been reported.

Acute toxicity is a minor risk from nickel or its compounds, with the exception of nickel carbonyl.

There is no convincing evidence that nickel salts produce point mutations in bacterial systems. However, some nickel salts are clastogenic *in vitro*, producing chromosome aberrations, (transformation) and sister chromatid exchanges in mammalian cells.

There is a lack of evidence of a carcinogenic risk from oral exposure to nickel, but the possibility that it acts as a promoter has been raised.

There is evidence of a carcinogenic risk through the inhalation of nickel metal dusts and some nickel compounds.

Very high concentrations of nickel are required to produce teratogenic or genotoxic effects.

The consequences of the effects of nickel on the immune system are not clear.

10.3 Environmental effects

Nickel is accumulated by plants. Growth retardation has been reported in some species, at high nickel concentrations.

There is no evidence that nickel may undergo biotransformation, though it does undergo complexation.

Nickel has been shown to be essential for the nutrition of many microorganisms, a variety of plants, and for some vertebrates.

11. RECOMMENDATIONS

1. The nomenclature for nickel compounds should be further standardized.
2. Analytical methods should be developed and standardized to facilitate speciation of nickel compounds in environmental media, biological materials, the workplace, and atmospheric emissions.
3. Studies are recommended on the kinetics and mass transfer of nickel, in order to elucidate the biogeochemical cycle on a global scale and its potential for long-range transport.
4. The use of nickel in consumer products that may release nickel when in contact with skin should be regulated. The specification and testing requirements should be standardized.
5. Studies should be performed on the absorption and cellular uptake, transport, and metabolism of well characterized nickel species, following different routes and types of administration.
6. Nickel compounds, to which human beings are exposed, should be tested for carcinogenicity in adequate, long-term, inhalation studies on animals.
7. Further large-scale cohort studies with well characterized exposures, including biological monitoring, are needed in all sectors of industry, to establish more accurate upper limits of cancer risk.
8. Priority should be given to improving industrial hygiene in occupations where exposure to high levels of soluble nickel compounds may occur.

12. PREVIOUS EVALUATIONS BY INTERNATIONAL BODIES

The carcinogenic risk of nickel and inorganic nickel compounds has been evaluated by the International Agency for Research on Cancer (IARC). An International Working Group evaluated epidemiological and experimental carcinogenicity data on nickel and certain nickel compounds encountered in the nickel refining industry (IARC, 1990). The evaluations are summarized below:

	Degree of evidence for carcinogenicity [a]		Overall evaluation
	Human	Animal	
Nickel and nickel compounds			
Nickel compounds			1
Nickel salts		L	
Nickel sulfate	S		
Combination of nickel oxides and sulfides encountered in the nickel refining industry	S		
Nickel monoxides		S	
Nickel trioxides		I	
Nickel sulfide, amorphous		I	
Nickel sulfides, crystalline		S	
Nickel antimonide		L	
Nickel arsenides		L	
Nickel carbonyl		L	
Nickel hydroxides		S	
Nickelocene		L	
Nickel selenides		L	
Nickel telluride		L	
Nickel titanate		I	
Metallic nickel	I	S	2B
Nickel alloys	I	L	

[a] S : Sufficient evidence of carcinogenicity.
L : Limited evidence of carcinogenicity.
I : Inadequate evidence of carcinogenicity.
1 : The agent is carcinogenic for human beings.
2B: The agent is possibly carcinogenic for human beings.

In the *Guidelines for drinking-water quality* (WHO, 1984), no guideline value was set for nickel in drinking-water, as the toxicological data available at the time indicated that a guideline value was not necessary.

In the *Air quality guidelines for Europe* (WHO, 1988), no safe level was recommended for nickel, because of its carcinogenic properties. At an air concentration of nickel dust of 1 µg/m^3, a conservative estimate of the lifetime risk is 4×10^{-4}.

REFERENCES

ABDULWAJID, A.W. & SARKAR, B. (1983) Nickel-sequestering renal glycoprotein. Proc. Natl Acad. Sci. (USA), 80(14): 4509-4512.

ABO-RADY, M.D. (1979a) Determination of heavy metals in two biological and two geological standard materials by atomic absorption spectroscopy. Fresenius Z. anal. Chem., 286: 380-382.

ABO-RADY, M.D. (1979b) The levels of heavy metals (Cd, Cu, Hg, Ni, Pb, Zn) in brook trout from the river Leine in the area of Göttingen (West Germany). Z. Lebensm. Unters. Forsch., 168(4): 259-263.

ABRAHAM, T.J., SALIH, K.Y., & CHACKO, J. (1986) Effects of heavy metals on the filtration rate of bivalve *Villorita cyprinoides* (Hanley) var. cochinensis. Indian J. mar. Sci., 15(3): 195-196.

ACHESON, E.D., COWDELL, R.H., & RANG, E.H. (1981) Nasal cancer in England and Wales: an occupational survey. Medicine, 38(3): 218-224.

ADALIS, D., GARDNER, D.E. & MILLER, F.J. (1978) Cytotoxic effects of nickel on ciliated epithelium. Am. Rev. respir. Dis., 118(2): 347-354.

ADAMSSON, E., LIND, B., NILSSON, B., & PISCATOR, M. (1980) Urinary and fecal elimination of nickel in relation to air-borne nickel in a battery factory. In: Brown, S.S. & Sunderman, F.W. Jr, ed. Nickel toxicology: Proceedings of the 2nd International Conference on Nickel Toxicology, Swansea, 3-5 September, 1980, London, New York, Academic Press, pp. 103-106.

ADKINS, B., RICHARDS, J.H., & GARDNER, D.E. (1979) Enhancement of experimental respiratory infection following nickel inhalation. Environ. Res., 20(1): 33-42.

AGGARWAL, S.K., KINTER, M., WILLS, M.R. SAVORY, J., & HEROLD, D.A. (1988) Isotope dilution gas chromatography - mass spectrometric method for the determination of nickel in biological materials. In: Fourth International Conference on Nickel Metabolism and Toxicology: Abstracts, Espoo, Finland, 5-9 September, 1988, Helsinki, Institute of Occupational Health (Abstract II.1).

AITIO, A. (1984) Biological monitoring of occupational exposure to nickel. In: Nickel in the human environment. Proceedings of a Joint Symposium, Lyon, 8-11 March, 1983, Lyon, International Agency for Research on Cancer, pp. 497-505 (IARC Scientific Publications No. 53).

AJMAL, M. & KHAN, A.U. (1985) Effect of electroplating factory effluent on germination and growth of hyacinth bean and mustard. Environ. Res., 38(2): 248-255.

AJMAL, M., KHAN, M.A., & NOMANI, A.A. (1985) Distribution of heavy metals in plants and fish of the Yamuna river (India). Environ. monit. Assess., 5(4): 361-367.

AKAGI, T., FUWA, K., & HARAGUCHI, H. (1985a) Simultaneous multi-element determination of trace metals in seawater by inductively coupled plasma atomic emission spectrometry after coprecipitation with gallium. Anal. Chim. Acta, 177: 139-155.

AKAGI, T., NOJIMI, Y., MATSUI, M., & HARAGUCHI, H. (1985b) Zirconium coprecipitation for simultaneous multi-element determination of trace metals in seawater by inductively coupled plasma atomic emission spectrometry. Appl. Spectroscop., 39(4): 662-667.

ALBERT, D.M., GONDER, J.R., PAPALE, J., CRAFT, J.L., DOHLMAN, H.G., REID, M.C., & SUNDERMAN, F.W. Jr (1980) Induction of ocular neoplasms in Fischer rats by intraocular injection of nickel subsulfide. In: Brown, S.S. & Sunderman, F.W. Jr, ed. Nickel toxicology. Proceedings of the 2nd International Conference on Nickel Toxicology, Swansea, 3-5 September, London, New York, Academic Press, pp. 55-58.

ALBERT, D.M., GONDER, J.R., PAPALE, J., CRAFT, J.L., DOHLMAN, H.G., REID, M.C., & SUNDERMAN, F.W. Jr (1982) Induction of ocular neoplasms in Fischer rats intraocular injection of nickel subsulfide. Invest. Opthalmol. vis. Sci., 22(6): 768-782.

ALBRACHT, S.P.J., GRAF, E.G., & THAUER, R.K. (1982) The ERP properties of nickel in hydrogenase from *Methanobacterium thermoautotrophicum*. FEBS Lett., 140: 311-313.

ALEXANDER, A.J., GOGGIN, P.L., & COOKE, M. (1983) A Fourier-transform infrared spectrometic study of the pyrosynthesis of nickel tetracarbonyl and iron trentacarbonyl by combustion of tobacco. Analyt. chim. Acta, 151: 1-12.

ALI, M., BISWAS, S.K., AKHTER, S., & KHAN, A.H. (1985) Multi-element analysis of water residue: a PIXE measurement. Fresenius Z. anal. Chem., 322(8): 755-760.

ALLOWAY, B.J. & MORGAN, H. (1986) The behaviour and availability of cadmium, nickel, and lead in polluted soils. In: Assink, J.W. & van den Brink, W.J., ed. Contaminated soil. Dordrecht, Boston, Lancaster, Martinus Nifhoff Publishers, pp. 101-113.

AMACHER, D. & PAILLET, S. (1980) Induction of trifluorothymidine-resistant mutants by metal ions in L5178Y/TK$^{+/-}$ cells. Mutat. Res., 78: 279-288.

AMBROSE, A.M., LARSON P.S., BORZELLECA, J.R., & HENNIGA, G.R. Jr (1976) Long-term toxicologic assessment of nickel in rats and dogs. J. food Sci. Technol., 13: 181-187.

AMLACHER, E. & RUDOLPH, CH. (1981) The thymidine incorporation inhibiting screening system (TSS) to test carcinogenic substances (a nuclear DNA synthesis suppressive short-term test). Arch. Geschwulstforsch., 51(7): 605-610.

ANDERSEN, A. (1988) Recent follow-up of respiratory cancer in a Norwegian nickel refinery. In: Fourth International Conference on Nickel Metabolism and

References

Toxicology: Abstracts, Espoo, Finland, 5-9 September, 1988, Helsinki, Institute of Occupational Health, p.49.

ANDERSEN, J.R., GAMMELGAARD, B., & REIMERT, S. (1986a) Direct determination of nickel in human plasma by Zeeman-corrected atomic absorption spectrometry. Analyst, 111(6): 721-722.

ANDERSEN, K.E., NIELSEN, G.D., FLYVHOLM, M.A., FREGERT, S., & GRUVBERGE, B. (1983) Nickel in tap water. Contact Dermatit., 9(2): 140-143.

ANDERSEN, O. (1983) Effects of coal combustion products and metal compounds on sister chromatid exchange (SCE) in a macrophage-like cell line. Environ. Health Perspect., 47: 239-253.

ANDERSEN, O. (1985) Evaluation of the spindle-inhibiting effect of Ni^{++} by quantitation of chromosomal super-contraction. Res. Commun. chem. Pathol. Pharmacol., 50: 379-386.

ANDERSON, B. G. (1950) The apparent thresholds of toxicity to Daphnia magna for chlorides of various metals when added to Lake Erie water. Serv. works J., 16: 96-113.

ANDERSSON, K., ELINDER, C.G., HOGSTEDT, C., KJELLSTROM, T., & SPANG, G. (1984) Mortality among cadmium and nickel-exposed workers in a Swedish battery factory. Toxicol. environ. Chem., 9(1): 53-62.

ANKE, M. (1974) [The significance of micronutrients for animal performance.] Akad. Landwirtschaftswiss. DDR Tagungsber., 132: 197-218 (in German).

ANKE, M., GRUN, M., DITTRICH, G., GROPPEL, B., & HENNIG, A. (1974) Low nickel rations for growth and reproduction in pigs. In: Hoekstra, W.G., ed. Trace element metabolism in animals, Baltimore, Maryland, University Park Press, Vol. 2, pp. 715-718.

ANKE, M., GRUN, M., & PARTSCHEFELD, M. (1976) [Nickel, a new essential trace element.] Arch. Tierernähr., 26: 740-741 (in German).

ANKE, M., HENNIG, A., GRUN, M., PARTSCHEFELD, M., GROPPEL, B., & LUDKE, H. (1977) [Nickel, an essential trace element. 1st communication: The supply of nickel as affecting the live weight gains, food consumption and body composition of growing pigs and goats.] Arch. Tierernähr., 27: 25-38 (in German).

ANKE, M., PARTSCHEFELD, M., GRUN, M., & GROPPEL, B. (1978) [Nickel - an essential trace element. 3rd communication: The influence of nickel content on the reproductive performance of female animals.] Arch. Tierernähr., 28: 83-90 (in German).

ANKE, M., KRONEMANN, H., GROPPEL, B., HENNIG, A., MEISSNER, D., & SCHNEIDER, H.-J. (1980) The influence of nickel deficiency on growth, reproduction, longevity and different biochemical parameters of goats. In Anke, M., Schneider, H.-J. & Brückner, Chr., ed. [3. Trace Element Symposium: Nickel, Jena, 7-11 July, 1980, Jena, Friedr.-Schiller Univ., pp. 3-10.

ANKE, M., GRUN, M., HOFFMANN, G., GROPPEL. B., GRUHN, K., & FAUST, H. (1981) Zinc metabolism in ruminants suffering from nickel deficiency. Mengen-Spurenelemente, 1: 189-196.

ANKE, M., GROPPEL, B., KRONEMANN, H., & GRUN, M. (1984) Nickel - an essential element. In: Nickel in the human environment, Proceedings of a Joint Symposium, Lyon, France, 8-11 March, 1983, Lyon, International Agency for Research on Cancer, pp. 339-365 (IARC Scientific Publications No. 53).

ARANYI, C., MILLER, F.J., ANDRES, S., EHRLICH, R., FENTERS, J., GARDNER, F.E., & WATERS, M.D. (1979) Cytotoxicity to alveolar macrophages of trace metals adsorbed to fly fish. Environ. Res., 20: 14-23.

ARDEN, G.P. & BYWATERS, E.G.L. (1978) Tissue reaction. In: Arden, G.B. & Ansel, B.M., ed. Surgical management of juvenile chronic polyarthritis. London, New York, Academic Press, pp. 263-275.

ARILLO, A., MARGIOCCO, C. MELODIA, F., & MENSI, P. (1982) Biochemical effects of long-term exposure to Cr. Cd, Ni on rainbow trout (*Salmo gairdneri* Rich): influence of sex and season. Chemosphere, 11:47-57.

ARLAUSKAS, A., BAKER, R., BONIN, A.M., TANDON, R.K., CRISP, P.T., & ELLIS, J. (1985) Mutagenicity of metal ions in bacteria. Environ. Res., 36(2): 379-388.

ARMIT, H.W. (1908) The toxicology of nickel carbonyl, Part II. J. Hyg., 8: 565-600.

ASATO, N., SOESTBERGEN, M.V., & SUNDERMAN F.W. Jr (1975) Binding of 63Ni (II) to ultrafiltrable constituents of rabbit serum *in vivo* and *in vitro*. Clin. Chem., 21(4): 521-527.

ASHRAF, M. & SYBERS, H.D. (1974) Lysis of pancreatic exocrine cells and other lesions in rats fed nickel acetate. Am. J. Pathol., 74: 199.

BABICH, H. & STOTZKY, G. (1982a) Nickel toxicity to microbes: effect of pH and implications for acid rain. Environ. Res., 29: 335-350.

BABICH, H. & STOTZKY, G. (1982b) Nickel toxicity to fungi: Influence of some environmental factors. Ecotoxicol. environ. Saf., 6: 577-589.

BABICH, H. & STOTZKY, G. (1983) Temperature, pH, salinity, hardness, and particulates mediate nickel toxicity to eubacteria, an actinomycete, and yeasts in lake, simulated estuarine, and sea waters. Aquat. Toxicol., 3: 195-208.

BAGO-GRANELL, J., AGUIRRE-CANYADELL, M., NARDI, J., & TALLADA, N. (1984) Malignant fibrous histiocytoma of bone at the site of a total hip arthroplasty. J. bone joint Surg. B., 66: 38-40.

BALOGH, I., SOMOGYI, E., SOTONYI, P., POGATSA, G., RUBANYI, G., & BELLUS, E. (1983) Electron-cytochemical detection of endogenous nickel in the myocardium in acute carbon monoxide poisoning. Applicability of a new cytochemical technique in forensic medicine. Z. Rechtsmed., 90(1): 7-14.

BARBEAU, C., DUPUIS, M., & ROY, J.C. (1985) Metallic elements in crude and milled chrysotile asbestos from Quebec. Environ. Res., 38(2): 275-282.

BARNES, J.M. & DENZ (1951) The effects of 2,3-dimercapto-propanol (BAL) on experimental nickel carbonyl poisoning. Br. J. ind. Med., 8: 117-126.

BARRIE, L.A. (1981) Atmospheric nickel in Canada. In: Effects of nickel in the Canadian environment, Ottawa, National Research Council of Canada, pp. 55-76 (Publication No. 18568).

References

BASRUR, P.K. & GILMAN, J.P.W. (1967) Morphologic and synthetic response of normal and tumor muscle cultures to nickel sulfide. Cancer Res., 27: 1168-1177.

BAUDOUIN, M. F. & SCOPPA, P. (1974) Acute toxicity of various metals to freshwater zooplankton. Bull. environ. Contam. Toxicol., 121(6):745-751.

BAUMGARDT, B., JACKWERTH, E., OTTO, H., & TOLG, G. (1986) Contribution to the trace analytical investigation of human lung tissue. Fresenius Z. anal. Chem., 323(5): 481-486.

BAYER, O. (1939) [Toxicology, pathology and clinical aspects of nickel carbonyl poisoning.] Arch. Gewerbepathol. Gewerbehyg., 9: 592-606 (in German).

BECKER, P., DORSTELMANN, K., FROMBERGER, W., & FORTH, W. (1980) [On the absorption of cobalt(II) and nickel(II) ions by isolated intestinal segments *in vitro* of rats.] In: Anke, M., Schneider, H.-J., & Brückner, Chr., ed. [3. Trace Element Symposium: Nickel, Jena, German Democratic Republic, 7-11 July, 1980,] Jena, Friedrich-Schiller University, pp. 79-85 (in German).

BEEFTINK, W.G. & NIEUWENHUIZE, J. (1982) Heavy-metal accumulation in salt marshes from the Western and Eastern Scheldt. Sci. total Environ., 25: 199-223.

BEIJER, K. & JERNELOV, A. (1986) Sources, transport and transformation of metals in the environment. In: Friberg, L., Nordberg, G.F., & Vouk, V.B., ed. Handbook on the toxicology of metals, Amsterdam, New York, Oxford, Elsevier Science Publishers, Vol. I, pp. 68-84.

BENCKO, V., WAGNER, V., WAGNEROVA, M., & REICHRTOVA, E. (1983) Immuno- biochemical findings in groups of individuals occupationally and non-occupationally exposed to immissions containing nickel and cobalt. J. Hyg. Epidemiol. Microbiol. Immunol., 27: 387-394

BENCKO, V., GEIST, T., ARBETOVA, D., DHARMADIKARI, D.M., & SVANDOVA, E. (1986) Biological monitoring of environmental pollution and human exposure to some trace elements. J. Hyg. Epidemiol. Microbiol. Immunol., 30: 1-10.

BENNETT, B.G. (1984) Environmental nickel pathways to man. In: Nickel in the human environment, Proceedings of a Joint Symposium, Lyon, 8-11 March, 1983, Lyon, International Agency for Research on Cancer, pp. 487-495 (IARC Scientific Publications No. 53).

BENSON, J.M., HENDERSON, R.F., & MACLELLAN, R.O. (1986) Comparative toxicity of four nickel compounds to canine and rodent alocolar macrophages *in vitro*. J. Toxicol. environ. Health, 19: 105-110.

BENSON, J.M., CARPENTER, R.L., HAHN, F.F., HALEY, P.J., HANSON, R.L., HOBBS, C.H., PICKRELL, J.A., & DUNNICK, J.K. (1987) Comparative inhalation toxicity of nickel subsulfide to F344/N rats and B6C3F1 mice exposed for twelve days. Fundam. appl. Toxicol., 9: 251-265.

BENSON, J.M., BURT, D.G., CARPENTER, R.L., EIDSON, A.F., HAHN, F.F., HALEY, P.J., HANSON, R.L., HOBBS, C.H., PICKRELL, J.A., & DUNNICK, J.K. (1988) Comparative inhalation toxicity of nickel subsulfide to F344/N rats and B6C3F1 mice exposed for twelve days. Fundam. appl. Toxicol., 10: 164-178.

BERGMAN, B., BERGMAN, M., MAGNUSSON, B., SOREMARK, R., & TODA, Y. (1980a) The distribution of nickel in mice. An autoradiographic study. J. oral Rehabil., 7(4): 319-324.

BERGMAN, M., BERGMAN, B., & SOREMARK, R. (1980b) Tissue accumulation of nickel released due to electrochemical corrosion of non-precious dental casting alloys. J. oral Rehabil., 7(4): 325-330.

BERMAN, E. & REHNBERG, B. (1983) Fetotoxic effects of nickel in drinking water in mice. Washington, DC, US Environmental Protection Agency, 11 pp (EPA-600/1-83-007).

BERNACKI, E.J., PARSONS, G.E., ROY, B.R., MIKAC-DEVIC, M., KENNEDY, C.D., & SUNDERMAN, F.W. Jr (1978a) Urine nickel concentrations in nickel-exposed workers. Ann. clin. lab. Sci., 8(3): 184-189.

BERNACKI, E.J., PARSONS, G.E., & SUNDERMAN, F.W. Jr (1978b) Investigation of exposure to nickel and lung cancer mortality: case control study at aircraft engine factory. Ann. clin. lab. Sci., 8(3): 190-194.

BERNACKI, E.J., ZYGOWICZ, E., & SUNDERMAN, F.W. Jr (1980) Fluctuations of nickel concentrations in urine of electroplating workers. Ann. clin. lab. Sci., 10(1): 33-39.

BERROW, M.C. & BURRIDGE, J.C. (1981) Persistence of metals in available form in sewage sludge-treated soils under field conditions. In: Proceedings of International Conference on Heavy Metals in the Environment, Amsterdam, Sept. 1981, Edinburgh, CEP Consultants Ltd., pp. 202-205

BERTRAND, G. & MACHEBOEUF, M. (1926a) Chimie biologique. Influence du nickel et du cobalt sur l'action exercée par l'insuline chez le lapin. C. R. Séances Acad. Sci., 182: 1504-1507.

BERTRAND, G. & MACHEBOEUF, M. (1926b) Chimie biologique. Influence du nickel et du cobalt sur l'action exercée par l'insuline chez le chien. C. R. Séances Acad. Sci., 183: 5-9.

BEVERIDGE, M.C., STAFFORD, E., & COUTTS, R. (1985) Metal concentrations in the commercially exploited fishes of an endorheic saline lake in the tin-silver province of Bolivia. Aquacult. Fish. Manage., 16(1): 41-53.

BIEDERMANN, K. & LANDOLPH, J.R. (1986) Induction of anchorage independence in human diploid foreskin fibroblasts by metal salts. Carcinogenesis, 542: 137.

BIESINGER, K.E. & CHRISTENSEN, G.M. (1972) Effects of various metals on survival, growth, reproduction, and metabolism of *Daphnia magna*D. J. Fish Res. Board. Can., 29(12): 1691-1700.

BIGGART, N.W. & MURPHY, E.C. Jr (1988) Analysis of metal-induced mutations altering the expression or structure of a retroviral gene in a mammalian cell line. Mutat. Res., 198: 115-129.

BINGHAM, E.W., BARKLEY, W., ZERWAS, M., STEMMER, K., & TAYLOR, P. (1972) Responses of alveolar macrophages to metals. I. Inhalation of lead and nickel. Arch. environ. Health, 25: 406-414.

References

BIRGE, W.J. (1978) Embryo-larval bioassays on inorganic coal elements and in situ Biomonitoring of coal-waste effluents. In: Surface mining and fish/wildlife needs in the eastern USA. Proceedings of a Symposium, December 1978. Washington, DC, US Fish and Wildlife Service, pp. 97-108 (9098.910 FWS/ OBS-78/81).

BIRGE, W.J. & BLACK, J.A. (1980) Aquatic toxicology of nickel. In: Nriagu, J.O., ed. Nickel in the environment, New York, Chichester, Brisbane, Toronto. John Wiley and Sons, pp. 349-406.

BLAYLOCK, B.G. & FRANK, M.L. (1979) A comparison of the toxicity of nickel to the developing eggs and larvae of carp (*Cyprinus carpio*). Bull. environ. Contam. Toxicol., 21:604-611.

BLANKENSTEIN, K. & STARCK, H.C. (1979) [Nickel compounds.] In: Bartholomé, E., Biekert, E., Hellmann, H., Ley, H., Weigert, W., & Weise, E., ed. [Ullmanns encyclopaedia of technical chemistry,] Weinheim, New York, Verlag Chemie, pp. 293-302 (in German).

BLOCK, G.T. & YEUNG, M. (1982) Asthma induced by nickel. J. Am. Med. Assoc., 247(11): 1600-1602.

BOLDT, J.R. & QUENEAU, P. (1967) The winning of nickel, its geology, mining, and extractive metallurgy, London, Methuen & Co. Ltd., 465 pp.

BOND, A.M. & WALLACE, G.G. (1983) Automated determination of nickel and copper by liquid chromatography with electrochemical and spectrophotometric detection. Anal. Chem., 55: 718-723.

BOND, A.M., KNIGHT, R.W., REUST, J.J.B., TUCKER, D.J., & WALLACE, G.G. (1986) Determination of metals in urine by direct injection of sample, high-performance liquid chromatography and electrochemical or spectrophotometric detection. Anal. chim. Acta, 182: 47-59.

BOUTAYRE, P., PIZZORNI, L., LIQUIER, J., TABOURY, J., & TAILLANDIER, E. (1984) Z-form induction in DNA by carcinogenic nickel compounds: an optical spectroscopy study. In: Nickel in the human environment. Proceedings of a Joint Symposium, Lyon, 8-11 March 1983, Lyon, International Agency for Research on Cancer, pp. 227-234 (IARC Scientific Publications No. 53).

BOYER, K.W. & HOWITZ, W. (1986) Special considerations in trace element analysis of food and biological materials. In: Environmental carcinogens - selected methods of analysis, Vol. 8: Some metals: As, Be, Cd, Cr, Ni, Pb, Se, Zn, Lyon, International Agency for Research on Cancer, pp. 191-220 (IARC Scientific Publications No. 71).

BOYLE, R.W. (1981) Geochemistry of nickel. In: Effects of nickel in the Canadian environment, Ottawa, National Research Council of Canada, pp. 31-44 (Publication No. NRCC 18568).

BOYSEN, M., WAKSVIK, H., SOLBERG, L.A., REITH, A., & HOGETVEIT, A.C. (1980) Histopathological follow-up studies and chromosome analysis in nickel workers. In: Brown, S.S. & Sunderman, F.W. Jr, ed. Nickel Toxicology. Proceedings of the 2nd International Conference on Nickel Toxicology, Swansea, 3-5 September 1980, London, New York, Academic Press, pp. 35-38.

BOYSEN, M., SOLBERG, L.A., ANDERSEN, I., HOGETVEIT, A.C., & TORJUSSEN, W. (1982) Nasal histology and nickel concentration in plasma and urine after improvements in the work environment at a nickel refinery in Norway. Scand. J. Work Environ. Health, 8(4): 283-290.

BOYSEN, M., SOLBERG, L.A., TORJUSSEN, W., POPPE, S., & HOGETVEIT, A.C. (1984) Histological changes, rhinoscopical findings and nickel concentration in plasma and urine in retired nickel workers. Acta otolaryngol., 97(1-2): 105-115.

BRADLEY, R.W. & MORRIS, J.R. (1986) Heavy metals in fish from a series of metal-contaminated lakes near Sudbury, Ontario. Water Air Soil Pollut., 27(3-4): 341-354.

BRAJTER, K. & SLONAWSKA. K. (1986) The efficiency of Cellex-P for the preconcentration of lead and other trace metals from waters. Anal. Chim. Acta, 185: 271-277.

BRANDES, W.W. (1934) Nickel carbonyl poisoning. J. Am. Med. Assoc., 102: 1204-1206.

BRINGMANN, G., KUHN, R., & WINTER, A. (1980) [Determination of the biological effect of water pollutants on protozoa. 3. Saprozoic flagellates.] Z. Wasser Abwasser Forsch., 13(5):170-173 (in German with English summary).

BRKOVIC-POPOVIC, L. & POPOVIC, M. (1977) Effects of heavy metals on survival and respiration rate of tubificid worms: Part 1. Effects on survival. Environ. Pollut., 13: 65-72.

BROOKS, R.R. (1980) Accumulation of nickel by terrestrial plants. In: Nriagu, J.O., ed. Nickel in the environment, New York, Chichester, Brisbane, Toronto, John Wiley and Sons, pp. 407-429.

BROWN, K.W., THOMAS, J.C., & SLOWEY, J.F. (1983a) The movement of metals applied to soils in sewage effluent. water air soil Pollut., 19(1): 43-54.

BROWN, K.W., THOMAS, J.C., & SLOWEY, J.F. (1983b) Metal accumulation by Bermuda grass grown on four diverse soils amended with secondarily treated sewage effluent. Water air soil Pollut., 20(4): 431-446.

BROWN, V. M. (1968) The calculation of the acute toxicity of mixtures of poisons to rainbow trout. Water Res., 2: 723-733.

BRULAND, K.W., FRANKS, R.P., KNAUER, G.A., & MARTIN, J.H. (1979) Sampling and analytical methods for the determination of copper, cadmium, zinc, and nickel at the nanogram per liter level in seawater. Anal. chim. Acta, 105: 233-245.

BRUNE, D. (1986) Metal release from dental biomaterials. Biomaterials, 7(3): 163-175.

BRYANT, V., NEWBERY, D.M., MCCLUSKY, D.S., & CAMPBELL, R. (1985) Effect of temperature and salinity on the toxicity of nickel and zinc to two estuarine invertebrates (*Corophium volutator, Macoma balthica*). Mar. Ecol. Prog. Ser., 24(1-2): 139-153.

BUHLER, E.V. (1965) Delayed contact hypersensitivity in the guinea-pig. Arch. Dermatol., 91: 171-175.

BURBA, P. & WILLMER, P.G. (1985) Analytical multielement preconcentration on metal hydroxide-coated cellulose. Fresenius Z. anal. Chem., 321(2): 109-118.

BURCH, J.D, HOWE, G.R., MILLER, A.B., & SEMENCIW, R. (1981) Tobacco, alcohol, asbestos, and nickel in the etiolgy of cancer of the larynx: a case-control study. J. Natl Cancer Inst., 67(6): 1219-1224.

BURGES, D. (1980) Mortality study of nickel platers. In: Brown, S.S. & Sunderman, F.J. Jr, ed. Nickel toxicology. Proceedings of the 2nd International Conference on Nickel Toxicology, Swansea, 3-5 September, 1980, London, New York, Academic Press, pp. 15-18.

BURROWS, D., CRESWELL, S., & MERRETT, J.D. (1981) Nickel, hands and hip prostheses. Br. J. Dermatol., 105(4): 437-443.

BUSELMAIER, W., ROHRBORN, G., & PROPPING, P. (1972) [Mutagenicity investigations with pesticides in the host-mediated assay and the dominant lethal test in mice.] Biol. Zbl., 91: 311-325 (in German).

BUTZ, I. (1984) [Effect of dilution water on the acute toxicity of nickel sulfate to rainbow trout.] Wasser Abwasser, 28: 41-56 (in German).

BYRNE, C.J. & DELEON, I.R. (1986) Trace metal residues in biota and sediments from Lake Pontchartrain, Lousiana, USA. Bull. environ. Contam. Toxicol., 37(1): 151-158.

CALAMARI, D., GAGGINO, G.F., & PACCHETTI, G. (1982) Toxokinetics of low levels of Cd, Cr, Ni and their mixture in long-term treatment of *Salmo gairdneri* Rich. Chemosphere, 11(1): 59-70.

CALAMARI, D., LLOYD, R., SOLBE, J.F., & SOLBE. L.G. (1984) Water quality criteria for European freshwater fish: Report on nickel and freshwater fish. Rome, Food and Agriculture Organization of the United Nations 20 pp (European Inland Fisheries Advisory Commission (EIFAC)) Technical.Paper 45).

CALLAN, W.M. & SUNDERMAN, F.W. Jr (1973) Species variations in binding of ^{63}NiCl$_2$ by serum albumin. Res. Commun. chem. Pathol. Pharmacol., 5: 459-472.

CALNAN, C.D. (1956) Nickel dermatitis. Br. J. Dermatol., 68: 229-236.

CAMNER, P., JOHANSSON, A., & LUNDBERG, M. (1978) Alveolar macrophages in rabbits exposed to nickel dust. Environ. Res., 16: 226-235.

CARLSON, H.E. (1984) Inhibition of prolactin and growth hormone secretion by nickel. Life Sci., 35(17): 1747-1754.

CARVAJAL, N.J. & ZIENIUS, R.H. (1986) Gas chromatographic analysis of trace metals isolated from aqueous solutions as diethyldithiocarbamates. J. Chromatogr., 355: 107-116.

CARVALHO, S.M. & ZIEMER, P.L. (1982) Distribution and clearance of ^{63}Ni administered as ^{63}NiCl$_2$ in the rat: intratracheal study. Arch. environ. Contam. Toxicol., 11: 245-248.

CASARETT-BRUCE, M., CAMNER, P., & CURSTEDT. T. (1981) Changes in pulmonary lipid composition of rabbits exposed to nickel dust. Environ. Res., 26(2): 353-362.

CASEY, C.E. & ROBINSON, M.F. (1978) Copper, manganese, zinc, nickel, cadmium and lead in human foetal tissues. Brit. J. Nutr., **39**: 639-646.

CASS, G.R. & MCRAE, G.J. (1983) Source-receptor reconciliation of routine air monitoring data for trace metals: An emission inventory-assisted approach. Environ. Sci. Technol., **17**(3): 129-139.

CASTLEMAN, B. & MCNEELY, B.U. (1965) Case records of the Massachusetts General Hospital, Case 38-1965. New Engl. J. Med., **273**, 494-504.

CASTRANOVA, V., BOWMAN, L., REASOR, M.J., & MILES, P.R. (1980) Effects of heavy metal ions on selected oxidative metabolic processes in rat alveolar macrophages. Toxicol. appl. Pharmacol., **53**(1): 14-23.

CATALANATTO, F.A., SUNDERMAN, F.W. Jr, & MCINTOSH, T.R. (1977) Nickel concentrations in human parotid saliva. Ann. clin. lab. Sci., **7**(2): 146-151.

CHAN, W.H. & LUSIS, M.A. (1986) Smelting operations and trace metals in air and precipitation in the Sudbury Basin. Adv. environ. Sci. Technol., **17**: 113-143.

CHANG, A.C., PAGE, A.L., & BINGHAM, F.T. (1982) Heavy metal absorption by winter wheat following termination of cropland sludge applications. J. environ. Qual., **11**(4): 705-708.

CHANG, A.C., WARNECKE, J.E., LUND, L.J., & PAGE, A.L. (1984) Accumulation of heavy metals in sewage sludge-related soils. J. environ. Qual., **13**(1): 87-91.

CHANG, C.C., TATUM. H.J., & KINCL, F.A. (1970) The effect of intrauterine copper and other metals on implantation in rats and hamsters. Fertil. Steril., **21**: 274-278.

CHAUSMER, A.B. (1976) Measurement of exchangeable nickel in the rat. Nutr. Rep. Int., **14**: 323-326.

CHEN, J.R., FRANCISCO, R.B., & MILLER, T.E. (1977) Légionnaires' disease: nickel levels. Science, **196**: 906-908.

CHEN, K.Y., YOUNG, C.S., JAN, T.K., & ROHATGI, N. (1974) Trace metals in wastewater effluents. J. Water Pollut. Fed., **46**(12): 2663-2675.

CHIAUDANI, G. & VIGHI, M. (1978) The use of *Selenastrum capri-cornutum* batch cultures in toxicity studies. Mitt. Int. Ver. Limnol., **21**: 316- 329.

CHORVATOVICOVA, D. (1983) The effect of nickel chloride on the level of chromosome aberrations in Chinese hamster *Cricetulus griseus*. Biologica, **38**: 1107-1112.

CHOVIL, A., SUTHERLAND, R.B., & HALLIDAY, M. (1981) Respiratory cancer in a cohort of nickel sinter plant workers. Br. J. ind. Med., **38**(4): 327-333.

CHRISTENSEN, J.M. & PEDERSEN, L.M. (1986) Enzymatic digestion of whole blood for improved determination of cadmium, nickel and chromium by electrothermal atomic absorption spectrophotometry: measurements in patients with rheumatoid arthritis and in normal humans. Acta pharmacol. toxicol., **59**(7): 399-402.

CHRISTENSEN, O.B & KRISTENSEN, M. (1982) Treatment with disulfiram in chronic nickel hand dermatitis. Contact Dermatit., 8(1): 59-63.

CHRISTENSEN, O.B. & MOELLER, H. (1975a) Nickel allergy and hand eczema. Contact Dermatit., 1(3): 129-135.

CHRISTENSEN, O.B. & MOELLER. H. (1975b) External and internal exposure to the antigen in the hand eczema of nickel allergy. Contact dermatit., 1(3): 136-141.

CHRISTENSEN, O. B., MOELLER, H., ANDRASKO, K., & LAGESSON, V. (1979) Nickel concentration of blood, urine and sweat after oral administration. Contact Dermatit., 5: 312-316.

CHRISTIE, N.T. & TUMMOLO, D. (1988) The role of Ni (II) in mutation. In: Fourth International Conference on Nickel Metabolism and Toxicology: Abstracts, Espoo, 5-9 September 1988, Helsinki, Institute of Occupational Health, p. 33.

CHRISTIE, N.T., TUMMOLO, D.M., BIGGART, N.W., & MURPHY, E.C. Jr (1988) Chromosomal changes in cell lines from mouse tumors induced by nickel sulfide and methylcholanthre. Cell Biol. Toxicol., 4: 427.

CHURCH, S.E. (1981) Multi-element analysis of fifty-four geochemical reference samples using inductively coupled plasma atomic-emission spectrometry. Geostand. Newsl., V: 133-160.

CICCARELLI, R.B. & WETTERHAHN, K.E. (1982) Nickel distribution and DNA lesions induced in rat tissues by the carcinogen nickel carbonate. Cancer Res., 42: 3544-3549.

CICCARELLI, R.B. & WETTERHAHN, K.E. (1985) *In vitro* interaction of ^{63}nickel (II) with chromatin and DNA from rat kidney and liver nuclei. Chem.-biol. Interact., 52(3): 347-360.

CICCARELLI, R.B., HAMPTON, T.H., & JENNETTE. K.W. (1981) Nickel carbonate induces DNA-protein crosslinks and DNA strand breaks in rat kidney. Cancer Lett., 12(4): 349-354.

CIRLA, A.M., BERNABEO, F., OTTOBONI, F., & RATTI, R. (1985) Nickel-induced occupational asthma - immunological and clinical aspects. In: Brown, S.S. & Sunderman, F.W. Jr, ed. Progress in Nickel toxicology. Proceedings of the 3rd International Congress on Nickel Metabolism and Toxicology, Paris, 4-7 September, 1984, Oxford, Blackwell Scientific Publications, pp. 165-168.

CLARK, J.R., VAN HASSEL, J.H., NICHOLSEN, R.B., CHERRY, D.S., & CAIRNS, J. (1981) Accumulation and depuration of metals by duckweed (*Lemna perpusilla*). Ecotoxicol. environ. Safety, 5: 87-96.

CLARY, J.J. (1975) Nickel chloride-induced metabolic changes in the rat and guinea-pig. Toxicol. appl. Pharmacol., 31: 55-65.

CLARY, J.J. & VIGNATI, I. (1973) Nickel chloride-induced changes in metabolism in the rat. Toxicol. appl. Pharmacol., 25: 467-468.

CLEMENTE, G.F., CIGNA ROSSI, L., & SANTARONI, G.P. (1980) Nickel in foods and dietary intake of nickel. In: Nriagu, J.O., ed. Nickel in the environment, New York, Chichester, Brisbane, Toronto, John Wiley and Sons, pp. 493-498.

CLEMMENSEN, O.J., MENNE, T., KAABER, K., & SOLGAARD, P. (1981) Exposure of nickel and the relevance of nickel sensitivity among hospital cleaners. Contact Dermatit., 7(1): 14-18.

CLEMONS, G.K. & GARCIA, J.F. (1981) Neuroendocrine effects of acute nickel chloride administration in rats. Toxicol. appl. Pharmacol., 61: 343-348.

CLOUTIER, N.R., CLULOW, F.V., LIM, T.P., & DAVE, N.K. (1986) Metal (copper, iron, cobalt, zinc, lead) and radium-226 levels in tissues of meadow voles *Microtus pennsylvanicus* living on nickel and uranium mine tailings in Ontario, Canada: Site, sex age and season effects with calculation of average skeletal radiation dose. Environ. Pollut., A41(4): 295-314.

COENEN, W., GROTHE, I., & KUHNEN, G. (1986) Exposure to welding fumes in the workplace with regard to nickel and chromates. Int. Congr. Ser. Exerpta Med., 676: 149-152.

COKER, E.G., & MATTHEWS, P.J. (1983) Metals in sewage sludge and their potential effects in agriculture. Water Sci. Technol., 15: 209-217.

CONWAY, K., & COSTA, M. (1989) The involvement of hereto-chromatic damage in nickel-induced transformation. In: Costa, M., & Rossmann, T.G., ed. Proceedings of the First International Meeting on the Molecular Mechanisms of Metal Toxicity and Carcinogenicity, Urbino, Italy, September 1988. Clifton, New Jersey, Humana Press Inc., pp.437-444 (Biological Trace Element Research, Vol. 21).

CORBETT, T.H., HEIDELBERGER, C., & DOVE, W.F. (1970) Determination of the mutagenic activity to bacteriophage T4 of carcinogenic and noncarcinogenic compounds. Mol. Pharmacol., 6: 667-679.

CORNELL, R.G. (1984) Mortality patterns among stainless-steel workers. In: Nickel in the human environment, Proceedings of a Joint Symposium, Lyon, 8-11 March, 1983, Lyon, International Agency for Research on Cancer, pp. 65-71 (IARC Scientific Publications No. 53).

CORNELL, R.G., & LANDIS, J.R. (1984) Mortality patterns among nickel/chromium alloy foundry workers: In: Nickel in the human environment, Proceedings of a Joint Symposium, Lyon, 8-11 March, 1983, Lyon, International Agency for Research on Cancer, pp. 87-93 (IARC Scientific Publications No. 53).

CORRICK, J.D. (1977) Mineral commodity profiles: Nickel, Washington, DC, US Department of the Interior, Bureau of Mines, 19 pp (MCP-4).

COSTA, M., & MOLLENHAUER, H.H. (1980a) Carcinogenic activity of particulate nickel compounds is proportional to their cellular uptake. Science, 209(4455): 515-517.

COSTA, M., & MOLLENHAUER, H.H. (1980b) Phagocytosis of nickel subsulfide particles during the early stage of neoplastic transformation in tissue culture. Cancer Res., 40: 2688-2694.

COSTA, M., NYE, J.S., SUNDERMAN, F.W. Jr, ALLPASS. P.R., & GONDOS, B. (1979) Induction of sarcomas in nude mice by implantation of Syrian hamster fetal cells exposed *in vitro* to nickel subsulfide. Cancer Res., 39: 3591-3597.

COSTA, M., JONES, M.K., & LINDBERG, O. (1980) Metal carcinogenesis in tissue culture. In: Martell, A.E., ed. Inorganic chemistry in biology and medecine. Washington, DC, pp. 45-73 (ACS Symp. Series 140).

COSTA, M., ABBRACCHIO, M.P., & SIMMONS-HANSEN, J. (1981a) Factors influencing the phagocytosis, neoplastic transformation, and cytoxicity of particulate nickel compounds in tissue culture systems. Toxicol. appl. Pharmacol., 60: 313-313.

COSTA, M., SIMMONS-HANSEN, J., BEDROSSIAN, C.W., BONURA, J., & CAPRIOLI, R.M. (1981b) Phagocytosis, cellular distribution, and carcinogenic activity of particulate nickel compounds in tissue culture. Cancer Res., 41(7): 2868-2876.

COSTA, M., HECK, J.D., & ROBINSON, S.H. (1982) Selective phagocytosis of crystalline nickel sulfide particles and DNA strandbreaks as a mechanism for the induction of cellular transformation. Cancer Res., 42: 2757-2763.

COWGILL, U.M. (1976) The chemical composition of two species of *Daphnia*, their algal food and their environment. Sci. total Environ., 6: 79-102.

COX, J.E., DOLL, R., SCOTT, W.A., & SMITH, S. (1981) Mortality of nickel workers: experience of men working with metallic nickel. Br. J. ind. Med., 38(3): 235-239.

CRAGLE, D.L., HOLLIS, D.R., NEWPORT, T.H., & SHY, C.M. (1984) A retrospective cohort mortality study among workers occupationally exposed to metallic nickel powder at the Oak-Ridge Tennessee USA gaseous diffusion plant. In: Nickel in the human environment, Proceedings of a Joint Symposium, Lyon, 8-11 March, 1983, Lyon, International Agency for Research on Cancer, pp. 57-64 (IARC Scientific Publications No. 53).

CRONIN, E. (1980) Contact dermatitis, Edinburgh, London, New York, Churchill Livingstone, pp. 279-390.

CRONIN, E., DI MICHIEL, A.D., & BROWN, S.S. (1980) Oral challenge in nickel-sensitive women with hand eczema. In: Brown, S.S., & Sunderman, F.W. Jr, ed. Nickel toxicology. Proceedings of the 2nd International Conference on Nickel Toxicology, Swansea, 3-5 September, 1980, London, New York, Academic Press, pp. 149-152.

CROSSMANN, G. (1988) [The cycle in the system soil-plant-animal on locations with extremely high soil contaminations by cadmium and nickel caused by sewage-sludge.] Berlin, Federal Agency of Environmental Protection, 95 pp (Research report No. 106 07 043) (in German).

CUCKLE, H., DOLL, R., & MORGAN, L.G. (1980) Mortality study of men working with soluble nickel compounds. In: Brown, S.S. & Sunderman. F.W. Jr, ed. Nickel toxicology. Proceedings of the 2nd International Conference on Nickel Toxicology, Swansea, 3-5 September, 1980, London, New York, Academic Press, pp. 11-14.

CURSTEDT, T., HAGMAN, M., ROBERTSON, B., & CAMNER, P. (1983) Rabbit lungs after long-term exposure to low nickel dust concentration. Environ. Res., 30: 89-94.

CURSTEDT, T., CASARETT-BRUCE, M., & CANNER, P. (1984) Changes in glycerophosphatides and their ether analogs in lung lavage of rabbits exposed to nickel dust. Exp. mol. Pathol., 41: 26-34

CUSTER, T.W., FRANSON, J.C., MOORE, J.F., & MYERS, J.E. (1986) Reproductive success and heavy metal contamination in Rhode Island common terns. Environ. Pollut., A 41(1): 33-52.

DALDRUP, T., HAARHOFF, K., & SZATHMARY, S.C. (1983) [Fatal nickel sulfate poisoning.] Beitr. gerichtl. Med., 41: 141-144 (in German).

DALLINGER, R., & KAUTZKY, H. (1985) The importance of contaminated food for the uptake of heavy metals by rainbow trout (*Salmo gairdneri*): A field study. Oecologia, 67(1): 82-89.

D'ALONZO, C.A., & PELL, S. (1963) A study of trace metals in myocardial infarction. Arch. environ. Health, 6: 381-385.

DAMJANOV, I., SUNDERMAN, F.W. Jr, MITCHELL, J.M., & ALLPASS, P.R. (1978) The induction of testicular sarcomas in Fischer rats by intratesticular injection of nickel subsulfide. Cancer Res., 38: 268-276.

DANIELSSON, L.-G., MAGNUSSON, B., & WESTERLUND, S. (1978) An improved metal extraction procedure for the determination of trace metals in seawater by atomic absorption spectrometry with electrothermal atomization. Anal. Chim. Acta, 98: 47-57.

DAVYDOVA, V.I., NEIZVESTNOVA, E.M., BLOKHIN, V.A., & SERGOVA, N.V. (1981) Toxcological evaluation of manganese, chromium and nickel. Gig. i Sanit., 7: 20-22.

DECATANZARO, J.B. & HUTCHINSON, T.C. (1985a) Effects of nickel addition on nitrogen mineralization, nitrification, and nitrogen leaching in some boreal forest soils. Water air soil Pollut., 24(2): 153-164.

DeCATANZARO, J.B., & HUTCHINSON, T.C. (1985b) Leaching and distribution of nitrogen and nickel in nickel-perturbed jack pine forest microcosms. Water air soil Pollut., 26(3): 281-292.

DECHENG, C., MING, J., LING, H., SHAN,. W., ZIQING, X., & XINSHUI, Z. (1987) Cytogenic analysis in workers occupationally exposed to nickel carbonyl. Mutat. Res., 188: 149-152.

DECSY, M.I., & SUNDERMAN, F.W. Jr. (1974) Binding of 63Ni to rabbit serum alpha-macroglobulin *in vivo* and *in vitro*. Bioinorg. Chem., 3(2): 95-105.

DEFLORA, S., ZANACCI, P., CAMOIRANO, A., BENICELLI, C., & BADOLATI, G.S. (1984) Genotoxic activity and potency of 135 compounds in the Ames reversion test and in a bacterial DNA-repair test. Mutat. Res., 133: 161-198.

DEKNUDT, G., & LEONARD, A. (1982) Mutagenicity tests with nickel salts in the male mouse. Toxicology, 25(4): 289-292.

DELGADO, E.A. (1958) Sarcoma following a surgically treated fractured tibia. Clin. Orthop., 12: 315-318.

DENG, C., & QU, B. (1981) The cytogenetic effects of nickel sulphate. Acta genet. sin., 8: 212-215.

DENG, C., QU, B., HUANG, J., ZHUO, Z.L., XIAN, H., YAO, M., CHEN, M.Y., LI, Z.X., SHENG, S.Y., & YEI, Z.F. (1983) [Cytogenetic effects of electroplating workers.] Acta scientiae circumstantiae, 3: 267-271 (in Chinese with English abstract).

DENG, C., LEE, H.H., XIAN, H., YAO, M., HUANG, J., & QU, B. (1988) Chromosomal aberrations and sister chromatid exchanges of peripheral blood lymphocytes in Chinese electroplating workers. J. trace Elem. exp. Med., 1: 57-62.

DERMOTT, R.M. & LUM, K.R. (1986) Metal concentrations in the annual shell layers of the bivalve *Elliptio complanata*. Environ. Pollut., B 12(2): 131-144.

DEUTMAN, R., MULDER, T.J., BRIAN, R., & NATER, J.P. (1977) Metal sensitivity before and after total hip arthroplasty. J. bone joint Surg., 59(7): 862-865.

DEWLING, R., MANGANELLI, R., & BAER, G. (1980) Fate and behaviour of selected heavy metals in incinerated sludge. J. Water Pollut. Control Fed., 52: 2552-2557.

DIEKERT, G. & RITTER, M. (1982) Nickel requirement of *Acetobacterium woodii*. J.Bacteriol., 151: 1043-1045.

DIEKERT, G. & THAUER, R.K. (1980) The effect of nickel on carbon monoxide dehydrogenase formation in *Clostridium thermoaceticum* and *Clostridium formicoaceticum*. FEMS microbiol. Lett., 7: 187-189.

DIETZ, R.N. & WIESER, R.F. (1983) Sulfate formation in oil-fired power plant plumes. Vol. 1: Parameters affecting primary sulfate emissions and a model for predicting emissions and plume opacity, Upton, NY, Brookhaven National Laboratories (Report No. EA-3231).

DIPAOLO, J.A. & CASTO, B.C. (1979) Quantitative studies of *in* vitro morphological transformation of Syrian golden hamster cells by inorganic metal salts. Cancer Res., 39: 1008-1013.

DITORO, D.M., MAHONY, J.D., KRICHGRABER, P.R., O'BYRNE, A.L., & PASQUALE, L.R. (1986) Effect of nonreversibility, particle concentration and ionic strength on heavy metal sorption. Environ. Sci. Technol., 20(1): 55-61.

DIXON, N.E., GAZZOLA, C., BEAKELEY, R.L., & ZERNER, B. (1975) Jack bean urease (EC 3.5.1.5). A metalloenzyme. A simple biological role for nickel? J. Am. Chem. Soc., 97: 4130-4133.

DJACHENKO, O.Z (in press) Effects of nickel and manganese on the chromosome aberrations and sister chromatid exchanges in human lymphocytes *in vitro*. In: Domnin, S.G., & Shcherbakov, S.V., ed. Problems of labour hygiene in steel and coloured metals industry, Moscow, Erisman Institute of Hygiene.

DOLL, R. (1958) Cancer of the lung and nose in nickel workers. Br. J. ind. Med., 15: 217-223.

DOLL, R., MORGAN, L.G., & SPEIZER, F.E. (1970) Cancers of the lung and nasal sinuses in nickel workers. Br. J. Cancer, 24(4): 623-632.

DOLL, R., MATTHEWS, J.D., & MORGAN, L.G. (1977) Cancers of the lung and nasal sinuses in nickel workers: A reassessment of the period of risk. Br. J. ind. Med., 34(2): 102-105.

DOLOVICH, J., EVANS, S.L., & NIEBOER, E. (1984) Occupational asthma from nickel sensitivity: I. Humam serum albumin in the antigenic determinant. Br. J. ind. Med., 41(1): 51-55.

DOOMS-GOOSSENS, A., CEUTERICK, A., VANMAELE, N., & DEGREEF, H. (1980) Follow-up study of patients with contact dermatitis caused by chromates, nickel and cobalt. Dermatologica, 160(4): 249-260.

DORMER, R.L., KERBEY, A.L., MCPHERSON, M., MANLEY, S., ASHCROFT, S.J.H., SCHOFIELD, J.G., & RANDLE, P.J. (1973) The effect of nickel on secretory systems. Studies on the release of amylase, insulin and growth hormone. Biochem. J., 140: 135-142.

DORNEMANN, A., & KLEIST, H. (1980) Nickel in liver and kidney samples. In: Brown, S.S., & Sunderman, F.W. Jr., ed. Nickel toxicology, Proceedings of the 2nd International Conference on Nickel Toxicology, Swansea, 3-5 September, 1980, London, New York, Academic Press, pp. 175-178.

DOSTAL, L.A., HOPFER, S.M., LIN, S.M., & SUNDERMAN, F.W. Jr (1989) Effects of nickel chloride on lactating rats and their suckling poups, and the transfer of nickel through rat milk. Toxicol. appl. Pharmacol., 101: 220-231.

DRAKE, H.L. (1982). Occurrence of nickel in carbon monoxide dehydrogenase from *Clostridium thermoaceticum*. J. Bacteriol., 149: 561-566.

DRINKER, K.R., FIARHALL, J.T., RAY, G.B., & DRINKER, C.K. (1924) The hygienic significance of nickel. J. ind. Hyg., 6: 307-360.

DUBE, V.E. & FISHER, D.E. (1972) Hemangioendothelioma of the leg following metallic fixation of the tibia. Cancer, 30: 1260-1266.

DUBINS, J.S., & LAVELLE, J.M. (1986) Nickel (II), genotoxicity: potentiation of mutagenesis of simple alkylating agents. Mutat. Res., 162(2): 187-199.

DUBREUIL, A., BOULEY, G., DURET, S., MESTRE, J.C., & BOUDENE, C. (1984) *In vitro* cytotoxicity of nickel chloride on a human pulmonary epithelial cell line. Arch. Toxicol. Suppl. 7: 391-393.

DUKE, J.M. (1980) Nickel in rocks and ores. In: Nriagu, J.O., ed. Nickel in the environment, New York, Chichester, Brisbane, Toronto, John Wiley and Sons, pp. 27-50.

DUNNICK, J.K., BENSON, J.M., HOBB, C.H., HAHN, F.F., CHENG, Y.S., & EIDSON, A.F. (1988) Comparative toxicity of nickel oxide, nickel sulfate hexahydrate, and nickel subsulfide after 12 days of inhalation exposure to F344/N rats and B6C3F1 mice. Toxicology, 50: 145-156.

DUYEVA, L.A. (1983) [Allergenicity of metals.] In: [Metals. Hygienic aspects of environmental evaluation and improvement]. Moscow, Meditsina, pp. 28-41 (in Russian).

References

EDMAN, B. & MOELLER, H. (1982) Trends and forecasts for standard allergens in a 12-year patch test material. Contact Dermatit., 8(2): 95-104.

EGEDAHL, R., & RICE, E. (1984) Cancer incidence at a hydrometallurgical nickel refinery. In: Nickel in the human environment, Proceedings of a Joint Symposium, Lyon, 8-11 March, 1983, Lyon, International Agency for Research on Cancer, pp. 47-56 (IARC Scientific Publications No. 53).

EGILSSON, V., EVANS, I.H., & WILKIE, D. (1979) Toxic and mutagenic effects of carcinogens on the mitochondria of *Saccharomyces cerevisiae*. Molec. gen. Genet., 174: 39-46.

EILERTSEN, E., SKAUG, V., & NORSETH, T. (1988) Tumor induction in rats after intrapleural instillation of various nickel oxides. In: Fourth International Conference on Nickel Metabolism and Toxicology, Abstracts, Espoo, Finland, 5-9 September 1988. Helsinki, Institute of Occupational Health, p. 43.

EKER, P. & SANNER, T. (1983) Assay for initiators and promoters of carcinogenesis based on attachment-independent survival of cells in aggregates. Cancer Res., 43: 320-323.

ELAKHOVSKAYA, N.P. (1972) [Metabolism of nickel entering the body with drinking water.] Gig.i Sanit., 37(6): 20-22 (in Russian).

ELLEN, G., VAN DEN BOSCH-TIBBESMA, G., & DOUMA, F.F. (1978) Nickel content in various Dutch foodstuffs. Z. Lebensm. Unters. Forsch., 166(3): 145-147.

ENGLISH, J.C., PARKER, R.D., SHARMA, R.P., & OBERG, S.G. (1981) Toxicokinetics of nickel in rats after intratracheal administration of a soluble and insoluble form. Am. Ind. Hyg. Assoc. J., 42(7): 486-492.

ENTERLINE, P.E., & MARSH, G.M. (1982) Mortality among workers in a nickel refinery and alloy manufacturing plant in West Virginia. J. Natl Cancer Inst., 68: 925-933.

EVANS, R.M., DAVIES, P.J.A., & COSTA, M. (1982) Video time-lapse microscopy of phagocytosis and intracellular fate of crystalline nickel sulfide particles in cultured mammalian cells. Cancer Res., 42: 2829-2735.

EVANS, R.M., YANO, E., MORGAN, L.G., & URANO, N. (1988) Chemiluminescent detection of free oxygen radical generation by stimulated PMN *in vitro*: effects of nickel compounds. In: Fourth International Conference on Nickel Metabolism and Toxicology, Abstracts, Espoo, 5-9 September 1988, Helsinki, Institute of Occupational Health, p. 37.

EVANS, W.H., READ, J.I., & LUCAS, B.E. (1978) Evaluation of a method for the determination of total cadmium, lead and nickel in foodstuffs using measurement by flame atomic-absorption spectrophotometry. Analyst, 103(1227): 580-594.

FALANDYSZ, J. (1985) Trace metals in flatfish from the southern Baltic, 1983. Z. Lebensm. Unters. Forsch., 181(2): 117-120.

FALANDYSZ, J. (1986a) Trace metals in cod from the southern Baltic, 1983. Z. Lebensm. Unters. Forsch., 182(3): 228-231.

FALANDYSZ, J. (1986b) Trace metals in herring from the southern Baltic, 1983. Z. Lebensm. Unters. Forsch., 182(1): 36-39.

FARRELL, R.L. & DAVIS, G.W. (1974) The effects of particulates on respiratory carcinogenesis by diethylnitrosamine. In: Karbe, E., & Park, I.F., ed. Experimental lung cancer, carcinogenesis, and bioassays, Berlin, Heidelberg, New York, Springer Verlag, pp. 219-233.

FEREN. K. & REITH, A. (1988). Phagocytosis and transformation testing of Ni_3S_2 using primary epithelial respiratory cells. Fourth International Conference on Nickel Metabolism and Toxicology, Abstracts, Espoo, Finland, 5-9 September 1988, Helsinki, Institute of Occupational Health, p.64.

FERM, V.H. (1972) The teratogenic effects of metals on mammalian embryos. Adv. Teratol., 5: 51-75.

FEZY, J.S., SPENCER, D.F., & GREENE, R.W. (1979) The effect of nickel on the growth of the freshwater diatom Navicula pelliculosa. Environ. Pollut., 20:131-137.

FIGONI, R., & TREAGAN, L. (1975) Inhibition effect of nickel and chromium upon antibody response of rats immunization with T-1 phage. Res. Commun. chem. Pathol. Pharmacol., 11: 335-338.

FILKOVA, L. & JAGER, J. (1986) Nonoccupational exposure to nickel tetracarbonyl. Cesk. Hyg., 31(5): 255-259.

FISHBEIN, L. (1981) Sources, transport and alterations of metal compounds: An overview. I. Arsenic, beryllium, cadmium, chromium, and nickel. Environ. Health Perspect., 40: 43-64.

FISCHER, T., FREGERT, S., GREENBERGER, B., & RYSTEDT, I. (1984) Contact sensitivity to nickel in white gold. Contact Dermatit., 10: 213-24.

FISHELSON, Z., PANBURN, M.K., & MüLLER-EBERHARD, H.J. (1983) C3 convertase of the alternative complement pathwaqy. Demonstration of an active, stable C3b,Bb(Ni) complex. J. biol. Chem., 258(12): 7411-7415.

FISHER, A.A. (1977) Allergic dermatitis presumably due to metallic foreign bodies containing nickel or cobalt. Cutis, 19(3): 285-286.

FISHER, A.A. (1985) Nickel dermatitis in men. Cutis, 35(5): 424-426.

FISHER, A.A. & SHAPIRO, A. (1956) Allergic eczematous contact dermatitis due to metallic nickel. J. Am. Med. Assoc., 161: 717-721.

FISHER, G.L., CHRISP, C.E., MCNEILL, K.L., MCNEILL, D.A., DEMOCKO, C., & FINCH, G.L. (1984) Mechanistic evaluations of the pulmonary toxicology of nickel subsulfide. Adv. mod. environ. Toxicol., 6: 49-60.

FISHER, G.L., CRISP, C.E., & MCNEILL, D.A. (1986) Lifetime effects of intertracheally instilled nickel subsulfide on B6C3F1 mice. Environ. Res, 40: 313-320.

FISHER, N.S. (1986) On the reactivity of metals for marine phytoplankton. Limnol. Oceanogr., 31(2): 443-449.

FLORA, C.J., & NIEBOER, E. (1980) Determination of nickel by differential pulse polarography at a dropping mercury electrode. Anal. Chem., 52: 1013-1020.

FORNACE, A.J. Jr (1982) Detection of DNA single-strand breaks produced during the repair of damage by DNA-protein cross-linking agents. Cancer Res., 42: 145-149.

FORTH, W. & RUMMEL, W. (1971) Absorption of iron and chemically related metals *in vitro* and *in vivo*: specificity of the iron-binding system in mucosa of the jejunum. In: Shoryna, S.C., & Waldron-Edward, D., ed. Intestinal absorption of metal ions, trace elements and radionuclides, New York, Pergamon Press, pp. 173-191.

FOULKES, E.C., & BLANCK, S. (1984) The selective action of nickel on tubule function in rabbit kidneys. Toxicology, 33: 245-249.

FOULKES, E.C., & McMULLEN, D.M. (1986) On the mechanism of nickel absorption in the rat jejunum. Toxicology, 38: 35-42.

FRACHE, R., BAFFI, F., DADONE, A., SCARPONI, G., & DAGNINO, I. (1980) The determination of heavy metals in the Ligurian sea. II. The geographical and vertical distribution of Cd, Cu, and Ni. Deep-Sea Res., 27A: 1079-1082.

FRANCIS, C.W., DAVIS, E.C., & GOYERT, J.C. (1985) Plant uptake of trace elements from coal gasification ashes. J. environ. Qual., 14(4): 561-569.

FREEMAN, B.M. & LANGSLOW, D.R. (1973) Response of plasma glucose, free fatty acids and glucagon to cobalt and nickel chlorides by *Gallus domesticus*. Comp. Biochem. Physiol., 46A: 427-436.

FRIBERG, L. (1950) Health hazards in the manufacture of alcaline accumulation. Acta med. Scand., 138(240): 124.

FRIEDLAND, A.J., JOHNSON, A.H., & SICCAMA, T.G. (1986) Zinc, Cu, Ni and Cd in the forest floor in the Northeastern United States. Water air soil Pollut., 29: 233-243

FRIEDRICH, B., HEINE, E., FINK, A., & FRIEDRICH, C.G. (1981) Nickel requirement for active hydrogenase formation in *Alcaligenes eutrophus*. J. Bacteriol., 145: 1144-1149.

FROSCH, P.J. & KLIGMAN, A.M., (1976) The chamber-scarification test for irritancy. Contact Dermatit., 2: 314-324.

FUKUNAGA, M., KURACHI, Y., & MIZYGUCHI, Y. (1982) Action of some metal ions on yeast chromosomes. Chem. pharm. Bull. (Tokyo), 30: 3017-3019.

FULLERTON, A., ANDERSEN, J.R., HOELGAARD, A., & MENNE, T. (1986) Permeation of nickel salts through human skin *in vitro*. Contact Dermatit., 15: 183-177.

FURST, A. & AL-MAHROUQ, H. (1981) Excretion of nickel following intratracheal administration of the carbonate. Proc. West. Pharmacol. Soc., 24: 119-121.

FURST, A. & CASSETTA, D. (1973) Carcinogenicity of nickel by different routes. Proc. Am. Assoc. Cancer Res., 121: 31.

FURST, A., & SCHLAUDER, M.C. (1971) The hamster as a model for metal carcinogenesis. Proc. West. Pharmacol. Soc., 14: 68-71.

FURST, A., CASSETTA, D.M., & SASMORE, D.P. (1973) Rapid induction of pleural mesotheliomas in the rat. Proc. West. Pharmacol. Soc., 16: 150-153.

GAINER, J.H. (1977) Effects of heavy metals and/or of deficiency of zinc on mortality rates in mice infected with encephalomyocarditis virus. Am. J. vet. Res., 38: 869-872.

GAWKRODGER, D.J., LLOYD, M., & HUNTER, J. (1986a) Occupational skin disease in hospital cleaning and kitchen workers. Contact Dermatit., 15(3): 132-135.

GAWKRODGER, D.J., COOK, S.W., FELL, G.S., & HUNTER, J.A.A. (1986b) Nickel dermatitis: the reaction to oral nickel challenge. Br. J. Dermatol., 115: 33-38.

GEMMER-COLOS, U., TUSS, H., SAUR, D., & NEEB, R. (1981) [Polarographic determination of Nickel in the ppb-range after extracting nickel as dioxime.] Fresenius Z. anal. Chem., 307: 347-351 (in German).

GENDREAU, R.M., JAKOBSEN, R.J., & HENRY, W.M. (1980) Fourier transform infrared spectroscopy for inorganic compound speciation. Environ. Sci. Technol., 14: 990-995.

GERIN, M., SIEMIATYCKI, J., RICHARDSON, L., PELLERIN, J., LAKHANI, R., & DEWAR, R. (1984) Nickel and cancer associations from a multicancer occupation exposure case-referent study - preliminary findings. In: Nickel in the Human Environment, Proceedings of a Joint Symposium, Lyon, 8-11 March, 1983, Lyon, International Agency for Research on Cancer, pp. 105-115 (IARC Scientific Publications No. 53).

GERSTLE, R.W., & ALBRINCK, D.N. (1982) Atmospheric emissions of metals from sewage sludge incineration. J. Air Pollut. Control Assoc., 32(11): 1119-1123.

GHIRINGHELLI, L. (1957) [Therapeutic utility of 2-3 dimercaptopropanol and thiotic acid in rats poisoned with nickel carbonyl.] Attiv. Soc. Lomb. Med. Biol., 12: 24-26 (in Italian).

GHIRINGHELLI, J. & AGAMENNONE, M. (1957) [The metabolism of nickel in animals experimentally poisoned with nickel carbonyl.] Med. Lav., 48: 187-194 (in Italian).

GIGNAC, L.D., & BECKETT, P.J. (1986) The effect of smelting operations on peatlands near Sudbury, Ontario, Canada. Can. J. Bot., 64(6): 1138-1147.

GILANI, S.H., & MARANO, M. (1980) Congenital abnormalities in nickel poisoning in chick embryos. Arch. environ. Contam. Toxicol., 9(1): 17-22.

GILMAN, J.P.W. (1962) Metal carcinogenesis. II. A study of the carcinogenic activity of cobalt, copper, iron and nickel compounds. Cancer Res., 22: 158-162.

GILMAN, J.P.W., & HERCHEN, H. (1963). The effect of physical form of implant on nickel sulphide tumourigenesis in the rat. Acta Union int. Cancer, 19: 615-619.

GILMAN, J.P.W. & RUCKERBAUER, G.M. (1962) Metal carcinogenesis. I. Observations on the carcinogenicity of a refinery dust, cobalt oxide, and colloidal thorium dioxide. Cancer Res., 22: 152-157.

References

GILMAN, J.P.W. & YAMASHIRO, S. (1985) Muscle tumorigenesis by nickel compounds. In: Brown, S.S., & Sunderman, F.W. Jr, ed. Progress in Nickel toxicology. Proceedings of the 3rd International Conference on Nickel Metabolism and Toxicology, Paris, 4-7 September, 1984, Oxford, Blackwell Scientific Publications, pp. 9-22.

GITLITZ, P.H., SUNDERMAN, F.W., & GOLDBLATT, P.J. (1975) Aminoaciduria and proteinuria in rats after single intraperitoneal injection of Ni(II). Toxicol. appl. Pharmacol., 34: 430-440.

GLAESS, E. (1956a) [Distribution of fragmentations and achromatic areas on chromosomes of *Vicia faba* following treatment with heavy metal salts]. Chromosome, 8: 260-284 (in German).

GLAESS, E. (1956b) [Investigations on the effects of heavy metal salts on root top mitosis of *Vicia faba*.] Ztschr. Botanik, 44: 1-58 (in German).

GLASER, U., HOCHRAINER, D., OLDIGES, H., & TAKENAKA, S. (1986) Long-term inhalation studies with NiO and As_2O_3 aerosols in Wistar rats. In: Stern, R.M., Berlin, A., Fletcher, A.C., & Jarvisalo, J., ed. Health hazards and biological effects of welding fumes and gases. Proceedings of The International Conference on Health Hazards and Biological Effects of Welding Fumes and Gases, Copenhagen, 18-21 February, 1985, Amsterdam, New York, Oxford, Elsevier Science Publishers, pp. 325-328.

GLENNON, J.D., & SARKAR, B. (1982) Nickel(II) transport in human blood serum. Studies of nickel(II) binding to human albumin and to native-sequence peptide, and ternary-complex formation with L-histidine. Biochem. J., 203(1): 15-23.

GODBOLD, J.H. Jr & TOMPKINS, E.A. (1979) A long-term mortality study of workers occupationally exposed to metallic nickel at the Oak Ridge Gaseous Diffusion Plant. J. occup. Med., 21(12): 799-806.

GOLDBERG, M., FUHRER, R., BRODEUR, J.-M., GOLDBERG, P., SEGNAN, N., LECLERC, A., CHASTANG, J.-F., BOURBONNAIS, R., & FRANCOIS, D. (1985a) Cancer des voies respiratoires parmi les travailleurs d' une entreprise d' extraction et du raffinage du nickel en Nouvelle Calédonie: Une étude cas-témoins au sein d' une cohorte. In: Brown, S.S. & Sunderman, J.W. Jr, ed. Progress in Nickel toxicology, Proceedings of the 3rd International Conference on Nickel Metabolism and Toxicology, Paris, 4-7 September, 1984, Oxford, Blackwell Scientific Publications, pp. 215-218.

GOLDBERG, P., BLANC, M., FUHRER, R., SEGNAN, N., BRODEUR, J.-M., LECLERC, A., FRANCOIS, D., DAB, W., GODARD, C., & GOLDBERG, M. (1985b) Incidence des cancers respiratoires chez les travailleurs d' une entreprise d' extraction et de raffinage du nickel et dans la population générale en Nouvelle Calédonie. In: Brown, S.S. & Sunderman, F.W. Jr, ed. Progress in Nickel toxicology. Proceedings of the 3rd International Conference on Nickel Metabolism and Toxicology, Paris, 4-7 September, 1984, Oxford, Blackwell Scientific Publications, pp. 211-214.

GOLDBERG, M., GOLDBERG, P., LECLERC, A., CHASTANG, J.F., MARNE, M.J., GUEZIEC, J., LAVIGNE, F., DUBOURDIEU, D., & HUERRE, M. (1988) A 7-year survey of respiratory cancers among nickel workers in New

Caledonia (1978-1984). In: Fourth International Conference on Nickel Metabolism and Toxicology, Abstracts, Espoo, Finland, 5-9 September 1988, Helsinki, Institute of Occupational Health, p.42.

GORDON, C.J. (1989) Effect of nickel chloride on body temperature and behavioral thermoregulation in the rat. Neurotoxicol. Teratol, 11: 317-320.

GORDON, C.J. & STEAD, A.G. (1986) Effect of nickel and cadmium chloride on autonomic and behavioral thermoregulation in mice. Neurotoxicology, 7: 97-106.

GORDON, C.J., FOGELSON, L., & STEAD, A.G. (1989) Temperature regulation following nickel intoxication in the mouse: effect of ambient temperature. Comp. Biochem. Physiol., 92: 73-76.

GORDYNIA, R.I. (1969) [Effect of a ration containing a nickel salt additive on carbohydrate metabolism in experimental animals.] Vopr. Racion. Pitan., 5: 167-170 (in Russian).

GRAHAM, J.A., GARDNER, D.E., WATERS, M.D., & COFFIN, D.L. (1975a) Effect of trace metals on phagocytosis by alveolar macrophages. Infect. Immun., 11: 1278-1283.

GRAHAM, J.A., GARDNER, D.E., MILLER. F.J., DANIELS, M.J., & COFFIN, D.L. (1975b) Effect of nickel chloride on primary antibody production in the spleen. Environ. Health Perspect., 12: 109-113.

GRAHAM, J.A., MILLER, F.J., DANIELS, M.J., PAYNE, E.A., & GARDNER, D.E. (1978) Influence of cadmium, nickel, and chromium on primary immunity in mice. Environ. Res., 16: 77-87.

GRANDJEAN, P. (1984) Human exposure to nickel. In: Nickel in the human environment. Proceedings of a Joint Symposium, Lyon, 8-11 March, 1983, Lyon, International Agency for Research on Cancer, pp. 469-485 (IARC Scientific Publications No. 53).

GRANDJEAN, P., SELIKOFF, I.J., SHEN, S.K., SUNDERMAN, F.W. Jr (1980) Nickel concentrations in plasma and urine of shipyard workers. Am J. ind. Med., 1(2): 181-189.

GRANDJEAN, P., ANDERSEN, O., & NIELSEN, G.D. (1988) Nickel. In: Alession, L., Berlin, A., Boris, M., & Roi, R., ed. Biological indicators for the assessment of human exposure to industrial chemicals. ISPRA (Varese), IPSRA Establishment, Joint Research Centre, Commission of the European Communities, pp. 56-80.

GREBNEV, V.L., & YELNICHNYKH, L.N. (1983) Experimental study of the action of nickel-containing aerosols in preparatory workshops of nickel production on parenchymour organs. Non-specific action of harmful occupation factors on workers' organism, Sverdlovsk, Medical College, pp. 90-95.

GREEN, M.H.L., MURIEL, W.J., & BRIDGES, B.A. (1976). Use of a simplified fluctuation test to detect low levels of mutagens. Mutat. Res., 38: 33-42.

GROPPEL, B., ANKE, M., RIEDEL, E., & GRUN, M. (1980) [The nickel supply and status of wild ruminants in the GDR.] In: Anke, M., Schneider, H.-J., & Brückner, Chr., ed. [3. Trace Element Symposium: Nickel, Jena, German

References

Democratic Republic, 7-11 July, 1980, Jena, Friedrich-Schiller University pp. 269-276(in German).

GROSS, P.R., KATZ, S.A., & SAMITZ, M.H. (1968) Sensitization of guinea-pigs to chromium salts. J. invest. Dermatol., 50: 424-427.

GUTENMANN, W.H., BACHE, C.A., LISK, D.J., HOFFMANN, D., ADAMS, J.D., & ELFVING, D.C. (1982) Cadmium and nickel in smoke of cigarettes prepared from tobacco cultured on municipal sludge-amended soil. J. Toxicol. environ. Health, 10(3): 423-431.

HACKETT, R.L., & SUNDERMAN, F.W. Jr (1967) Acute pathological reactions to administration of nickel carbonyl. Arch. environ. Health, 14(4): 604-613.

HACKETT, R.L., & SUNDERMAN, F.W. Jr (1968) Pulmonary alveolar reaction to nickel carbonyl: Ultrastructural and histochemical studies. Arch. environ. Health, 16(3): 349-362.

HACKETT, R.L., & SUNDERMAN, F.W. Jr (1969) Nickel carbonyl. Effects upon the ultrastructure of hepatic parenchymal cells. Arch. environ. Health, 19: 337-343.

HAGEDORN-GOTZ, H., KUPPERS, G., & STOPPLER, M. (1977) On nickel contents in urine and hair in a case of exposure to nickel carbonyl. Arch. Toxicol., 38(4): 275-285.

HALE, J.G. (1977) Toxicity of metal mining wastes. Bull. environ. Contam. Toxicol., 17:66-73.

HALEY, P.Y., BICE, D.E., MUGGENBURG, B.A., HAHN, F.F., & BENJAMIN, S.A. (1987) Immunopathologic effects of nickel subsulfide on the primate pulmonary system. Toxicol. appl. Pharmacol., 88: 1-12

HALL, T.M. (1982) Free ionic nickel accumulation and localization in the freshwater zooplankter *Daphnia magna*. Limnol. Oceanogr., 27(4): 718-727.

HANSEN, K., & STERN, R.M. (1983) *In vitro* toxicity and transformation potency of nickel compounds. Environ. Health Perspect., 51: 223-226.

HANSEN, K. & STERN, R.M. (1984) Toxicity and transformation potency of nickel compounds in BHK cells *in vitro*. In: Nickel in the human environment. Proceedings of a Joint Symposium, Lyon, 8-11 March, 1983, Lyon, International Agency for Research on Cancer, pp. 193-200 (IARC Scientific Publications No. 53).

HANSEN, L.D. & FISHER, G.L. (1980) Elemental distribution in coal fly ash particles. Environ. Sci. Technol., 14(9): 1111-1117.

HANSEN, L.D., SILBERMAN, D., FISHER, G.L., & EATOUGH, D.J. (1984) Chemical speciation of elements in stack-collected, respirable-size coal fly ash. Environ. Sci. Technol., 18: 181-186.

HARO, R.T., FURST, A., & FALK, H.L. (1968) Studies on the acute toxicity of nickelocene. Proc. West. Pharmacol. Soc., 11: 39-42.

HARNETT, P.B., ROBISON, S.H., SWARTZENDRUBER, D.E., & COSTA, M. (1982) Comparison of protein, RNA, and DNA binding and cell-cycle-specific growth inhibitory effects of nickel compounds in cultured cells. Toxicol. appl. Pharmacol., 64: 20-30.

HARTENSTEIN, R., NEUHAUSER, E.F., & NARAHARA, A. (1981) Effects of heavy metal and other elemental additives to activated sludge on growth of *Eisenia foetida*. J. environ. Qual., 10(3): 372-377.

HARTWIG, A. & BEYERSMANN, D. (1989) Enhancement of UV-induced mutagenesis and sister-chromatid exchanges by nickel ions in V79 cells: evidence for inhibition of DNA repair. Mutat. Res., 217: 65-73.

HAWLEY, J.E. (1955) Spectographic study of some Nova Scotia coals. Can. min. metall. Bull., 48: 712-726.

HAWORTH, S., LAWLOR, T., MORTELMANS, K., SPECK, W., & ZEIGER, E. (1983) *Salmonella* mutagenicity test results in 250 chemicals. Environ. mol. Mutagen., 5: 3-142.

HAYASHI, YL, TAKAHASHI, M., MAEKAWA, A., KUROKAWA, Y., & KOLUBO, T. (1985) Screening of environmental pollutants for promotion effects on carcinogenesis. Annu. Rep. Minist. Health Welfare Jpn 1982-1984, 20: 1-10.

HEATH, J.C. & DANIEL, M.R. (1964) The production of malignant tumors by nickel in the rat. Br. J. Cancer, 18: 261-264.

HECK, J.D. & COSTA, M. (1982) Extracellular requirements for the endocytosis of carinogenic crystalline nickel sulfide particles by facultative phagocytes. Toxicol. Lett., 12: 243-250.

HEINRICHS, H., & R. MAYER, R. (1980) Distribution and cycling of nickel in forest ecosystems. In: Nriagu, J.O., ed. Nickel in the environment, New York, Chichester, Brisbane, Toronto, John Wiley and Sons, pp. 431-455.

HENDEL, R.C., & SUNDERMAN, F.W. Jr (1972) Species variations in the properties of ultrafiltrable and protein-bound serum nickel. Res. Commun. chem. Pathol. Pharmacol., 4(1): 141-146.

HENRY, W.M. & KNAPP, K.T. (1980) Compound forms of fossil fuel fly ash emissions. Environ. Sci. Technol., 14(4): 450-456.

HENRY, W.M., BARBOUR, R.L., JAKOBSEN, R.J., & SCHUMAKER, P.M. (1982) Inorganic compound identification of fly ash emissions from municipal incinerators. Final Report, Washington, DC, US Environmental Protection Agency, 32 pp (Report prepared by Battelle Columbus Laboratories, Ohio, under the EPA Contract 68-02-2296)(EPA-600/3-82-095. PB83-146175).

HERLANT-PEERS, M.C., HILDEBRAND, H.F., & BISERTE, G. (1982) ^{63}Ni (II) incorporation into lung and liver cytosol of Balb/C mouse. An *in vitro* and *in vivo* study. Zentralbl. Bakteriol. Mikrobiol. Hyg. (B), 176(4): 368-382.

HILDEBRAND, H.F. & BISERTE, G. (1979a) Cylindrical laminated bodies in nickel subsulphide-induced rhabdomyosarcoma in rabbits. Eur. J. cell Biol., 19: 276-280.

HILDEBRAND, H.F. & BISERTE, G. (1979b) Nickel sub-sulphide-induced leiomyosarcoma in rabbit white skeletal muscle. A light microscopical and ultrastructural study. Cancer, 43: 1358-1374.

HILDEBRAND, H.F., & TETAERT, D. (1981) Ni3S2-induced leiomyosarcomas in rabbit skeletal muscle: analysis of the tumoral myosin and its significance in the retrodifferentiation concept. Oncodev. Biol. Med., 2: 101-108.

HO, W., & FURST, A. (1973) Nickel excretion by rats following a single treatment. Proc. West. Pharmacol. Soc., 16: 245-248.

HOBBS, R.J. & STREIT, B. (1986) Heavy metal concentrations in plants growing on a copper mine spoil heap in the Grand Canyon, Arizona. Am. Mid. Nat., 115(2): 277-281.

HODGSON, G.W. (1954) Vanadium, nickel, and iron trace metals in crude oil of Western Canada. Bull. Am. Assoc. Pet. Geol., 38(12): 2537-2554.

HOEY, M.J. (1966) The effects of metallic salts on the histology and functioning of the rat testis. J. reprod. Fertil., 12: 461-471.

HOFF, R.M. & BARRIE, L.A. (1986) Air chemistry observations in the Canadian arctic. Water Sci. Tech., 18: 97-107.

HOGETVEIT, A.C., & BARTON, R.T. (1976) Preventive health program for nickel workers. J. occup. Med., 18(12): 805-808.

HOGETVEIT, A.C. & BARTON, R.T. (1977) Monitoring nickel exposure in refinery workers. In: Brown, S.S., ed. Clinical chemistry and chemical toxicology of metals, Amsterdam, Oxford, New York, Elsevier Science Publishers, Vol. 1, pp. 265-268.

HOGETVEIT, A.C., BARTON, R.T., & KOSTOL, L.C. (1978) Plasma nickel as a primary index of exposure in nickel refining. Ann. occup. Hyg., 21: 113-120.

HOGETVEIT, A.C., BARTON, R.T., & ANDERSEN, I. (1980) Variations of nickel in plasma and urine during the work period. J. occup. Med., 22: 597-600.

HOHNADEL, D.C., SUNDERMAN, F.W. Jr, NECHAY, M.W., & MCNEELY, M.D. (1973) Atomic absorption spectrometry of nickel, copper, zinc and lead in sweat collected from healthy subjects during sauna bathing. Clin. Chem., 19: 1288-1292.

HONDA, K., FUJISE, Y., TATSUKAWA, R., & ITANO, K. (1984) Composition of chemical components in bone of striped dolphin, *Stenella coeruleoalba*: Distribution characteristics of heavy metals in various bones. Agric. biol. Chem., 48(3): 677-680.

HONDA, K., MIN BYUNG, Y., & TATSUKAWA, R. (1985) Heavy metal distribution in organs and tissues of the eastern great white egret *Egretta alba modesta*. Bull. environ. Contam. Toxicol., 35(6): 781-790.

HOPFER, S.M. & SUNDERMAN, F.W. Jr (1988) Hypothermia and deranged circadian rhythm of core body temperature in nickel chloride-treated rats. Res. Commun. chem. Pathol. Pharmacol., 62: 495-505.

HOPFER, S.M., SUNDERMAN, F.W. Jr, FREDRICKSON, R.N., & MORSE, F.E. (1978) Nickel induced erythrocytosis: efficacies of nickel compounds and susceptabilities of rat strains. Ann. clin. lab. Sci., 8: 396-402.

HOPFER, S.M., LINDEN, J.V., CRISOSTOMO, C., CATALANATTO, F.A., GALEN, M., & SUNDERMAN, F.W. Jr (1985) Hypernickelemia in hemodialysis patients. In: Brown, S.S., & Sunderman, F.W. Jr, ed. Progress in Nickel toxicology. Proceedings of the 3rd International Conference on Nickel Metabolism and Toxicology, Paris, 4-7 September, 1984, Oxford, Blackwell Scientific Publications, pp. 133-136.

HOPFER, S.M., LINDEN, J.V., REZUKE, W.N., O'BRIEN, J.E., SMITH, L., WATTERS, F., & SUNDERMAN, F.W. Jr (1987) Increased nickel concentrations in body fluids of patients with chronic alcoholism during disulfiram therapy. Res. Commun. chem. Pathol. Pharmacol., **55**: 101-109.

HOPFER, S.M., FAY, W.P., & SUNDERMAN, F.W. Jr (1989) Serum nickel concentration in hemodialysis patients with environmental nickel exposure. Ann. clin. lab. Sci.

HORAK, E. & SUNDERMANN, F.W. Jr (1973) Fecal nickel excretion by healthy adults. Clin. Chem., **19**(4): 429-430.

HORAK, E. & SUNDERMAN, F.W. Jr (1975a) Effects of Ni(II), other divalent metal ions, and glucagon upon plasma glucose concentrations in normal, adrenalectomized and hypophysectomized rats. Toxicol. appl. Pharmacol., **32**: 316-329.

HORAK, E. & SUNDERMAN, F.W. Jr (1975b) Effects of Ni(II) upon plasma glucagon and glucose in rats. Toxicol. appl. Pharmacol., **33**: 388-391.

HORIE, A., HARATAKE, J., TANAKA, I., KODAMA, Y., & TSUCHIYA, K. (1985) Electron microscopical findings, with special reference to cancer in rats caused by inhalation of nickel oxide. Biol. trace Res., **7**: 223-239.

HOWARD, J.M. (1980) Serum nickel in myocardial infarction. Clin. Chem., **26**(10): 1515.

HRUDEY, S.E. (1985) Residues from hazardous waste treatment. Effluent Water Treat. J., **25**(1): 7-12.

HSI, A.W., O'NEILL, J., & SAN SEBASTIAN, J.R. (1979a) Quantitative mammalian cell genetic toxicology: study of the cytotoxicity and mutagenicity of 70 individual environmental agents related to energy technologies and 3 subfractions of a crude synthetic oil in the CHO/HGPRT system. Environ. Sci. Res., **15**: 291-315.

HSIE, A.W., JOHNSON, N.P., COUCH, D.B., SAN SEBASTIAN, I., O'NEILL, I.P., HOESCHELE, I.D., RAHN, R.O., & FORBES, N.L. (1979b) Quantitative mammalian cell mutagenesis and a preliminary study of the mutagenic potential of metallic compounds. In: Kharasch, N., ed. Trace metals in health and disease, New York, Raven Press, pp. 55-69.

HUANG, J., LU, Y., LIU, G., ZHENG, G., MEI, C., LI, Z., SHENG, S., & YE, Z. (1986) The distribution of trace elements in rats. Acta zool. sin., **32**: 35- 39.

HUEPER, W.C. (1952) Experimental studies in metal cancerogenesis. I. Nickel cancer in rats. Texas Rep. Biol. Med., **10**: 167-186.

HUEPER, W.C. (1955) Experimental studies in metal cancerigenesis. IV. Cancer produced by parenterally introduced metallic nickel. J. Natl Cancer Inst., **16**: 55-67.

HUEPER, W.C. (1958) Experimental studies in metal cancerigenesis. IX. Pulmonary lesions in guinea-pigs and rats exposed to prolonged inhalation of powdered metallic nickel. Arch. Pathol., **65**: 600-607.

HUEPER, W.C. & PAYNE, W.W. (1962) Experimental studies in metal carcinogenesis. Arch. environ. Health, **5**: 445-462.

HUGHES, G.M., PERRY, S.F., & BROWN, V.M. (1979) A morphometric study of effects of nickel, chromium and cadmium on the secondary lamellae of rainbow trout gills. Water Res., **13**:665-679.

HUGHES, A.W., SHERLOCK, D.A., HAMBLEN, D.L., & REID, R. (1987) Sarcoma at the site of a single hip screw. J. bone joint Surg., **B 69**: 470-472.

HUI, G. & SUNDERMAN, F.W. Jr (1980) Effects of nickel compounds on incorporation of the (^{3}H) thymidine into DNA in rat liver and kidney. Carcinogenesis, **1**: 297-304.

HULETT, L.D., WEINBURGER, A.J., NORTHCUTT, K.J., & FERGUSON, M. (1980) Chemical species in fly ash from coal-burning power plants. Science, **210**: 1356-1358.

HUNZIKER, N. (1960) De l'eczéma experimental. Quelques expériences concernant la sensibilisation du cobaye au nickel. Dermatologica, **121**: 307-312.

HUTCHINSON, T.C. (1973) Comparative studies of the toxicity of heavy metals to phytoplankton and their synergistic interactions. Water Pollut. Res. Can., **8**: 68-90.

HUTCHINSON, T.C. & CZYRSKA, H. (1975) Heavy metal toxicity and synergism in floating aquatic weeds. Verh. Int. Ver. Limnol., **19**: 2102-2111.

HUTCHINSON, T.C., & WHITBY, L.M. (1977) The effects of acid rainfall and heavy metal particulates on a boreal forest ecosystem near the Sudbury smelting region of Canada. Water air soil Pollut., **7**: 421-438.

HUTCHINSON, T.C., FEDORENKO, A., FITCHKO, J., KUJA, A., VANLOON, J., & LICHWA, J. (1975) Movement and compartmentation of nickel and copper in an aquatic ecosystem. In: Nriagu, J.O., ed. Environmental biogeochemistry, Ann Arbor, Michigan, Ann Arbor Science Publishers Inc., Vol. 2, pp. 565-585.

HUTCHINSON, T.C., FREEDMAN, B., & WHITBY, L. (1981) Nickel in Canadian soils and vegetation. In: Effects of nickel in the Canadian environment, Ottawa, National Research Council of Canada, pp. 119-157 (Publication No. NRCC 18568).

IARC (1976) Cadmium, nickel, some epoxides, miscellaneous industrial chemicals and general considerations on volatile anaesthetics. Lyon International Agency for Research on Cancer, pp.75-112 (IARC Monographs on the Evaluation of Carcinogenic Risk of Chemicals to Man, Vol. 11).

IARC (1989) Mortality and cancer incidence follow-up of an historical cohort of European welders. Lyon, International Agency for Research on Cancer (Internal Technical Report No. 89/003).

IARC (1990) Nickel and nickel compounds. In: Chromium, nickel and welding. Lyon, International Agency for Research on Cancer, pp. 257-445 (IARC Monographs on the Evaluation of Carcinogenic Risks to Humans, Vol. 49).

ICHINOKI, S. & YAMAZAKI, M. (1985) Simultaneous determination of nickel, lead, zinc, and copper in citrus leaves and rice flour by liquid chromatography with hexamethylene dithiocarbamate extraction. Anal. Chem., 57(12): 2219-2222.

IKEBE, K. & TANAKA, R. (1979) Determination of vanadium and nickel in marine samples by flameless and flame atomic absorption spectrophotometry. Bull. environ. Contam. Toxicol., 21(4-5): 526-532.

INOUE, S., & KAWANISHI, S. (1989) ESR evidence for superoxide, hydroxyl radicals and singlet oxygen produced from hydrogen peroxide and nickel(ii) complex of glycylglycyl-L-histidine. Biochem. Biophys. Res. Commun., 159(2): 445-451.

IZMEROV, N.F. (1983) [Occupational diseases: guidelines.] Moscow, Meditsina, pp. 319-329 (in Russian).

JACOBSEN, N., ALFHEIM, I., & JONSEN, J. (1978) Nickel and strontium distribution in some mouse tissues. Passage through placenta and mammary glands. Res. Commun. chem. Pathol. Pharmacol., 20(3): 571-584.

JACQUET, P., & MAYENCE, A. (1982) Application of the *in vitro* embryo culture to the study of the mutagenic effets of nickel in male germ cells. Toxicol. Lett., 11(1-2): 193-197.

JAFFRE, T., KERSTEN, W., BROOKS, R.R., & REEVES, R.D. (1979) Nickel uptake by Flacourtiaceae of New Caledonia. Proc. R. Soc. Lond. (Biol), 205(1160): 385-394.

JANSEN, L.H., BERRENS, L., & VAN DELDEN, J. (1964) Contact sensitivity to simple chemicals: The role of intermediates in the process of sensitization. Naturwissenschaften, 51: 387-388.

JARSTRAND, C., LUNDBORG, M., WIRNIK, A., & CAMNER, P. (1978) Alveolar macrophage function in nickel dust exposed rabbits. Toxicology, 11(4): 353-359.

JASMIN, G. & RIOPELLE, J.L. (1976) Renal carcinomas and erythrocytosis in rats following intrarenal injection of nickel subsulfide. Lab. Invest., 35: 71-78.

JENKINS, D.W. (1980a) Nickel accumulation in aquatic biota. In: Nriagu, J.O., ed. Nickel in the environment, New York, Chichester, Brisbane, Toronto, John Wiley and Sons, pp. 283-337.

JENKINS, D.W. (1980b) Nickel accumulation in terrestrial wildlife. In: Nriagu, J.O., ed. Nickel in the environment, New York, Chichester, Brisbane, Toronto, John Wiley and Sons, pp. 457-462.

JOHANSSON, A. & CAMNER, P. (1980) Effects of nickel dust on rabbit alveolar epithelium. Environ. Res., 22: 510-516.

JOHANSSON, A., CAMNER, P., JARSTRAND, C., & WIERNIK, A. (1980) Morphology and function of alveolar macrophages after longterm nickel exposure. Environ. Res., 23: 170-180.

JOHANSSON, A., CAMNER, P., & ROBERTSON, B. (1981) Effects of longterm nickel dust exposure on rabbit alveolar epithelium. Environ. Res., **25**: 391-403.

JOHANSSON, A., CAMNER, P., JARSTRAND, C., & WIERNIK, A. (1983a) Rabbit lungs after long-term exposure to low nickel dust concentration. II. Effects on morphology and function. Environ. Res., **30**: 142-151.

JOHANSSON, A., CURSTEDT, T., ROBERTSON, B., & CAMNER, P. (1983b) Rabbit lung after inhalation of soluble nickel. II. Effects on lung tissue and phospholipids. Environ. Res., **31**: 399-412.

JOHANSSON, A., WIERNIK, A., LUNDBORG, M., JARSTRAND, C., & CAMNER, P. (1988) Alveolar macrophages in rabbits after combined exposure to nickel and trivalent chromium. Environ. Res., **46**: 120-132.

JONES, J.G. & WARNER, C.G. (1972) Chronic exposure to iron oxide, chromium oxide, and nickel oxide fumes in metal dressers in a steelworks. Br. J. ind. Med., **29**: 168-177.

JORDAN, W.P. & KING, S.E. (1979) Nickel feeding in nickel-sensitive patients with hand eczema. J. Am. Acad. Dermatol., **1**(6): 506-508.

KAABER, K., VEIEN, N.K., & TJELL, J.C. (1978) Low nickel diet in the treatment of patients with chronic nickel dermatitis. Br. J. Dermatol., **98**: 197-201.

KADOTA, I., & KURITA, M. (1955) Hyperglycemia and islet cell damage caused by nickelous chloride. Metabolism, **4**: 337-342.

KALDOR, J., PETO, J., EASTON, D., DOLL, R., HERMON, C., & MORGAN, L. (1986) Models for respiratory cancer in nickel refinery workers. J. Natl Cancer Inst., **77**: 841-848.

KALIMO, K. & LAMMINTAUSTA, K. (1984) 24 and 48 h allergen exposure in patch testing. Contact Dermatit., **10**: 25-29.

KALLIOMAKI, P.L., RAHKONEN, E., VAARANEN, V., KALLIOMAKI, K., & AITTONIEMI, K. (1981) Lung-retained contaminants, urinary chromium and nickel among stainless steel welders. Int. Arch. occup. environ. Health, **49**(1): 67-75.

KANEMATSU, N., HARA, M., & KADA, T. (1980) REC assay and mutagenicity studies on metal compounds. Mutat. Res., **77**: 109-116.

KASPRZAK, K.S. (1974) An autoradiographic study of nickel carcinogenesis in rats following injection of $^{63}Ni_3S_2$ and $Ni_3^{35}S_2$. Res. Commun. chem. Pathol. Pharmacol., **8**(1): 141-150.

KASPRZAK, K.S. & POIRIER, L.A. (1983) Effects of calcium, magnesium and sodium acetates on tissue distribution of nickel (II) in Strain A mice. In: Brown, S.S. & Savory, J., ed. Chemical toxicology and clinical chemistry of metals. Proceedings of the 2nd International Conference, Montreal, 19-22 July, 1983, Oxford, Blackwell Scientific Publications, pp. 374-376.

KASPRZAK, K.S. & POIRIER, L.A. (1985) Effects of calcium and magnesium salts on nickel subsulfide carcinogenesis in Fisher rats. In: Brown, S.S. & Sunderman, F.W. Jr, ed. Progress in nickel toxicology. Proceedings of the 3rd

International Conference on Nickel Metabolism and Toxicology, Paris, 4-7 September, 1984, Oxford, Blackwell Scientific Publications, pp. 29-32.

KASPRZAK, K.S. & SUNDERMAN, F.W. Jr (1969) The metabolism of nickel carbonyl-^{14}C. Toxicol. appl. Pharmacol., 15(2): 295-303.

KASPRZAK, K.S. & SUNDERMAN, F.W. Jr (1977) Mechanisms of dissolution of nickel subsulfide in rat serum. Res. Commun. chem. Pathol. Pharmacol., 16(1): 95-108.

KASPRZAK, K.S., MARCHOW, L., & BREBOROWICZ, J. (1973) Pathological reactions in rat lungs following intratracheal injection of nickel subsulfide and 3,4-benzpyrene. Res. Commun. chem. Pathol. Pharmacol., 6(1): 237-245.

KASPRZAK, K.S., GABRYEL, P., & JARCZEWSKA, K. (1983) Carcinogenicity of nickel(II)hydroxides and nickel(II)sulfate in Wistar rats and its relation to the *in vitro* dissolution rates. Carcinogenesis, 4(3): 275-279.

KASPRZAK, K.S., WAALKES, M.P., & POIRIER, L.A. (1986) Antagonism by essential divalent metals and amino acides of nickel (II)-DNA binding *in vitro*. Toxicol. appl. Pharmacol., 82: 336-343.

KASPRZAK, K.S. & RODRIGUEZ, R.E. (1988) Nickel-iron interactions in carcinogenesis: dose effect and involvement of active oxygen and related enzymes. In: Fourth International Conference on Nickel Metabolism and Toxicology, Abstracts, Espoo, Finland, 5-9 September 1988, Helsinki, Institute of Occupational Health, p. 59.

KASPRZAK, K.S., KOVATCH, R.M., & POIRIER, L.A. (1988) Inhibitory effect of zinc on nickel subsulfide carcinogenesis in Fischer rats. Toxicology, 52: 153-262.

KASPRZAK, K.S., WARD, J.M., POIRIER, L.A., REICHARDT, D.A., DENN, A.C., III, & REYNOLDS, C.W. (1987) Nickel-magnesium interactions in carcinogenesis: dose effects and involvement of natural killer cells. Carcinogenesis, 8(7): 1005-1011.

KATZ, S.A. & SAMITZ, M.H. (1975) Leaching of nickel from stainless steel consumer commodities. Acta dermatol., 55: 113-115.

KEEFER, R.F., SINGH, R.N., & HORVATH, D.J. (1986) Chemical composition of vegetables grown on an agricultural soil amended with sewage sludges. Environ. Qual., 15(2): 146-152.

KELLER, W. & PITBLADO, J.R. (1986) Water quality changes in Sudbury area lakes: A comparison of synoptic surveys in 1974-1976 and 1981-1983. Water air soil Pollut., 29(3): 285-296.

KEMKA, R. (1971) [Determination of the nickel and cobalt contents in urine and the atmosphere.] Prac. Lek., 23: 80-85 (in Czech).

KESKINEN, H., KALLIOMAKI, P.L., & ALANKO, K. (1980) Occupational asthma due to stainless steel welding fume. Clin. Allergy, 10: 151-159.

KHAN, S.N., RAHMAN, M.A., & SAMAD, A. (1984) Trace elements in serum from Pakistani patients with acute and chronic ischemic heart disease and hypertension. Clin. Chem., 30: 644-648.

KIM, M.K., FISHER, A.M., & MACKAY, R.J. (1969) Pulmonary effects of metallic dusts - nickel and iron. Toronto, University of Toronto, School of Hygiene, Department of Physiological Hygiene, 24 pp.

KINCAID, J.F., STRONG, J.S., & SUNDERMAN, F.W. (1953) Nickel poisoning. I. Experimental study of the effects of acute and subacute exposure to nickel carbonyl. Arch. ind. Hyg. occup. Med., 8: 48-60.

KINCAID, J.F., STANLEY, E.L., BECKWORTH, C.H., & SUNDERMAN, F.W. (1956) Nickel poisoning. III. Procedures for detection, prevention, and treatment of nickel carbonyl exposure including a method for the determination of nickel in biologic materials. Am. J. clin. Pathol., 26: 107-119.

KING, M.M., LYNN, K.K., & HUANG, C.Y. (1985) Activation of the calmodulin-dependant phosphoprotein phosphatase by nickel ions In: Brown, S.S. & Sunderman, F.W. Jr, ed. Progress in nickel toxicology. Proceedings of the 3rd International Conference on Nickel Metabolism and Toxicology, Paris, 4-7 September, 1984, Oxford, Blackwell Scientific Publications, pp. 117-122.

KIRCHGESSNER, M. & SCHNEGG, A. (1979) [Activity of proteases leucinaryl-amidase and a-amylase in pancreas at nickel deficiency.] Nutr. Metab., 23: 62-64 (in German).

KIRCHGESSNER, M. & SCHNEGG, A. (1980a) Biochemical and physiological effects of nickel deficiency. In: Nriagu, J.O., ed. Nickel in the environment, New York, Chichester, Brisbane, Toronto, John Wiley and Sons, pp. 635-652.

KIRCHGESSNER, M. & SCHNEGG, A. (1980b) [Hydrocarbon metabolism in nickel deficiency.] In: Anke, M., Schneider, H.-J., & Brückner, Chr., ed. [3. Trace Element Symposium: Nickel, Jena, German Democratic Republic, 7-11 July, 1980, Jena, Friedrich-Schiller University, pp. 23-26 (in German).

KJELLSTROM, T., FRIBERG, A., & RAHNSTER, B. (1979) Mortality and cancer morbidity among cadmium exposed workers. Environ. Health Perspect., 28: 199-204.

KLEIN, L.A., LANG, M., NASH, N., & KIRSCHNER, S.L. (1974) Sources of metals in New York City wastewater. J. Water Pollut. Control Fed., 46(12): 2653-2662.

KNIGHT, Y.A., REZUKE, W.N., WONG, S.H-Y., HOPFER, S.M., ZAHARIA, O., & SUNDERMAN, F.W. Jr (1987) Acute thymic involution and increased lipoperoxides in thymus of nickel chloride-treated rats. Res. Commun. chem. Pathol. Pharmacol., 55: 291-302.

KODAMA, Y., TANAKA, I., MATSUNO, K., ISHIMATSU, S., HORIE, A., & TSUCHIYA, K. (1985) Pulmonary deposition and clearance of inhaled nickel oxide aerosol. In: Brown, S.S. & Sunderman, F.W. Jr, ed. Progress in nickel toxicology, Proceedings of the 3rd International Conference on Nickel Metabolism and Toxicology, Paris, 4-7 September 1984, Oxford, Blackwell Scientific Publications, pp. 81-84.

KOLLER, L.D. (1980) Immunotoxicology of heavy metals. Int. J. Immunopharmacol., 2: 269-279.

KOLPAKOV, F.I. (1963) [Permeability of skin to nickel compounds.] Arkh. Patol., 25(6): 38-45 (in Russian).

KOMCZYNSKI, L., NOWAK, H., & REJNIAK, L. (1963) Effect of cobalt, nickel and iron on mitosis in the roots of the broad bean (*Vicia faba*). Nature (Lond.), 198: 1016-1017.

KOPONEN, M., GUSTAFSSON, T., KALLIOMAKI, P.L., & PYY, L. (1981) Chromium and nickel aerosols in stainless steel manufacturing, grinding and welding. Am. Ind. Hyg. Assoc. J., 42(8): 596-601.

KOVACS, P. & DARVAS, Z. (1982) Studies on the Ni content of the centriole. Acta histochem., 71: 169-173.

KOWALCZYK, G.S., GORDON, G.E., & RHEINGROVER, S.W. (1982) Identification of atmospheric particulate sources in Washington, DC, using chemical element balances. Env. Sci. Technol., 16(2): 79-89.

KRISHNAN, E.R. & HELLWIG, G.V. (1982) Trace emissions from coal and oil combustion. Environ. Prog., 1(4): 290-295.

KUCHARIN, G.M. (1970) [Occupational disorders of the nose and nasal sinuses in workers in an electrolytic nickel refining plant.] Gig. Tr. prof. Zabol., 14: 38-40 (in Russian).

KUEHN, K. & SUNDERMAN, F.W. Jr (1982) Dissolution half-times of nickel compounds in water, rat serum, and renal cytosol. J. inorg. Biochem., 17(1): 29-39.

KUEHN, K., FRASER, O.B., & SUNDERMAN, F.W. Jr (1982) Phagocytosis of particulate nickel compounds by rat peritoneal macrophage *in vitro*. Carcinogenesis, 3(3): 321-326.

KUROKAWA, Y., MATSUSHIMA, M., IMAZAWA, T., TAKAMURA, N., TAKAHASHI, M., & HAYASHI, Y. (1985) Promoting effect of metal compounds on rat renal tumorigenesis. J. Am. Coll. Toxicol., 4: 321-330.

LABELLA, F., DULAR, R., LEMON, P., VIVIAN, S., & QUEEN, G. (1973a) Prolactin secretion is specifically inhibited by nickel. Nature (Lond.), 245: 331-332.

LABELLA, F., DULAR, R., VIVIAN, S., & QUEEN, G. (1973b) Pituitary hormone releasing or inhibiting activity of metal ions present in hypothalamic extracts. Biochem. Biophys. Res. Commun., 52: 786-791.

LANGARD, S. & STERN, R.M. (1984) Nickel in welding fumes - A cancer hazard to welders? A review of epidemiological studies on cancer in welders. In: Nickel in the human environment, Proceedings of a Joint Symposium, Lyon, 8-11 March, 1983, Lyon, International Agency for Research on Cancer, pp. 95-103 (IARC Scientific Publications No. 53).

LARRAMENDY, M.L., POPESCU, N.C., & DIPAOLO, J.A. (1981) Induction by inorganic metal salts of sister chromatid exchanges and chromosome aberrations in human and Syrian hamster cell strains. Environ. Mutagen., 3: 597-606.

LAU, T.J., HACKETT, R.L., & SUNDERMAN, F.W. Jr (1972) The carcinogenicity of intravenous nickel carbonyl in rats. Cancer Res., 32: 2253-2258.

LAVELLE, J.M. & WITMER, C.M. (1981) Mutagenicity of NiCl2 and the analysis of mutagenicity of metal ions in a bacterial fluctuation test. Environ. Mutagen., 3: 320-321.

LAZAREVA, L.P. (1985) Changes in biological characteristics of *Daphnia magna* from chronic action of copper and nickel at low concentrations. Gidrobiol. Zh., 21: 59-62.

LEACH, C.A. Jr & SUNDERMAN, F.W. Jr (1987) Hypernickelemia following coronary arteriography, caused by nickel in the radiographic contrast medium. Ann. clin. lab. Sci. 17(3): 137-143.

LEACH, C.A. Jr, LINDEN, J.V., HOPFER, S.M., CRISOSTOMO, M.C., & SUNDERMAN, F.W. Jr (1985) Nickel concentrations in serum of patients with acute myocardial infarction or unstable angina pectoris. Clin. Chem. 31(4): 556-560.

LECHNER, J.F., TOKIWA, T., MCCLENDON, I.A., & HAUGEN, A. (1984) Effects of nickel sulfate on growth and differentiation of normal human bronchial epithelial cells. Carcinogenesis, 5: 1697-1703.

LEE, R.E. & VON LEHMDEN, D.J. (1973) Trace metal pollution in the environment. J. Air. Pollut. Control Assoc., 23: 853-857.

LESLIE, A.C.D., WINCHESTER, J.W., LEYSIEFFER, F.W., & AHLBERG, M.S. (1976) Prediction of health effects of pollution aerosols. Trace Subst. environ. Health, 10: 497-504.

LESSARD, R., REED, D., MAHEUX, B., & LAMBERT, J. (1978) Lung cancer in New Caledonia, a nickel smelting island. J. occup. Med., 20(12): 815-817.

LESTROVOI, A.P., ITSKOVA, A.I., & ELISEEV, I.N. (1974) [Effect of nickel on the iodine fixation of the thyroid gland when administered perorally and by inhalation.] Gig. i Sanit., 10: 105-106 (in Russian).

LEVAN, A. (1945) Cytological reactions induced by inorganic salt solutions. Nature (Lond.), 3973: 751-751.

LEWIS, C.L. & OTT, W.L. (1970) Analytical chemistry of nickel, Oxford, New York, Pergamon Press, 232 pp.

LICHTY, P.D. & ZEY, J.N. (1985) Health Hazard Evaluation Report: General Motors Corporation, Dayton, Ohio. Cincinnati, Ohio, National Institute for Occupational Safety and Health, 37 pp (HETA 84-060-1645).

LIGETI, L., RUBANYI, G., KOLLER, A., & KOVACH, A.G.B. (1980) [Effect of nickel ions on hemodynamics, cardiac performance and coronary blood flow in anaesthetized dogs.] In: Anke, M., Schneider, H.-J., & Brückner, Chr., ed. [3. Trace Element Symposium Nickel, Jena, German Democratic Republic, 7-11 July, 1980], Jena, Friedrich-Schiller University, pp. 117-122 (in German).

LINDEN, J.V., HOPFER, S.M., GOSSLING, H.R., & SUNDERMAN, F.W. Jr (1985) Blood nickel concentrations in patients with stainless-steel hip prostheses. Ann. clin. lab. Sci., 15(6): 459-464.

LIU, Y. ET AL. (1983) [The role of nickel sulfate in inducing nasoharyngeal carcinoma (NPC) in rats]. Guangzhou, Cancer Institute of Zhongshan Medical

College, WHO Collaborating Centre for Research on Cancer, pp. 48-49 (Cancer Research Reports, Vol. 4)(in Chinese with English abstract).

LLOYD, G.K. (1980) Dermal absorption and conjugation of nickel in relation to the induction of allergic contact dermatitis - preliminary results. In: Brown, S.S. & Sunderman, F.W., ed. Nickel toxicology, Proceedings of the 2nd International Conference on Nickel Toxicology, Swansea, 3-5 September, 1980, London, New York, Academic Press, pp. 145-148.

LOKEN, A.C. (1950) [Carcinoma of the lung in nickel workers.] Tidskr. Nor. Laegeforen., 70: 376-378 (in Norwegian).

LONG-ZHU, J. & ZHE-MING, N. (1985) Determination of nickel in urine and other biological samples by graphite furnace atomic absorption spectrometry. Fresenius Z. anal. Chem., 321: 72-76.

LOW, K.S. & LEE, C.K. (1981) Copper, zinc, nickel and chromium uptake by "Kangkong Air" (*Ipomea acquatica* Forsk). Pertanika, 4(1):16-20.

LU, C.C., MATSUMOTO, N., & IIJIMA, S. (1979) Teratogenic effects of nickel chloride on embryonic mice and its transfer to embryonic mice. Teratology, 19(2): 137-142.

LU, C.C., MATSUMOTO, N., & IIJIMA, S. (1981) Placental transfer and body distribution of nickel chloride in pregnant mice. Toxicol. appl. Pharmacol., 59(3): 409-413.

LUCASSEN, M. & SARKAR, B. (1979) Nickel(II)-binding constituents of human blood serum. J. Toxicol. environ. Health, 5(5): 897-905.

LUDEWIGS, H.-J. & THIESS, A.M. (1970) [Knowledge in occupational medicine of nickel carbonyl poisoning.] Zentralbl. Arbeitsmed. Arbeitsschutz, 20: 329-339 (in German).

LUMB, G.D., SUNDERMAN, F.W. Jr, & SCHNEIDER, H.P. (1985) Nickel-induced malignant tumors. Ann. clin. lab. Sci., 15(5): 374-380.

LUNDBORG, M. & CAMNER, P. (1982) Decreased level of lysozyme in rabbit lung lavage fluid after inhalation of low nickel concentrations. Toxicology, 22(4): 353-358.

LUNDBORG, M. & CAMNER, P. (1984) Lysozyme levels in rabbit lung after inhalation of nickel, cadmium, cobalt, and copper chlorides. Environ. Res., 34: 335-342.

MACHELETT, B. & PODLESAK, W. (1980) [Nickel uptake by oat and mustard plants as a function of lime status of the soil.] In: Anke, M., Schneider, H.J., & Brückner, Chr. Jr ed. [3. Trace Element Symposium Nickel, Jena, German Democratic Republic, 7-11 July 1980, Jena, Friedrich-Schiller University, pp. 207-213 (in German).

MACKENZIE PEERS, A. (1986) Determination of arsenic, chromium, nickel, cadmium, lead, beryllium and selenium in air and airborne particulates by inductively coupled argon atomic emission spectroscopy. NIOSH 7300 adapted by Mackenzie Peers, A. In: Environmental carcinogens - selected methods of analysis. Vol. 8: Some metals: As, Be, Cd, Cr, Ni, Pb, Se, Zn, Lyon, International Agency for Research on Cancer, pp. 225-230 (IARC Scientific Publications No. 71).

MCCONNELL, L.H., FINK, J.N., SCHLUETER, D.P, & SCHMIDT, M.G. Jr (1973) Asthma caused by nickel sensitivity. Ann. intern. Med., 78(6): 888-890.

MCDONALD, I. (1981) Malignant lymphoma associated with internal fixation of a fractured tibia. Cancer, 48: 1009-1011.

MCDOUGALL, A (1956) Malignant tumor at site of bone plating. J. bone joint Surg. B,. 38: 709-713.

MCGREGOR, D.B., BROWN, A., CATTANACH, P., EDWARDS, I., McBRIDE, D., RIACH, C., & CASPARY, W.J. (1988) Responses of the L5178Y tK^+/tK^- mouse lymphoma cell forward mutation assay: III. 72 coded chemicals. Environ. molec. Mutagen., 12: 85-154.

MCLAREN, J.W., MYKYTIUK, A.P., WILLIE, S.N., & BERMAN, S.S. (1985) Determination of trace metals in seawater by inductively coupled plasma mass spectrometry with preconcentration on silica-immobilized 8-hydroxyquinoline. Anal. Chem., 57: 2907-2911.

MCLEAN, J.R., MCWILLIAMS, R.S., KAPLAN, J.G., & BIRNBOIM, H.C. (1982) Rapid detection of DNA strand breaks in human peripheral blood cells and animal organs following treatment with physical and chemical agents. In: Bora, K.C., Douglas, G.R., & Nestmann, E.R., ed. Chemical mutagenesis, human population monitoring and genetic risk assessment. Amsterdam, Oxford, New York, Elsevier Science Publishers, pp. 137-141 (Progress in Mutation Research, Vol. 3).

McNEELY, M.D., SUNDERMAN, F.W. Jr, NECHAY, M.W., & LEVINE, H. (1971a) Abnormal concentrations of nickel in serum in cases of myocardial infarction, stroke, burns, hepatic cirrhosis, and uremia. Clin. Chem., 17(11): 1123-1128.

MCNEELY, M.D., NECHAY, M.W., & SUNDERMAN, F.W. Jr (1971b) Measurement of nickel in serum and urine as indices of environmental exposure to nickel. Clin. Chem., 18: 992-995.

MAENZA, R.M,., PRADHAN, A.M., & SUNDERMAN, F.W. Jr (1971) Rapid induction of sarcomas in rats by combination of nickel sulfide and 3,4-benzypyrene. Cancer Res., 31: 2067-2071.

MAGNUS, K., ANDERSEN, A., & HOGETVEIT, A.C. (1982) Cancer of respiratory organs among workers at a nickel refinery in Norway. Int. J. Cancer, 30(6): 681-685.

MAGNUSSON, B. & KLIGMAN, A.M. (1970) Allergic contact dermatitis in the guinea pig. Springfield, Illinois, Charles C. Thomas, 141 pp.

MALO, J.L., CARTIER, A., DOEPNER, M., NIEBOER, E., EVANS, S., & DOLOVICH, J. (1982) Occupational asthma caused by nickel sulfate. J. Allergy clin. Immunol., 69: 55-59.

MAREK, M. & TREHARNE, R.W. (1982) An *in vitro* study of the release of nickel from two surgical implant alloys. Clin. Orthop., 167: 291-295.

MARZOUK, A. & SUNDERMAN, F.W. Jr (1985) Biliary excretion of nickel. Toxicol. Lett., 27: 65-71.

MAS, A., HOLT, D. & WEBB, M.C. (1985) The acute toxicology and teratogenicity of nickel in pregnant rats. Toxicology, 35: 47-57.

MASON, B. (1952) Principles of geochemistry, New York, John Wiley & Sons, London, Chapman & Hall, Limited, 276 pp.

MASTROMATTEO, E. (1967) Yant memorial lecture: Nickel: A review of its occupational health aspects. J. occup. Med., 9(3): 127-136.

MASTROMATTEO, E. (1986) Nickel. Am Ind. Hyg. Assoc. J., 47(10): 589-601.

MATHIS, B.J. & CUMMINGS, T.F. (1973) Selected metals in sediments, water, and biota in the Illionois River. J. Water Pollut. Control Fed., 45(7):1573- 1583.

MATHUR, A.K., DATTA, K.K., TANDON, S.K., & DIKSHITH, T.S. (1977) Effect of nickel sulphate on male rats. Bull. environ. Contam. Toxicol., 17: 241-247.

MATHUR, A.K., DIKSHITH, T.S., LAL, M.M., & TANDON, S.K. (1978) Distribution of nickel and cytogenetic changes in poisoned rats. Toxicology, 10(2): 105-113.

MEDINSKY, M.A., BENSON, J.M., & HOBBS, C.H. (1987) Lung clearance and disposition of the ^{63}Ni in F344/N rats after intratracheal instillation of nickel sulfate solutions. Environ. Res., 43: 168-178.

MEDVEDEVA, V.J. (1965) [Nickel content in blood of a new born baby and in venous and retroplacental blood and milk of the mother.] Vest. Akad. Navuk. Belarusk. SSR. Ser. biyal Navuk, 2: 114-115 (in Russian).

MENDEN, E.E., ELIA, V.J., MICHAEL, L.W., & PETERING, H.G. (1972) Distribution of cadmium and nickel of tobacco during cigarette smoking. Environ. Sci. Technol., 6: 830-832.

MENNE, T., BORGAN, O., & GREEN, A. (1982) Nickel allergy and hand dermatitis in a stratified sample of the Danish female population: an epidemiological study including a statistic appendix. Acta dermatovenereol., 62(1): 35-41.

MENNE, T., ANDERSEN, K.E., KAABER, K., OSMUNDSEN, P.E., ANDERSEN, J.R., YOING, F., & VALEUR, G. (1987) Evaluation of the dimethylglyoxime stick test for the detection of nickel. Derm. Beruf Umwelt, 35: 128-130.

MERCER, T.T. (1967) On the role of particle size in the dissolution of lung burdens. Health Phys., 13: 1211-1221.

MERTZ, W. (1981) The essential trace elements. Science, 213: 1332-1338.

MEYER, A. & NEEB, R. (1985) Determination of cobalt and nickel in some biological matrices - comparison of chelate gas-chromatography and adsorption voltammetry. Fresenius Z. anal. Chem., 321(3): 235-241.

MIKAC-DEVIC, D., SUNDERMAN, F.W. Jr, & NOMOTO, S. (1977) Furildioxime method for nickel analysis in serum and urine by electrothermal atomic absorption spectrometry. Clin. Chem., 23(6): 948-956.

MIKHEYEV, M.J. (1971) [Distribution and elimination of nickel carbonyl.] Gig. Tr. prof. Zabol., 15: 35-38 (in Russian).

MIKI, H., KASPRZAK, K.S., KENNEY, S., & HEINE, U.K. (1987) Inhibition of intercellular communication by nickel(II): antagonistic effect of magnesium. Carcinogenesis, 8(11): 1757-1760.

MILFORD, J.B. & DAVIDSON, C.I. (1985) The sizes of particulate trace elements in the atmosphere - a review. J. air Pollut., 35(12): 1249-1260.

MITSUMASA, O. (1987) Induction of ocular tumor by nickel subsulfide in the Japanese common newt, *Cynops pyrrhogaster*. Cancer Res., 47: 5213-5217.

MIYAKI, M., AKAMATSU, N., ONO, T., & KOYAMA, H. (1979) Mutagenicity of metal cations in cultured cells from Chinese hamsters. Mutat. Res., 68: 259-263.

MOELLER, H. (1984) Attempts to induce contact allergy to nickel in the mouse. Contact Dermatit., 10: 65-68.

MOFFA, I.P., ELLISON, J.E., & HAMILTON, J.C. (1983) Incidence of nickel sensitivity in dental patients. J. dent. Res., 62: 199.

MORGAN, J.G.A. (1958) Some observations on the incidence of respiratory cancer in nickel workers. Br. J. ind. Med., 15: 224-234.

MORGAN, J.G.A. (1960) A simplified method for the estimation of nickel in urine. Br. J. ind. Med., 17: 209-212.

MORGAN, L.G. & ROUGE, P.J. (1979) A study into the correlation between atmospheric and biological monitoring of nickel in nickel refinery workers. Ann. occup. Hyg., 22(3): 311-317.

MORITA, H., NODA, K., & UMEDA (1985) Mutagenicities of nickel and cobalt compounds in a mammalian cell line. Mutat. Res., 147: 265-266.

MORROW, P.E. (1970) Models for the study of particle retention and elimination in the lung. Inhalation carcinogenesis, Oake Ridge, Tennessee, US Atomic Energy Commission, Division of Technical Information, pp. 103-116 (AEC Symposium Series, No. 18).

MORSE, E.E., LEE, T.Y., REISS, R.F., & SUNDERMAN, F.W. Jr (1977) Dose-response and time-response study of erythrocytosis in rats after intrarenal injection of nickel subsulfide. Ann. clin. lab. Sci., 7: 17-24.

MORTIMER, D.C. (1985) Freshwater aquatic macrophytes as heavy metal monitors the Ottawa river experience Canada. Environ. monit. Assess., 5(3): 311-324.

MUHLE, H., BELLMANN, B., & TAKENAKA, S. (1988) Chronic effects of intratracheally instilled nickel containing particles in hamsters. In: Fourth International Conference on Nickel Metabolism and Toxicology: Abstracts, Espoo, Finland, 5-9 September 1988, Helsinki, Institute of Occupational Health, p. 41.

MUKUBO, K. (1978) Studies on experimental lung tumor by the chemical carcinogens and inorganic substances. III. Histopathological studies on lung tumor induced by pertracheal vinyl tube infusion of 20-methylcholanthrene combined with chromium and nickel powder. J. Nara Med. Ass., 29: 321-340.

MURTHY, R.C., BARKLEY, W., HOLLINGSWORTH. C., & BINGHAM, E. (1983) Enzymatic changes in alveolar macrophages of rats exposed to lead and nickel by inhalation. J. Am. Coll. Toxicol., 2: 193-199.

MYRON, D.R., ZIMMERMAN, T.J., SHULER, T.R., KLEVAY L.M., LEE, D.E., & NIELSEN, F.H. (1978) Intake of nickel and vanadium by humans. A survey of selected diets. Am. J. clin. Nutr., 31(3): 527-531.

NADEENKO, V.G., LENCHENKO, V.G., ARKHIPENKO, T.A., SAICHENKO, S.P., & PETROVA, N.N. (1979) [Embryotoxic effect of nickel entering the body via drinking water.] Gig. i Sanit., 6: 86-88 (in Russian).

NADEENKO, V.G., LECHENKO, V.G., & SAICHENKO, S.P. (2988) [Combined action of nickel and cobalt administered with drinking water.] In: Domnin, S.G. & Tartakovskaya, L.Ya., ed. [Combined effects of physical and chemical factors of industrial environment], Moscow, Erisman Institute of Hygiene, pp. 84-90 (in Russian).

NAS (1975) Nickel. Washington, DC, National Academy of Sciences, 277 pp.

NATUSCH, D.F.S., WALLACE, J.R., & EVANS, C.A. (1974) Toxic trace elements: preferential concentration in respirable particles. Science, 183: 202-204.

NEBEKER, A.V., STINCHFIELD, A., SAVONEN, C., & CHAPMAN, G.A. (1986) Effects of copper, nickel and zinc on three species of Oregon freshwater snails. Environ. Toxicol. Chem., 5(9): 807-811.

NECHAY, M.W. & SUNDERMAN, F.W. Jr (1973) Measurements of nickel in hair by atomic absorption spectrometry. Ann. clin. lab. Sci., 3(1): 30-35.

NEUHAUSER, E.F., LOEHR, R.C., & MALECKI, M.R. (1985) Contact and artificial soil tests using earthworms to evaluate the impact of wastes in soil. In: Petros, J.K., Lacy, W.J., & Conway, R.A., ed. Hazardous and industrial solid waste testing: Fourth Symposium, Philadelphia, American Society for Testing and Materials, pp. 192-203 (ASTM STP 886).

NEUMULLER. O.A. (1985) Römpp's chemical encyclopaedia, 8th ed., Stuttgart, Franck'sche Verlagsbuchhandlung, Vol. 4, 3230 pp.

NEWMAN, S.M., SUMMIT, R.L., & NUNEZ, L. J. (1982) Incidence of nickel-induced sister chromatid exchange. Mutat. Res., 101: 67-75.

NIEBOER, E. & JUSYS, A.A. (1983) Contamination control in routine ultratrace analysis of toxic metals. In: Brown, S.S. & Savory, J., ed. Chemical toxicology and clinical chemistry of metals, Proceedings of the 2nd International Conference, Montreal, 19-22 July, 1983, London, New York, Blackwell Scientific Publications, pp. 3-16.

NIEBOER, E., MAXWELL, R.J., ROSSETTO, F.E., STAFFORD, A.R., & STETSKO, P.J. (1986) Concepts in nickel carcinogenesis. In: Xavier, A.V., ed. Frontiers of bioinorganic chemistry. 2nd ed., Weinheim, VCH Verlag, pp. 142-151.

NIELSEN, F.H. (1971) Studies on the essentiality of nickel. In: Mertz, W. & Cornatzer, W.E., ed. New trace elements in nutrition, New York, Basel, Marcel Dekker, pp. 215- 253.

NIELSEN, F.H. (1974) Essentiality and function of nickel. In: Hoekstra, W.G., Suttie, J.W., Ganther, H.E., & Mertz, W., ed. Trace element metabolism in animals. Proceedings of the 2nd International Symposium on Trace Element Metabolism in Animals, Madison, Wisconsin, 18-22 June, 1973, Baltimore, Maryland, University Park Press, pp. 381-395.

NIELSEN, F.H. (1984) Ultratrace elements in nutrition. Annu. Rev. Nutr., 4: 21-41.

NIELSEN, F.H. & HIGGS, D.J. (1971) Further studies involving a nickel deficiency in chicks. Proc. trace Subst. environ. Health, 4: 241-246.

NIELSEN, F.H. & OLLERICH, D.A. (1974) Nickel: a new essential trace element. Fed. Proc., 33: 1767-1772.

NIELSEN, F.H. & SAUBERLICH, H.E. (1970) Evidence of a possible requirement for nickel by the chick. Proc. Soc. Exp. Biol. Med., 134(3): 845-849.

NIELSEN, F.H. & SHULER, T.R. (1979) Effect of dietary nickel and iron on the trace element content of rat liver. Biol. trace elem. Res., 1: 337-346.

NIELSEN, F.H. & SHULER, T.R. (1981) Effect of form of iron on nickel deprivation in the rat. Liver content of copper, manganese and zinc. Biol. trace elem., Res., 3: 245-256.

NIELSEN, F.H., OLLERICH, D.A., FOSMIRE, G.J., & SANDSTEAD, H.H. (1974) Nickel deficiency in chicks and rats: effects on liver morphology, function and polysomal integrity. Adv. exp. Med. Biol., 48: 389-403.

NIELSEN, F.H., MYRON, D.R., GIVAND, S.H., & OLLERICH, D. A. (1975a) Nickel deficiency and nickel-rhodium interaction in chicks. J. Nutr., 105: 1607-1619.

NIELSEN, F.H., MYRON, O.R., GIVAND, S.H., ZIMMRMANN, T.J., & OLLERICH, D.A. (1975b) Nickel deficiency in rats. J. Nutr., 105: 1620-1630.

NIELSEN, F.H., SHULER, T.R., ZIMMERMANN, T.J., COLLINGS, M.E., & UTHUS, E.O. (1979a) Interaction between nickel and iron in the rat. Biol. trace elem. Res., 1: 325-335.

NIELSEN, F.H., ZIMMERMAN, T.J., COLLINGS, M.E., & MYRON, D.R. (1979b) Nickel deprivation in rats: Nickel-iron interactions. J. Nutr., 109(9): 1623-1632.

NIELSEN, G.D. (in press) Oral challenge of nickel-allergic patients with hand eczema. In: Nieboer, E. & Aitio, A., ed. Nickel and human health: Current perspectives. New York, Chichester, Brisbane, Toronto, John Wiley & Sons (Advances in environmental science and technology Series).

NIELSEN, G.D. & FLYVHOLM, M. (1984) Risks of high nickel intake with diet. In: Nickel in the human environment. Proceedings of a Joint Symposium, Lyon, 8-11 March 1983, Lyon, International Agency for Research on Cancer, pp. 333-338 (IARC Scientific Publications No. 53).

NIELSEN, G.D., JORGENSEN, P.J., KEIDING, K., & GRANDJEAN, P. (1987a) Urinary nickel excretion before and after loading with naturally occurring nickel. In: Trace elements in human health and disease. Second Nordic Symposium,

Odense, Denmark, 17-21 August 1987: Symposium Abstracts, Odense, Odense University (Abstract C3).

NIELSEN, G.D., ENDERSEN, O., & GRANDJEAN, P. (1987b) Effect of diethyldithiocarbamate on toxicokinetics of ^{57}Ni in mice. In: Trace elements in human health and disease. Extended abstracts from the second Nordic Symposium, Odense, 17-21 August 1987, Copenhagen, World Health Organization, Regional Office for Europe, pp. 78-81 (Environmental Health Series No. 20).

NILZEN, A. & WIKSTROM, K. (1955) The influence of lauryl sulphate on the sensitization of guinea-pigs to chrome and nickel. Acta dermatovenereol., 35: 292-299.

NISHIMURA, M. & UMEDA, M. (1979) Induction of chromosomal aberrations in cultured mammalian cells by nickel compounds. Mutat. Res., 68: 337-349.

NISHIOKA, H. (1975) Mutagenic activities of metal compounds in bacteria. Mutat. Res., 31: 185-189.

NODIYA, P.J. (1972) [Cobalt and nickel balance in students of an occupational technical school.] Gig. i Sanit., 37: 108-109 (in Russian).

NOMOTO, S. & SUNDERMAN, F.W. (1970) Atomic absorption spectrometry of nickel in serum, urine, and other biological materials. Clin. Chem., 16: 477-485.

NOMOTO, S. & SUNDERMAN, F.W. Jr (1988) Presence of nickel in alpha-2-macroglobulin isolated from human serum by high performance liguid chromatography. Ann. clin. lab. Sci., 18: 78-84.

NOMOTO, S., HIRABAYASHI, T., & FUKUDA, T. (1983) Serum nickel concentrations in women during pregnancy, parturition, and post-partum. In: Brown, S.S. & Savory, J., ed. Chemical toxicology and clinical chemistry of metals, Proceedings of the 2nd International Conference, Montreal, 19-22 July, 1983, London, New York, Blackwell Scientific Publications, pp. 351- 352.

NOMOTO, S., MCNEELY, M.D., & SUNDERMAN, F.W. (1971) Isolation of a nickel a$_2$-macroglobulin from rabbit serum. Biochemistry, 10(9): 1647-1651.

NORGAARD, O. (1955) Investigations with radioactive Ni57 into the resorption of nickel through the skin in normal and in nickel-hypersensitive persons. Acta dermtovenereol., 35: 111-117.

NORGAARD, O. (1957) Investigations with radioactive nickel, cobalt and sodium on the resorption through the skin in rabbits, guinea-pigs and man. Acta dermatovenereol., 37: 440-445.

NORSETH, T. (1975) Urinary excretion of nickel as an index of nickel exposure in welders and nickel refinery workers. Int. Congr. Occup. Health, 18: 327.

NOVEY, H.S., HABIB, M., & WELLS, I.D. (1983) Asthma and IgE antibodies induced by chromium and nickel salts. J. Allergy clin. Immunol., 72: 407-412

NOZDRYUKHINA, L.R. (1978) Use of blood trace elements for diagnosis of heart and liver disease. In: Kirchgessner, M., ed. Trace element metabolism in man and animals. Proceedings of the 3rd International Symposium, Freising, Federal

Republic of Germany, July 1977, Freising-Weihenstephan, Technical University Munich, pp. 336-339.y

NRIAGU, J.O. (1979) Global inventory of natural and anthropogenic emissions of trace metals to the atmosphere. Nature (Lond.), 279(5712): 409-411.

NRIAGU, J.O. (1980) Global cycle and properties of nickel. In: Nriagu, J.O., ed. Nickel in the environment, New York, Chichester, Brisbane, Toronto, John Wiley and Sons, pp. 1-26.

OAKLEY, A.M.M., IVE, F.A., & CARR, M.M.C. (1987) Skin clips are contraindicated when there is nickel allergy. J. R. Soc. Med., 80: 290-291.

O'DELL, G.D., MILLER, W.J., MOORE, S.L., KING, W.A., ELLERS, J.C., & JURECEK, H. (1971) Effect of dietary nickel level on excretion and nickel content of tissues in male calves. J. anim. Sci., 32(4): 769-773.

OGAWA, H.I., TSURUTA, S., NIYITANI, Y., MINO, H., SAKATA, K., & KATO, Y. (1987) Mutagenesis of metal salts in combination with 9-amonoacridine in *Salmonella typhimurium*. Jpn. J. Genet., 62: 159.

OHNO, H., HANAOKA, F., & YAMADA, M.A. (1982) Inducibility of sister-chromatid exchanges by heavy-metal ions. Mutat. Res., 104: 141-145.

OLEJAR, S., OLEJAROVA, E., & VRABEL, K. (1982) [Lung tumors in workers in a nickel refinery.] Prac. Lek., 34(8): 80-282 (in Czech).

OLSEN, I. & JONSEN, J. (1979a) Whole-body autoradiography of 63Ni in mice throughout gestation. Toxicology, 12(2): 165-172.

OLSEN, I. & JONSEN, J. (1979b) Effect of cadmium acetate, copper sulfate and nickel chloride on organ cultures of mouse trachea. Acta pharmacol. toxicol., 44: 120-127.

OLSEN, J. & SABROE, S. (1984) Occupational causes of laryngeal cancer. J. Epidemiol. community Health, 38(2): 117-121.

ONKELINX, C. & SUNDERMAN, F.W. (1980) Modeling of nickel metabolism. In: Nriagu, J.O., ed. Nickel in the environment, New York, Chichester, Brisbane, Toronto, John Wiley and Sons, pp. 525-545.

ONKELINX, C., BECKER, J., & SUNDERMAN, F.W. (1973) Compartmental analysis of the metabolism of ^{63}Ni(II) in rats and rabbits. Res. Commun. chem. Pathol. Pharmacol., 6(2): 663-676.

OSKARSSON, A. & TJALVE, H. (1979a) The distribution and metabolism of nickel carbonyl in mice. Br. J. ind. Med., 36(4): 326-335.

OSKARSSON, A. & TJALVE, H. (1979b) An autoradiographic study on the distribution of ^{63}NiCl2 in mice. Ann. clin. lab. Sci., 9(1): 47-59.

OSKARSSON, A., ANDERSSON, Y., & TJALVE, H. (1979) Fate of nickel subsulfide during carcinogenesis studied by autoradiography and X-ray powder diffraction. Cancer Res., 39(10): 4175-4182.

OSTAPCZUK, P., VALENTA, P., STOEPPLER, M., & NURNBERG, H.W. (1983) Voltammetric determination of nickel and cobalt in body fluids and other biological materials. In: Brown, S.S. & Savory, J., ed. Chemical toxicology and clinical chemistry of metals, Proceedings of the 2nd International Conference, Montreal, 19-22 July, 1983; London, New York, Blackwell Scientific Publications, pp. 61-64.

OSTAPCZUK, P., FRONING, M., STOEPPLER, M., & NUERNBERG, H.W. (1985a) Square wave voltammetry: A new approach for the sensitive determination of nickel and cobalt in human samples. In: Brown, S.S. & Sunderman, F.W. Jr ed. Progress in Nickel toxicology. Proceedings of the 3rd International Conference on Nickel Metabolism and Toxicology, Paris, 4-7 September, 1984, Oxford, Blackwell Scientific Publications, pp. 129-132.

OSTAPCZUK, P., FRONING, M., STOEPPLER, M., & NüRNBERG, H.W. (1985b) [Determination of heavy metals in environmental samples and biotic material by means of square wave voltammetry.] Fresenius Z. anal. Chem., 320: 645 pp. (in German).

OSTERREICHISCHES FORSCHUNGSZENTRUM SEIBERSDORF GMBH (1983) [Assessment of the acute toxicity of nickel sulfate to rainbow trout.] Vienna, Austrian Research Centre, 21 pp. (Report No. 4213, BL-424/83)(in German).

OTTOLENGHI, A.D., HASEMAN, J.K., PAYNE, W.W., FALK, H.L., & MACFARLAND, H.N. (1974) Inhalation studies of nickel sulfide in pulmonary carcinogenesis of rats. J. Natl Cancer Inst., 54(5): 1165-1172.

OU, B., LIU, Y., HUANG, X., & FENG, G. (1981) [The promoting action of nickel in the induction of nasopharyngeal carcinoma in rats.] Guangzhou, Cancer Institute of Zhongshan Medical College, WHO Collaborating Centre for Research on Cancer, pp. 3-8 (Cancer Research Reports, Vol. 2)(in Chinese with English abstract).

OU, B., LIU, Y., & SHENG, G. (1983) [Tumor induction in next generation of dinitrosopiperazine-treated pregnant rats.] Guangzhou, Cancer Institute of Zgongshan Medical College, WHO Collaborating Centre for Research on Cancer pp. 44-45, (Cancer Research Reports, Vol. 4)(in Chinese with English abstract).

PACYNA, J.M., OTTAR, B., TOMZA, U., & MAENHAUT, W. (1985) Long-range transport of trace elements to Ny Alesund, Spitsbergen. Atmos. Environ., 19(16): 857-865.

PAINTER, R.B. & HOWARD, R. (1982) The HeLa DNA-synthesis inhibition test as a rapid screen for mutagenic carcinogens. Mutat. Res., 92: 427-437.

PALO, J. & SAVOLAINEN, H. (1973) Biochemical diagnosis of aspartyl-glycosaminuria. Ann. clin. Res., 5: 156-162.

PARKER, D. & TURK, J.L. (1978) Delay in the development of the allergic response to metals following intratracheal instillation. Int. Arch. Allergy appl. Immunol., 57: 289-293.

PARKER, G.H. (1985) Copper, nickel, and iron in plumage of three upland gamebird species from non-contaminated environments. Bull. environ. Contam. Toxicol., 35(6): 776-780.

PARKER, K. & SUNDERMAN, F.W. Jr (1974) Distribution of ^{63}Ni in rabbit tissues following intravenous injection of ^{63}NiCl$_2$. Res. Commun. chem. Pathol. Pharmacol., 7: 755-762.

PATIERNO, S.R. & COSTA, M. (1985) DNA-protein cross-links induced by nickel compounds in intact cultured mammalian cells Chem.-biol. Interact., 55: 75-91.

PATIERNO, S.R., SUGIYAMA, M., BASILIKON, J.P., & COSTA, M. (1985) Preferential DNA-protein cross-linking by $NiCl_2$ in magnesium-insoluble regions of fractionated Chinese hamster ovary cell chromatin. Cancer Res., 45(11): 5787-55794.

PATON, G.R. & ALLISON, A.C. (1972) Chromosome damage in human cell cultures induced by metal salts. Mutat. Res., 16: 332-336.

PAYNE, W.W. (1964) Carcinogenicity of nickel compounds in experimental animals. Proc. Am. Assoc. Cancer Res., 5: 50 (Abstract 197).

PEDERSEN, E., HOGETVEIT, A.C., & ANDERSEN, A. (1973) Cancer of respiratory organs among workers at a nickel refinery in Norway. Int. J. Cancer, 12(1): 32-41.

PELTONEN, L. (1979) Nickel sensitivity in the general population. Contact Dermatit., 5(1): 27-32.

PENMAN, H.G. & RING, P.A. (1984) Osteosarcoma in association with total hip replacement. J. bone joint Surg., B 66: 632-634.

PENNINGTON, J.A.T. & JONES, J.W. (1987) Molybdenum, nickel, cobalt, vanadium and strontium in total diets. J. Am. Diet. Assoc., 87(12): 1644-1650.

PETO, J. (1988) The international nickel speciation research project. In: Fourth International Conference on Nickel Metabolism and Toxicology: Abstracts, Espoo, Finland, 5-9 September 1988, Helsinki, Institute of Occupational Health, p. 50.

PETO, J., CUCKLE, H., DOLL, R., HERMON, C., & MORGAN, L.G. (1984) Respiratory cancer mortality of Welsh nickel refinery workers. In: Nickel in the human environment. Proceedings of a Joint Symposium, Lyon, 8-11 March, 1983, Lyon, International Agency for Research on Cancer, pp. 37-46 (IARC Scientific Publications No. 53).

PHATAK, S.S. & PATWARDHAN, V.N. (1950) Toxicity of nickel. J. Sci. ind. Res., 9B: 70-76.

PHATAK, S.S. & PATWARDHAN, V.N. (1952) Toxicity of nickel. Accumulation of nickel in rats fed on nickel-containing diets and its elimination. J. Sci. ind. Res., 11B: 173-176.

PHILLIPS, D.J.H. & MUTTARASIN, K. (1985) Trace metals in bivalve molluscs from Thailand. Mar. environ. Res., 15(3): 215-234.

PICKERING, Q.H. (1974) Chronic toxicity of nickel to the fathead minnow. J. Water Pollut. Control Fed., 46: 760-765.

PICKERING, Q.H. & HENDERSON, C. (1966) The acute toxicity of some heavy metals to different species of warm water fishes. Air Water Pollut. Int. J., 10: 453-463.

PIENTA, R.J., POILEY, J.A., & LEBHERZ, W.B. (1977) Morphological transformation of early passage golden hamster embryo cells derived from cryopreserved primary cultures as a reliable *in vitro* bioassay for identifying diverse carcinogens. Int. J. Cancer, 19: 642-655.

PIHLAR, B., VALENTA, P., & NURNBERG, H.W. (1981) New high-performance analytical procedure for the voltammetric determination of nickel in routine analysis of waters, biological materials and food. Fresenius Z. anal. Chem., 307: 337-346.

PIKALEK, P. & NECASEK, J. (1983) The mutagenic activity of nickel in Corynebacterium sp. Folia microbiol., 28: 17-21.

POIRIER, L.A., THEISS, J.C., ARNOLD, L.J., & SHIMKIN, M.B. (1984) Inhibition by magnesium and calcium acetates of lead subacetate- and nickel acetate-induced lung tumors in strain A mice. Cancer Res., 44: 1520-1522.

POLEDNAK, A.P. (1981) Mortality among welders, including a group exposed to nickel oxides. Arch. environ. Health, 36(5): 235-242.

POLEMIO, M., SENESI, N., & BUFO, S.A. (1982) Soil contamination by metals; a survey in industrial and rural areas of Southern Italy. Sci. total Environ., 25: 71-80.

PORT, C.D., FENTERS, J.D., EHRLICH, R., COFFIN, D.L., & GARDNER, D. (1975) Interaction of nickel oxide and influenza infection in the hamster. Environ. Health Perspect., 10: 268.

POTT, F., ZIEM, U., REIFFER, F.-J., HUTH, F., ERNST, H., & MOHR, U. (1987) Carcinogenicity studies on fibres, metal compounds, and some other dusts in rats. Exp. Pathol., 32: 129-152.

POTT, F., RIPPE, M., ROLLER, M., ROSENBRUCH, M., & HUTH, F. (1988) Carcinogenicity studies on nickel compounds and nickel alloys after intraperitoneal injection in rats. In: Fourth International Conference on Nickel Metabolism and Toxicology: Abstracts, Espoo, Finland, 5-9 September 1988, Helsinki, Institute of Occupational Health, p. 42.

POWLESLAND, C. & GEORGE, I.C. (1986) Acute and chronic toxicity of nickel to larvae of Chironomus riparis (Meigen). Environ. Pollut., 42(1): 47-64.

PROKIPCAK, B. & ORMROD, D.P. (1986) Visible injury and growth responses of tomato and soybean to combinations of nickel, copper and ozone. Water air soil Pollut., 27(3-4): 329-340.

PRYSTOWSKY, S.D., ALLEN, A.M., SMITH, R.W., NONOMURA, J.H., ODOM, R.B., & AKERS, W.A. (1979) Allergic contact hypersensivity to nickel, neomycin, ethylenediamine, and benzocaine. Relationships between age, sex, history of exposure, and reactivity to standard patch tests and use tests in a general population. Arch. Dermatol., 115(8): 959-965.

PUCHYR, R. & SHAPIRO, R. (1986) Determination of trace elements in food by HCl-HNO3 leaching and flame atomic absorption spectroscopy. J. Assoc. Off. Anal. Chem., 69(5): 868-870.

RADIAN CORPORATION (1984) Locating and estimating emissions from sources of nickel. Research Triangle Park, North Carolina, US Environmental Protection Agency, Office of Air Quality Planning and Standards, 188 pp (EPA-450/4-84-007F).

RAHKONEN, E., JUNTTILA, M.-L., KALLIOMAKI, P.-L., OLKINUORA, M., KOPONEN, M., & KALLIOMAKI, K. (1983) Evaluation of biological monitoring among stainless steel welders. Int. Arch. occup. environ. Health, 52: 243-255.

RAITHEL, H.J. (1987) [Research report nickel. Investigations on occupational nickel exposure and strain-situations of 837 persons.] Sankt Augustin, Federation of Industrial Cooperative Societies, 199 pp (in German).

RAITHEL, H.J., MAYER, P, SCHALLER, K.H., MOHRMANN, W., WELTLE, D., & VALENTIN, H. (1981) [Exposure to nickel in glass industry workers. I. analysis and quantification of external and internal nickel load.] Zentralbl. Arbeitsmed. Arbeitsschutz prophyl. Ergonomie, 31(8): 332-339 (in German).

RANTA, W.B., TOMASSINI, F.D., & NIEBOER, E. (1978) Elevation of copper and nickel levels in primaries from black and mallard ducks collected in the Sudbury district, Ontario. Can. J. Zool., 56(4): 581-586.

RASMUSON, A. (1985) Mutagenic effects of some water-soluble metal compounds in a somatic eye-color test system in *Drosophila melanogaster*. Mutat. Res., 157: 157-162.

REDDY, M.R. & DUNN, S.J. (1984) Accumulation of heavy metals by soybean from sludge-amended soil. Env. Pollution, B 7: 281-296.

REDMOND, C.K. (1984) Site-specific cancer mortality among workers involved in the production of high nickel alloys. Nickel in the human environment. Proceedings of a Joint Symposium, Lyon, 8-11 March, 1983, Lyon, International Agency for Research on Cancer, pp. 73-86 (IARC Scientific Publications No. 53).

REDMOND, C.K., LEGASSE, A.A., & BASS, G. (1983) Cancer mortality in workers in the high nickel alloy industries. Pittsburgh, Pennsylvania, University of Pittsburgh, Graduate School of Public Health, Department of Biostatistics, 145 pp.

REHWOLDT, R., BIDA, G., & NERRIE, B. (1971) Acute toxicity of copper, nickel and zinc ions to some Hudson river fish species. Bull. environ. Contam. Toxicol., 6: 445-448.

REHWOLDT, R., MENAPACE.L.W., NERRIE, B. & ALESSANDRELLO, D. (1972) The effect of increased temperature upon the acute toxicity of some heavy metal ions. Bull. environ. Contam. Toxicol., 8: 91-96.

REHWOLDT, R., LASKO, L., SHAW, C. & WIRHOWSKI, E. (1973) The acute toxicity of some heavy metal ions toward benthic organisms. Bull. environ. Contam. Toxicol., 10:291-294.

REICHRTOVA, E., KOVACIKOVA, Z., TAKAC, L., & ORAVEC, C. (1986a) The effect of metal particles from a nickel refinery dump on alveolar macrophages. Part. 1. Chamber exposure of Wistar rats. Environ. Pollut., A 40: 87-94.

REICHRTOVA, E., TAKAC, L., & KOVACIKOVA, Z. (1986b) The effect of metal particles from a nickel refinery dump on alveolar macrophages. Part 2. environmental exposure of rabbits. Environ. Pollut., A 40: 101-107.

REZUKE, W.N., KNIGHT, J.A., & SUNDERMAN, F.W. Jr (1987) Reference values for nickel concentrations in human tissue and bile. Ann. J. ind. Med., 11: 419-426.

RICHTER, O.R. & THEIS, T.L. (1980) Nickel speciation in a soil/water system. In: Nriagu, J.O., ed. Nickel in the environment, New York, Chichester, Brisbane, Toronto, John Wiley and Sons, pp. 189-202.

RIGAUT, J.-P. (1983) Preliminary report on nickel health criteria, Luxemburg, Commission of the European Communities, 1009 pp. (CCE/LUX/V/E/24/83)

RITTMANN, D., NOCHRAINER, D., OBERDORSTER, G., KORDEL, W., SPIEGELBERG, T., KONIG, H., FRITSCHE, V., KUHNEN, H., & FINGERHUT, R. (1981) [Inhalation experiment on rats and pathophysiological effect of nickel. Environmental research plan of the Federal Ministry of the Interior.] Berlin, Federal Ministry of the Interior (Research Report No. 107 06 00702)(in German).

RIVEDAL, E. & SANNER, T. (1980) Synergistic effect on morphological transformation of hamster embryo cells by nickel sulphate and benz(a)pyrene. Cancer Lett., 8(3): 203-208.

RIVEDAL, E. & SANNER, T. (1981) Metal salts as promoters of *in vitro* morphological transformation of hamster embryo cells initiated by benzo(*a*)pyrene. Cancer Res., 41: 2950-2953.

RIVEDAL, E., HEMSTAD, J., & SANNER, T. (1980) Synergistic effects of cigarette smoke extracts, benz(a)pyrene and nickel sulphate on morphological transformation of hamster embryo cells. In: Holmstedt, B., Lauwerys, R., Mercier, M., & Roberfroid, M., ed. Mechanisms of toxicity and hazard evaluation. Proceedings of the 2nd International Congress on Toxicology, Brussels, Belgium, 6-11 July 1980. Amsterdam, New York, Oxford, Elsevier/North Holland Biomedical Press, pp. 259-263 (Developments in Toxicology and Environmental Science, Vol. 8).

ROBERTS, R.S., JULIAN, J.A., MUIR, D.C., & SHANNON, H.S. (1984) Cancer mortality associated with the high temperature oxidation of nickel subsulfide. In: Nickel in the human environment, Proceedings of a Joint Symposium, Lyon, 8-11 March, 1983, Lyon, International Agency for Research on Cancer, pp. 23-35 (IARC Scientific Publications No. 53).

ROBISON, S.H. & COSTA, M. (1982) The induction of DNA strand breakage of nickel compounds in cultured Chinese hamster ovary cells. Cancer Lett., 15: 35-40.

ROBISON, S.H., CANTONI, O., & COSTA, M. (1982) Strand breakage and decreased molecular weight of DNA induced by specific metal compounds. Carcinogenesis, 3: 657-662.

ROCKSTROH, H. (1959) [On the etiology of bronchial cancer in arsenic processing nickel smelting plants.] Arch. Geschwulstforsch., 14: 151-162 (in German).

RODRIGUEZ-ARNAIZ, R. & RAMOS, P. (1986) Mutagenicity of nickel sulphate in *Drosophila melanogaster*. Mutat. Res., 170(3): 115-117.

ROSSMAN, T.G., MOLINA, M., & MEYER, L.W. (1984) The genetic toxicology of metal compounds: I. Induction of prophage in $E.$ coli WP2s. Environ. Mutagen., 6: 59-69.

ROUSH, C.C., MEIGS, I.W., KELLY, J.A., FLANNERY, I.T., & BURDO, H. (1980) Sinonasal cancer and occupation: a case-control study. Am. J. Epidemiol., 111: 183-193.

ROY, B.R. (1985) Nepera 82-06 sampling method for inorganic nickel. In: Brown, S.S. & Sunderman, F.W. Jr, ed. Progress in Nickel toxicology. Proceedings of the 3rd International Conference on Nickel Metabolism and Toxicology, Paris, 4-7 September, 1984, Oxford, Blackwell Scientific Publications, pp. 191-194.

RUBANYI, G. & INOVAY, J. (1982) Effect of nickel ions on spontaneous electrically and norepinephrine stimulated isometric contractions in the isolated portal vein of the rat. Acta physiol. Acad. Sci. Hung., 59: 181-186.

RUBANYI, G. & KOVACH, A.G.B. (1980) Cardiovascular actions of nickel ions. Acta physiol. Acad. Sci. Hung., 55: 345-353.

RUBANYI, G., BALOGH, I., KOVACH, A.G.B., SOMOGYI, E., & SOTONYI, P. (1980) Effect of nickel ions on ultrastructure of isolated perfused rat heart. J. mol. cell Cardiol., 12: 609-618.

RUBANYI, G., LIGETI, L., & KOLLER, A. (1981) Nickel is released from the ischemic myocardium and contracts coronary vessels by a Ca-dependent mechanism. J. mol. cell Cardiol., 13(11): 1023-1026.

RUBANYI, G., BIRTALAN, I., GERGELY, A., & KOVACH, A.G. (1982a) Serum nickel concentration in women during pregnancy, during parturition, and post partum. Am. J. Obstet. Gynecol., 143: 167-169.

RUBANYI, G., KALABAY, L., PATAKI, T., & HAJDU, K. (1982b) Nickel induces vasoconstriction in the isolated canine coronary artery by a tonic Ca^{2+}-activation mechanism. Acta physiol. Acad. Sci. Hung., 59: 155-159.

RUBANYI, G., BAKOS, M., HAJDU, K., & PATAKI, T. (1982c) Dependence of nickel induced coronary vasoconstriction on the activity of the electrogenic Na^+, K^+-pump. Acta physiol. Acad. Sci. Hung., 59: 169-174.

RUBANYI, G., HAJDU, K., PATAKI, T., & BAKOS, M. (1982d) The role of adrenergic receptors in Ni^{2+}-induced coronary vasoconstriction. Acta Physiol. Acad. Sci. Hung., 59: 161-167.

RUBANYI, G., LIGETI, L., KOLLER, A., & KOVACH, A.G. (1984) Possible role of nickel ions in the pathogenesis of ischemic coronary vasoconstriction in the dog heart. J. mol. cell Cardiol., 16: 533-546.

RUTHERFORD, G.K. & BRAY, C.R. (1979) Extent and distribution of soil heavy metal contamination near a nickel smelter at Coniston, Ontario. J. environ. Qual., 8(2): 219-222.

RYU, R.K.N., BOVILL, E.G. Jr, SKINNER, H.B., & MURREY, W.R. (1987) Soft tissue sarcoma associated with aluminum oxide ceramic total hip arthroplasty. Clin. Orthop. relat. Res., 216: 207-212.

SADIQ, M., ZAIDI, T.H., UL-HODA, A., & MIAN, A.A. (1982) Heavy metal concentrations in shrimp, crab, and sediment obtained from Ad-Dammam sewage outfall area. Bull. environ. Contam. Toxicol., 29(3): 313-320.

SAICHENKO, S.P. (1985) Experimental evaluation of genetic danger of metals administered with drinking water. In: Domnin, S.C. & Kogan, F.V., ed. Problems of occupational pathology and toxicology in mining and metallurgy. Moscow, Erisman Institute of Hygiene, pp. 75-80.

SAICHENKO, S.P. & SHARAPOVA, N.Z (1987) About the mutagenicity of some metals for *Salmonella typhimurium*. In: Domnin, S.C. & Shcherbakov S.V., ed. Problems of hygiene and occupational pathology in colored metals and steel industry. Moscow, Erisman Institute of Hygiene, pp. 97-101.

SAKNYN, A.V. & BLOKHIN, V.A. (1978) [Development of malignant tumors in rats under the influence of nickel-containing aerosols.] Vopr. Onkol., 24(4): 44-48 (in Russian).

SAKNYN, A.V. & SHABYNINA, N.K. (1970) [Some statistical material on carcinogenic danger in the production of nickel from oxidized ores.] Gig. Tr. prof. Zabol., 14(11): 10-13 (in Russian).

SAKNYN, A.V. & SHABYNINA, N.K. (1973) [Epidemiology of malignant neoplasms in nickel plants.] Gig. Tr. prof. Zabol., 9: 25-28 (in Russian).

SAKNYN, A.V., SIPATOV, G.YA., YELNICHUYKH, L.N., & STARKOV, P.S. (1978) [Hygienic characterization of dusts in primary nickel production.] In: Domnin, S.G., ed. Occupational disease of dust aetiology. Erisman Institute of Hygiene, Moscow, Vol. 4, pp. 18-23.

SALTZMANN, B.E., CHOLAK, J., SCHAFER, L.J., YEAGER, D.W., MEINERS, B.O., SVETLIK, J., & ZWILLENBERG, M.L. (1985) Concentrations of six metals in the air of eight cities. Environ. Sci. Technol., 19(4): 328-333.

SAMELA, D., TSOUMPAS, G.M., & WALSHANS, G.K. (1986) Environmental aspects of the combustion of sewage sludge in a utility boiler. Environ. Prog., 5(2): 110-115.

SAMITZ, M.H. & KATZ, S.A. (1975) Nickel dermatitis hazards from prostheses. *In vivo* and *in vitro* solubilization studies. Br. J. Dermatol., 92: 287.

SAMITZ, M.H. & KATZ, S.A. (1976) Nickel-epidermal interactions: diffusion and binding. Environ. Res., 11(1): 34-39.

SAMITZ, M.H. & POMERANTZ, H. (1958) Studies of the effects on the skin of nickel and chromium salts. Arch. ind. Health, 18: 473-479.

SAMITZ, M.H., KATZ, S.A., SCHEINER, D.M., & LEWIS, J.E. (1975) Attempts to induce sensitization in guinea-pigs with nickel complexes. Acta dermtovenereol., 55: 475-480.

SANDE, J.H. VAN DE, MACINTOSH, L.P., & JOVIN, T.M. (1982) Mn^{2+} and other transition metals at low concentration induce the right-to-left helical transformation of poly (d(G-C)). EMBO J., 1: 777-782.

SANDERS, J.R., MCGRATH, S.P., & ADAMS, T.M. (1986) Zink, copper and nickel concentrations in ryegrass grown on sewage sludge-contaminated soils of different pH. J. Sci. Food Agric., 37: 961-968.

SANFORD, W.E., NIEBOER, E., BACH, P., STACE, B., GREGG, N., & DOBROTA, M. (1988) The renal clearance and toxicity of nickel. In: Fourth International Conference on Nickel Metabolism and Toxicology, Abstracts, Espoo, Finland, 5-9 September 1988, Helsinki, Institute of Occupational Health, p.17.

SANINA, J.P. (1968) [Toxicology of nickel carbonyl.] Toxikol. Nov. Prom. Khim. Veshchestv., 10: 144-149 (in Russian).

SANOTSKII, J.V. (1955) [Action mechanism of nickel carbonyl.] Farmakol. Toksikol., **18**(2): 48-50 (in Russian).

SANTUCCI, B., CRISTAUDO, A., CANNISTRACI, C., & PICARDO, M. (1988) Nickel sensitivity: effects of prolonged oral intake of the element. Contact Dermatit., **19**: 202-295.

SANZOLONE, R.F., CHAO, T.T., & CRENSHAW, G.L. (1979) Atomic absorption spectrometric determination of cobalt, nickel, and copper in geological materials with matrix marking and chelation-extraction. Anal. chim. Acta, **105**: 247-253.

SAPEK, B. & SAPEK, A. (1980) [Nickel content in the profiles of organic soils.] In: Anke, M., Schneider, H.-J., & Brückner, Chr., ed. [3. Trace Element Symposium: Nickel, Jena, German Democratic Republic, 7-11 July, 1980, Jena, Friedrich-Schiller University, pp. 185-191 (in German).

SARKAR, B. (1980) Nickel in blood and kidney. In: Brown, S.S. & Sunderman, F.W. Jr, ed. Nickel toxicology, Proceedings of the 2nd International Conference on Nickel Toxicology, Swansea, 3-5 September, 1980, London, New York, Academic Press, pp. 81-84.

SAXHOLM, H.J.K., REITH, A., & BROEGGER, A. (1981) Oncogenic transformation and cell lysis in C3H/10T 1/2 cells and increased sister chromatid exchange in human lymphocytes by nickel subsulfide. Cancer Res., **41**(10): 4136-4139.

SCHMIDT, J.A. & ANDREN, A.W. (1980) The atmospheric chemistry of nickel. In: Nriagu, J.O., ed. Nickel in the environment, New York, Chichester, Brisbane, Toronto, John Wiley and Sons, pp. 93-135.

SCHMIDT, W. & DIETL, F. (1981) [Determination of nickel and cobalt in digested soils by means of flameless atomic absorption with zirconium coated graphite tubes.] Fresenius Z. anal. Chem., **308**: 129-132 (in German with English summary).

SCHNEGG, A. & KIRCHGESSNER, M. (1975a) [The essentiality of nickel for the growth of animals.] Z. Tierphysiol., Tierernähr., Futtermittelkd., **36**: 63-74 (in German).

SCHNEGG, A. & KIRCHGESSNER, M. (1975b) [Changes in the hemoglobin content, erythrocyte content and hematocrit in nickel deficiency.] Nutr. Metabol., **19**: 268-278 (in German).

SCHNEGG, A. & KIRCHGESSNER, M. (1976a) [Interaction of nickel with iron, copper and zinc.]. Arch. Tierernähr., **26**(8): 543-549 (in German with English summary).

SCHNEGG, A. & KIRCHGESSNER, M. (1976b) [Absorption and metabolic efficiency of iron during nickel deficiency.] Int. Z. Vitam .Forsch., **46**: 96-99 (in German with English summary).

SCHNEGG, A. & KIRCHGESSNER, M. (1977a) [Changes in enzyme activities in liver and kidney with nickel and iron deficiency, respectively.] Z. Tierphysiol., Tierernährg., Futtermittelkd., **38**: 200-205 (in German with English summary).

SCHNEGG, A. & KIRCHGESSNER, M. (1977b) [Alkaline and acid phosphatase activity in liver and serum during nickel versus Fe-deficiency.] Int. Z. Vitam. Forsch., **47**: 274-276 (in German with English summary).

SCHNEGG, A. & KIRCHGESSNER, M. (1977c) [Changes of various substrate concentrations in the serum and liver in Ni- or Fe-deficiency.] Z. Tierphysiol. Tierernährg. Futtermittelkd, 39(5-6): 247-251 (in German with English summary).

SCHNEGG, A. & KIRCHGESSNER, M. (1977d) [Differential diagnosis of Fe and Ni deficiency by determination of some enzyme activities.] Zbl. Vet. Med., 24A: 242-247 (in German).

SCHNEGG, A. & KIRCHGESSNER, M. (1980) [On the essentiallity of nickel for animal growth.] In: Anke, M., Schneider, H.-J. & Brückner, Chr., ed. [3. Trace Element Symposium: Nickel,] Jena, German Democratic Republic, 7-11 July, 1980, Jena, Friedrich-Schiller University, pp. 11-16 (in German).

SCHNEIDER, H.-J., ANKE, M., & KLINGER, G. (1980)[The nickel-status of human beings.] In: Anke, M., Schneider, H.-J. & Brückner, Chr., ed. [3. Trace Element Symposium: Nickel, Jena, German Democratic Republic, 7-11 July, 1980, Jena, Friedrich-Schiller-University, pp. 277-283 (in German).

SCHROEDER, H.A. (1968) Serum cholesterol levels in rats fed thirteen trace elements. J. Nutr., 94: 475-480.

SCHROEDER, H.A. (1970) A sensible look at air pollution by metals. Arch. environ. Health, 21: 798-806.

SCHROEDER, H.A. & MITCHENER, M. (1971) Toxic effects of trace elements on the reproduction of mice and rats. Arch. environ. Health, 23: 102-106.

SCHROEDER, H.A. & MITCHENER, M. (1975) Life time effects of mercury, methyl mercury and nine other trace metals on mice. J. Nutr., 105: 452-458.

SCHROEDER, H.A., BALASSA, J.J., & TIPTON, J.H. (1962) Abnormal trace metals in man - nickel. J. chron. Dis., 15: 51-65.

SCHROEDER, H.A., BALASSA, J.J., & VINTON, W.H. Jr (1964) Chromium, lead, cadmium, nickel and titanium in mice: effect on mortality, tumors and tissue levels. J. Nutr., 83: 239-250.

SCHROEDER, H.A., MITCHENER, M., & NASON, A.P. (1974) Life-term effects of nickel in rats: survival, tumors, interactions with trace elements and tissue levels. J. Nutr., 104: 239-243.

SCHUMANN, H. (1980) [Nickel contamination of surface waters and drinking water and some influencing factors.] In: Anke, M., Schneider, H.-J., & Brückner, Chr., ed. [3. Trace Element Symposium Nickel, Jena, German Democratic Republic, 7-11 July, 1980.] Jena, Friedrich-Schiller University, pp. 155-161 (in German).

SEEMANN, J., WITTIG, P., KOLLMEIER, H., & ROTHE, G. (1985) Analytical measurements of Cd, Pb, Zn, Cr and Ni in human tissues. Lab. Med., 9: 294-299.

SEN, P. & COSTA, M. (1985) Induction of chromosomal damage in Chinese hamster ovary cells by soluble and particulate nickel compounds: preferential fragmentation of the heterochromatic long arm of the X-chromosome by carcinogenic crystalline NiS particles. Cancer Res., 45(5): 2320-2325.

SEN, P. & COSTA, M. (1986) Pathway of nickel uptake influences its interaction with heterochromatic DNA. Toxicol. appl. Pharmacol., 84(2): 278-285.

SHANNON, H.S., JULIAN, J.A., MUIR, D.C.F., ROBERTS, R.S., & CECUTTI, A.C. (1984) A mortality study of Falconbridge workers. In: Nickel in the human environment, Proceedings of a Joint Symposium, Lyon, 8-11 March, 1983, Lyon, International Agency for Research on Cancer, pp. 117-124 (IARC Scientific Publications No. 53).

SHAW, T.L. & BROWN, V.M. (1971) Heavy metals and the fertilization of rainbow trout eggs. Nature (Lond.), 230: 251.

SHI, Z. (1986) Acute nickel carbonyl poisoning: a report of 179 cases. Br. J. ind. Med., 43: 422-424.

SHI, Z., LATA, A., & HAN, Y. (1986) A study of serum monoamine oxidase (MAO) activity and EEG in nickel carbonyl workers. Br. J. ind. Med., 43: 425-426.

SHIBATA, M., IZUMI, K., SANO, N., AKAGI, A., & OTSUKA, H. (1989) Induction of soft tissue tumours in F344 rats by subcutaneous, intramuscular, intra-articular, and retroperitoneal injection of nickel sulphide (Ni_3S_2). J. Pathol., 157: 263-274.

SILVERSTEIN, M., MIRER, F., KOTELCHUCK, D., SILVERSTEIN, B., & BENNETT, M. (1981) Mortality among workers in a diecasting and electroplating plant. Scand. J. Work Environ. Health, 7: 156-165.

SINA., J.F., BEAN, C.L., DYSART, G.R., TAYLOR, V.I., & BRADLEY, M.O. (1983) Evaluation of the alkaline elution/rat hepatocyte assay as a predictor of carcinogenic/mutagenic potential. Mutat. Res., 113: 357-391.

SINGH, I. (1984) Induction of gene conversion and reverse mutation by manganese sulphate and nickel sulphate in Saccharomyces cerevisiae. Mutat. Res., 137: 47-49.

SIROVER, M.A. & LOEB, L.A. (1977) On the fidelity of DNA replication. Effect of metal activators during synthesis with avian myeloblastosis virus DNA polymerase. J. biol. Chem., 252: 3605-3610.

SJOVALL, P., CHRISTENSEN, O.B., & MOLLER, H. (1987) Oral hyposensitization in nickel allergy. J. Am. Acad. Dermatol., 17: 774-778.

SKAUG, V., GYLSETH, B., REISS, A.P., & NORSETH, T. (1985) Tumor induction in rats after intrapleural injection of nickel subsulfide and nickel oxide. In: Brown, S.S. & Sunderman, F.W. Jr, ed. Nickel toxicology, Proceedings of the 2nd International Conference on Nickel Toxicology, Swansea, 3-5 September, 1980, London, New York, Academic Press, pp. 37-40.

SMART, G.A. & SHERLOCK, J.C. (1987) Nickel in foods and the diet. Food Addit. Contam., 4(1): 61-71.

SMIALOWICZ, R.J. (1985) The effect of nickel and manganese on natural killer cell activity. In: Brown, S.S. & Sunderman, F.W. Jr, ed. Progress in Nickel toxicology. Proceedings of the 3rd International Conference on Nickel Metabolism and Toxicology, Paris, France, 4-7 September, 1984, Oxford, Blackwell Scientific Publications, pp. 161-164.

SMIALOWICZ, R.J., ROGERS, R.R., RIDDLE, M.M., & STOTT, G.A. (1984) Immunologic effects of nickel: 1. Suppression of cellular and human immunity. Environ. Res., 33: 413-427.

SMIALOWICZ, R.J., RODGERS, R.R., RIDDLE, M.M., GARNER, R.J., ROWE, D.G., & LUEBKE, R.W. (1985) Immunologic effects of nickel. II. Suppression of natural killer cell activity. Environ. Res., 36: 56-66.

SMIALOWICZ, R.J., ROGERS, R.R., RIDDLE, M.M., ROWE, D.G., & LUEBKE, R.W. (1986) Immunological studies in mice following in utero exposure to NiCl$_2$. Toxicology, 38: 293-303.

SMIALOWICZ, R.J., ROGERS, R.R., RIDDLE, M.M., LUEBKE, R.W., FOGELSON, L.D., & ROWE, D.G. (1987) Effects of manganese, calcium, magnesium, and zinc on nickel-induced suppression of murine natural killer cell activity. J. Toxicol. environ. Health, 20(1-2): 67-80.

SMITH, J.C. (1969) A controlled environment system for trace element deficiency studies. Trace Subst. environ. Health, 2: 223-242.

SMITHE, J.C. & HACKLEY, B. (1968) Distribution and excretion of nickel-63 administered intravenously to rats. J. Nutr., 95: 541-546.

SMITH-SONNEBORN, J., PALIZZI, R.A., McCANN, E.A., & FISHER, G.L. (1983) Bioassay of genotoxic effects of environmental particles in a feeding ciliate. Environ. Res., 32: 474-479.

SMITH-SONNEBORN, J., LEIBOVITZ, B., DONATHAN, R., & FISHER, G.L. (1986) Bioassay of environmental nickel dusts in a particle feeding ciliate. Environ. Mutagen., 8(4): 621-626.

SNODGRAS, W. (1980) Distribution and behaviour of nickel in the aquatic environment. In: Nriagu, J.O., ed. Nickel in the environment, New York, Chichester, Brisbane, Toronto, John Wiley and Sons, pp. 203-274.

SOLOMONS, N.W., VITERI, F., SHULER, T.R., & NIELSEN, F.H. (1982) Bioavailability of nickel in man: effects of foods and chemically-defined dietary constituents on the absorption of inorganic nickel. J. Nutr., 112(1): 39-50.

SONNENFELD, G., STREIPS, U.N., & COSTA, M. (1983) Differential effects of amorphous and crystalline nickel suflide on murine α/β interferon production. Environ. Res., 32: 474-479.

SORAHAN, T. & WATERHOUSE, J.A.W. (1983) Mortality study of nickel-cadmium battery workers by the method of regression models in life tables. Br. J. ind. Med., 40: 293-300.

SORAHAN, T., BURGES, D.C.L., & WATERHOUSE, J.A.H. (1987) A mortality study of nickel/chromium platers. Br. J. ind. Med., 44: 250-258.

SORINSON, S.N., KORNILOVA, A.P., & ARTEMEVA, A.M. (1958) [Concentrations of nickel in blood and urine of workers in the nickel carbonyl industry.] Gig. i Sanit., 23(9): 69-72 (in Russian).

SPEARS, J.W. (1984) Nickel as a "newer trace element" in the nutrition of domestic animals. J. anim. Sci., 59: 823-835.

SPEARS, J.W. & HATFIELD, E.E. (1977) Role of nickel in animal nutrition. Feedstuffs, 49: 24-28.

SPEARS, J.W., HATFIELD, E.E., FORBES, R.M., & KOENIG, S.E. (1978) Studies on the role of nickel in the ruminant. J. Nutr., 108: 313-320.

SPEARS, J.W., JONES, E.E., SAMSELL, L.J., & ARMSTRONG, W.D. (1984) Effect of dietary nickel on growth, urease activity, blood parameters and tissue mineral concentrations in the neonatal pig. J. Nutr., 114: 845-854.

SPENCER, D.F. (1980) Nickel and aquatic algae. In: Nriagu, J.O., ed. Nickel in the environment, New York, Chichester, Brisbane, Toronto, John Wiley and Sons, pp. 339-347.

SPENCER, D.F. & GREENE, R.W. (1981) Effects of nickel on seven species of freshwater algae. Environ. Pollut., A 25: 241-247.

SPIEGELBERG, T., KORDEL, W., & HOCHRAINER, D. (1984) Effects of NiO inhalation on alveolar macrophages and the humoral immune systems of rats. Ecotoxicol. environ. Safety, 8: 516-525.

SPRUIT, D., MALI, J.W.H., & DE GROOT, N. (1965) The interaction of nickel ions with human cadaverous dermis. J. Invest. Dermatol., 44: 103-106.

STACK, M.V., BURKITT, A.J., & NICKLESS, G. (1976) Trace metals in teeth at birth (1957-1963 and 1972-1973). Bull. environ. Contam. Toxicol., 16: 764-766.

STAHLY, E.E. (1973) Some considerations of metal carbonyls in tobacco smoke. Chem. Ind., 12: 620-623.

STEDMAN, D.H. (1986a) Method 5 - determination of nickel carbonyl in air colorimetry. In: O'Neill, I.K., Schuller, P., & Fishbein, L., ed. Environmental carcinogens - selected methods of analysis. Vol. 8: Some metals: As, Be, Cd, Cr, Ni, Pb, Se, Zn. Lyon, International Agency for Research on Cancer, pp. 261-267 (IARC Scientific Publications No. 71).

STEDMAN, D.H. (1986b) Method 6 - determination of nickel carbonyl in air by chemiluminescence. In: O'Neill, J.K., Schuller, P., & Fishbein, L., ed. Environmental carcinogens - selected methods of analysis. Vol. 8: Some metals: As, Be, Cd, Cr, Ni, Pb, Se, Zn. Lyon, International Agency for Research on Cancer, pp. 269-273 (IARC Scientific Publications No. 71).

STEDMAN, D.HJ. & HIKADE, D.A. (1980) The rate of decay of traces of nickel carbonyl in air. In: Brown, S.S. & Sunderman, F.W. Jr, ed. Nickel toxicology, Proceedings of the 2nd International Conference on Nickel Toxicology, Swansea, 3-5 September, 1980, London, New York, Academic Press, pp. 183-186.

STERN, A.C., BOUBEL, R.W., TURNER, D.B., & FOX, D.L. (1984) Fundamentals of air pollution, 2nd ed., Orlando, San Diego, San Francisco, Academic Press, pp. 103-113.

STERN, R.M. (1983) Assessment of risk of lung cancer for welders. Arch. env. Health, 38(3): 148-156.

STERN, R.M., HANSEN, K., & THOMSEN, E. (1985) An *in vitro* genotoxicity assay to complement *in vivo* inhalation studies and for occupational monitoring of metallic aerosols and welding fumes. In: Brown, S.S. & Sunderman, F.W. Jr, ed. Progress in nickel toxicology. Proceedings of the 3rd International Conference on Nickel Metabolism and Toxicology, Paris, 4-7 September, 1984, Oxford, Blackwell Scientific Publications, pp. 77-80.

STEWART, S.G. & CROMIA, F.E. (1934) Experimental nickel dermatitis. J. Allergy, 5: 575-582.

STOEPPLER, M. (1980) Analysis of nickel in biological materials and natural waters. In: Nriagu, J.O. ed. Nickel in the environment, New York, Chichester, Brisbane, Toronto, John Wiley & Sons, pp. 661-821.

STOKES, P. (1975) Uptake and accumulation of copper and nickel by metal-tolerant strains of *Scenedesmus*. Int. Ver. Limnol., 19:128-137.

STOKES, P.M. (1981) Nickel in aquatic systems. In: Effects of nickel in the Canadian environment, Ottawa, National Research Council of Canada, pp. 77-117 (Publication No. NRCC-18568).

STONER, G.O., SHIMKIN, M.B., TROXELL, M.C., THOMPSON, T.L., & TERRY, L.S. (1976) Test for carcinogenicity of metallic compounds by the pulmonary tumor response in strain of mice. Cancer Res., 36: 1744-1747.

STORENG, R. & JONSEN, J. (1980) Effect of nickel chloride and cadmium acetate on the development of preimplantation mouse embryos *in vitro*. Toxicology, 17: 183-187.

STORENG, R. & JONSEN, J. (1981) Nickel toxicity in early embryogenesis in mice. Toxicology, 20(1): 45-51.

STOVBUN, A.T., YATSYUK, M.D., POMARENKO, U.I., & YAKOVLEVA, L.S. (1962) [Data on the trace element composition of human milk and various modifications of cow's milk.] Nauk. Zap. Ivano.-Frankivskii Derzk. Med. Inst., 5: 38-40 (in Russian).

STRAIN, W.H., VARNES, A.W., DAVIS, B.R., & KARK, E.C. (1980) [Nickel in drinking and household water.] In: Anke, M., Schneider, H.-J., & Brückner, Chr., ed. [3. Trace Element Symposium: Nickel, Jena, German Democratic Republic, 7-11 July, 1980] Jena, Friedrich-Schiller University, pp. 149-154 (in German).

STRATTON, G.W. & CORKE, C.T. (1979) The effect of nickel on the growth, photosynthesis and nitrogenase activity of Anabaena inequalis. Can. J. Microbiol., 25: 1094-1099.

STURGEON, R.E., BERMAN, S.S., & WILLIE, S.N. (1982) Concentration of trace metals from sea-water by complexation with 8-hydroxyquinoline and adsorption in C_{18}-bonded silica gel. Talanta, 29: 167-171.

SUMINO, K., HAYAKAWA, K., SHIBATA, T., & KITAMURA, S. (1975) Heavy metals in normal Japanese tissues. Arch. environ. Health, 30: 487-494.

SUNDERMAN, F.W. (1964) Nickel and copper mobilization by sodium diethyldithiocarbamate. J. New Drugs, 4: 154-161.

SUNDERMAN, F.W. (1970) Nickel poisoning. In: Sunderman, F.W. & Sunderman, F.W. Jr, ed. Laboratory diagnosis of diseases caused by toxic agents, St. Louis, Missouri, Warren H. Green Inc., pp. 387-396.

SUNDERMAN, F.W. (1971) The treatment of acute nickel carbonyl poisoning with sodium diethyldithiocarbamate. Ann. clin. Res., 3(3): 182-185.

SUNDERMAN, F.W. Jr (1977) The metabolism and toxicology of nickel. In: Brown, S.S. & Savory, J., ed. Chemical toxicology and clinical chemistry of metals. Proceedings of the 2nd International Conference, Montreal, 19-22 July, 1983, London, New York, Blackwell Scientific Publications, pp. 231-259.

References

SUNDERMAN, F.W. Jr (1980) Analytical biochemistry of nickel. International Union of Pure and Applied Chemistry, Subcommittee on environmental and Occupational Toxicology of Nickel. Pure appl. Chem., 52: 527-544.

SUNDERMAN, F.W. Jr (1983a) Potential toxicity from nickel contamination of intravenous fluids. Ann. clin. lab. Sci., 13(1): 1-4.

SUNDERMAN, F.W. Jr (1983b) Organs and species specificity in nickel subsulfide carcinogenesis. In: Langenbach, R., Nesnow, S., & Rice, J.M., ed. Organ and species specificity in chemical carcinogenesis, New York, Plenum Press pp. 107-127 (Basic life sciences. Vol. 24).

SUNDERMAN, F.W. Jr (1984) Carcinogenicity of nickel compounds in animals. In: Nickel in the human environment. Proceedings of a Joint Symposium, Lyon, 8-11 March, 1983, Lyon, International Agency for Research on Cancer, pp. 127-142 (IARC Scientific Publications No. 53).

SUNDERMAN, F.W. Jr (1986) Iatrogenic exposures to toxic metals. 9th Annual meeting of the National Academy of Clinical Biochemistry on Metal Metabolism and Disease, Atlanta, TA, USA, July 9-20, 1985. Clin. Physiol. Biochem., 4(1): 112.

SUNDERMAN, F.W. Jr (1987) Lipid peroxidation as a mechanism of acute nickel toxicity. Toxicol. environ. Chem., 15: 59-69.

SUNDERMAN, F.W. Jr (1989a) Mechanisms of nickel carcinogensis. Scand. J. Work Environ. Health, 15: 1-12.

SUNDERMAN, F.W. Jr (1989b) Carcinogencity of metal alloys in orthopaedic prostheses: clinical and experimental studies. Fundam. appl. Toxicol., 13: 1-12.

SUNDERMAN, F.W. Jr & BARBER, A. M. (1988) Finger loops, ongogens, and metals. Ann. clin. lab. Sci., 18: 267-288.

SUNDERMAN, F.W. Jr & DONELLY, A.J. (1965) Studies of nickel carcinogenesis metastasizing pulmonary tumors in rats induced by the inhalation of nickel carbonyl. Am. J. Pathol., 46: 1027-1041.

SUNDERMAN, F.W. Jr & HOPFER, S.M. (1983) Correlation between the carcinogenic activities of nickel compounds and their potencies for stimulating erythropoiesis in rats. In: Saker, B., ed. Biological aspects of metals and metal-related diseases, New York, Raven Press, pp. 171-181.

SUNDERMAN, F.W. & KINCAID, J.F. (1954) Nickel poisoning II. Studies on patients suffering from acute exposure to vapors of nickel carbonyl. J. Am. Med. Assoc., 155(10): 889-894.

SUNDERMAN, F.W. Jr & MCCULLY, K.S. (1983) Effects of manganese compounds on carcinogenicity of nickel subsulfide in rats. Carcinogenesis, 4(4): 461-465.

SUNDERMAN, F.W. Jr & MAENZA, R.M. (1976) Comparisons of carcinogenicities of nickel compounds in rats. Res. Commun. chem. Pathol. Pharmacol., 14: 319-330.

SUNDERMAN, F.W. Jr & OSKARSSON, A. (1988) Nickel. In: Merian, ed. Metals and their compounds in the environment, Weinheim, VCH Verlagsgesellschaft, pp. 1-19.

SUNDERMAN, F.W. Jr & SELIN, C.E. (1968) The metabolism of nickel-63 carbonyl. Toxicol. appl. Pharmacol., **12**: 207-217.

SUNDERMAN, F.W. & SUNDERMAN, F.W. Jr (1961a) Nickel poisoning. XI. Implication of nickel as a pulmonary carcinogen in tobacco smoke. Am. J. clin. Pathol., **35**: 203-209.

SUNDERMAN, F.W. & SUNDERMAN, F.W. Jr (1961b) Löffler's syndrome associated with nickel sensitivity. Arch. int. Med., **107**: 405-408.

SUNDERMAN, F.W. Jr, KINCAID, J.F., DONNELLY, A.J., & WEST, B. (1957) Nickel poisoning. IV. Chronic exposure of rats to nickel carbonyl: a report after one year of observation. Arch. ind. Health, **16**: 480-485.

SUNDERMAN, F.W., DONELLY, A.J., WEST, B., & J.F. KINCAID (1959) Nickel poisoning. IX. Carcinogenesis in rats exposed to nickel carbonyl. Arch. ind. Health, **20**: 36-41.

SUNDERMAN, F.W., RANGE, C.L., SUNDERMAN, F.W. JR., DONELLY, A.J., & LUCYSZYN, G.W. (1961) Nickel poisoning. XII. Metabolic and pathologic changes in acute pneumonitis from nickel carbonyl. Am. J. clin. Pathol., **36**: 477-491.

SUNDERMAN, F.W. Jr, WHITE, J.C., & SUNDERMAN, F.W. (1963) Metabolic balance studies in hepatolenticular degeneration treated with diethyldithiocarbamate. Am. J. Med., **34**: 875-888.

SUNDERMAN, F.W. Jr, ROSZEL, N.O., & CLARK, R.J. (1968) Gas chromatography of nickel carbonyl in blood and breath. Arch. environ. Health, **16**(6): 836-843.

SUNDERMAN, F.W. Jr, NOMOTO, S., PRADHAN, A.M., LEVINE, H., BERNSTEIN, S.H., & HIRSCH, R. (1970) Increased concentrations of serum nickel after acute myocardial infarction. J. Med., **283**(17): 896-899.

SUNDERMAN, F.W. Jr, NOMOTO, S., & NECHAY, M. (1971) Nickel metabolism in myocardial infarction. II. Measurements of nickel in human tissues. Trace Subst. environ. Health, **4**: 352-356.

SUNDERMAN, F.W. Jr, NOMOTO, S., MORANG, R., NECHAY, M.W., BURKE, C.N., & NIELSEN, S.W. (1972) Nickel deprivation in chicks. J. Nutr., **102**: 259-268.

SUNDERMAN, F.W. Jr, KASPRZAK, K., HORAK, E., GITLITZ, P., & ONKELINX, C. (1976a) Effects of triethylenetetramin upon the metabolism and toxicity of $63NiCl_2$ in rats. Toxicol. appl. Pharmacol., **38**: 177-188.

SUNDERMAN, F.W. Jr, KASPRZAK, K.S., LAU, T.J., MINGHETTI, P.P. MAECIZA, R.M., BECKER, N., ONKELINX, C., & GOLDBLATT, P.J. (1976b) Effects of manganese on carcinogenicity and metabolism of nickel subsulfide. Cancer Res., **36**: 1790-1800.

SUNDERMAN, F.W. Jr, SHEN, S.K., MITCHELL, J.M., ALLPASS, P.R., & DAMJANOV, I. (1978a) Embryotoxicity and fetal toxicity of nickel in rats. Toxicol. appl. Pharmacol., **43**(2): 381-390.

SUNDERMAN, F.W. Jr, ALLPASS, P., & MITCHELL, J. (1978b) Recent progress in nickel carcinogenesis. Ann. Ist. Super. Sanita, **22**(2): 669-679.

SUNDERMAN, F.W. Jr, ALLPASS, P., & MITCHELL, J. (1978c) Ophthalmic malformations in rats following prenatal exposure to inhalation of nickel carbonyl. Ann. clin. lab. Sci., 8: 499-500.

SUNDERMAN, F.W. Jr, MITCHELL, J., ALLPASS, P., & BASELT, R. (1978d) Embryotoxicity and teratogenicity of nickel carbonyl in rats. Toxicol. appl. Pharmacol., 45: 345.

SUNDERMAN, F.W. Jr, MAENZA, R.M., ALPASS, P.R., MITCHELL, J.M., DAMJANOV, I., & GOLDBLATT, P.J. (1978e) Carcinogenicity of nickel subsulfide in Fischer rats and Syrian hamsters after administration by various routes. Adv. exp. Med. Biol., 91: 57-67.

SUNDERMAN, F.W. Jr, ALLPASS, P.R., MITCHELL, J.M., BASELT, R.C., & ALBERT, D.M. (1979a) Eye malformations in rats: induction by prenatal exposure to nickel carbonyl. Science, 203: 550-553.

SUNDERMAN, F.W. Jr, MAENZA, R.M., HOPFER, S.M., MITCHELL, J.M., ALLPASS, P.R. & DAMJANOV, I. (1979b) Induction of renal cancers in rats by intrarenal injection of nickel subsulfide. J. environ. Pathol. Toxicol., 2(6): 1511-1527.

SUNDERMAN, F.W. Jr, SHEN, S.K., REID, M.C., & ALLPASS, P.R. (1980) Teratogenicity and embryotoxicity of nickel carbonyl in Syrian hamsters. Teratog. Carcinog. Mutagen., 1(2): 223-233.

SUNDERMAN, F.W. Jr, MCCULLY, K.S., & RINEHIMER, L.A. (1981) Negative test for transplacental carcinogenicity of nickel subsulfide in Fischer rats. Res. Commun. chem. Pathol. Pharmacol., 31(3): 545-554.

SUNDERMAN, F.W. Jr, REID. M.C., SHEN, S.K., & KEVORKIAN, C.B. (1983) Embryotoxicity and teratogenicity of nickel compounds. In: Clarkson, T.W., Nordberg, G.F., & Sager, P.R., ed. Reproductive and developmental toxicity of metals. New York, London, Plenum Press, pp. 399-416.

SUNDERMAN, F.W. Jr, CRISOSTOMO, M.C., REID, M.C., HOPFER, S.M., & NOMOTO, S. (1984a) Rapid analysis of nickel in serum and whole blood by electrothermal atomic absorption spectrophotometry. Ann. clin. lab. Sci., 14(3): 232-241.

SUNDERMAN, F.W. Jr, MACCULLY, K.S., & HOPFER, S.M. (1984b) Association between erythrocytosis and renal cancers in rats following intrarenal injection of nickel compounds. Carcinogenesis, 5: 1511-1517.

SUNDERMAN, F.W. Jr, MARZOUK, A., CRISOSTOMO, M.C., & WEATHERBY, D.R. (1985) Electrothermal atomic absorption spectrophotometry of nickel in tissue homogenates. Ann. clin. lab. Sci., 15(4): 299-307.

SUNDERMAN, F.W. Jr, HOPFER, S.M., CRISOSTOMO, M.C., & STOEPPLER, M. (1986a) Rapid analysis of nickel in urine by electrothermal atomic absorption spectrophotometry. Ann. clin. lab. Sci., 16(3): 219-230.

SUNDERMAN, F.W. Jr, AITIO, A., MORGAN, L.G., & NORSETH, T. (1986b) Biological monitoring of nickel. Toxicol. ind. Health, 2(1): 17-78.

SUNDERMAN, F.W. Jr, HOPFER, S.M., KNIGHT, J.A., McCEILLY, K.S., CECUTTI, A.G., THORNHILL, A.G., CONWAY, K., MILLER, C., PATIERNO, S.R., & COSTA, M. (1987a) Physiochemical characteristics and biological effects of nickel oxides. Carcinogenesis, 8: 305-313.

SUNDERMAN, F.W. Jr, AN, L.L., ZAHARIA, O., WONG, S.H.-Y., & HOPFER, S.M. (1987b) Acute depletion of pulmonary lavage cells and enhanced lipid peroxidation in alveolar macrophages following parenteral administration of nickel chloride to rats. In: Rand, N.Y., & Raper, C., ed. Pharmacology, Amsterdam, New York, Oxford, Elsevier Science Publishers, pp. 817-820.

SUNDERMAN, F.W. Jr, HOPFER, S.M., & KNIGHT, J.A. (1988a) Biological reactivity of nickel oxides and related compounds. In: Fourth International Conference on Nickel Metabolism and Toxicology: Abstracts, Espoo, Finland, 5-9 September 1988, Helsinki, Institute of Occupational Health, p. 15.

SUNDERMAN, F.W. Jr, DINGLE, B., HOPFER, S.M., & SWIFT, T. (1988b) Acute nickel toxicity in electroplating workers who accidentally ingested a solution of nickel sulfate and nickel chloride. Ann. J. ind. Med., 14: 257-266.

SUNDERMAN, F.W. Jr, HOPFER, S.M., SWEENEY, K.C., MARCUS, A.H., MOST, B.M., & CREASON, J. (1989a) Nickel absorption and kinetics in human volunteers. Proc. Soc. exp. Biol. Med., 191: 5-11.

SUNDERMAN, F.W. Jr, HOPFER, S.M., SWIFT, T., REZUKE, W.N., ZIEBKA, L., HIGHAM, P., EDWARDS, B., FOLCIK, M., & GOSSLING, H.R. (1989b) Cobalt, chrominium and nickel concentrations in body fluids of patients with porous-coated knee or hip prostheses. J. Orthop. Res., 7(3): 307-315.

SUNDERMAN, F.W. Jr, HOPFER, F.W. Jr, LIN, S.-M., PLOWMAN, M.C., STOJANOVIC, T., WONG, S.H.-Y., ZAHARIA, O., & ZIEBKA, L. (1989c) Toxicity to alveolar macrophages in rats following parenteral injection of nickel chloride. Toxicol. appl. Pharmacol., 100: 107-118.

SUNG, J.F.C., NEVISSI, A.E., & DEWALLE, F.B. (1986) Concentration and removal efficiency of major and trace elements in municipal wastewater. J. environ. Sci. Health, A 21(5): 435-448.

SUSHENKO, O.V., & RAFIKOVA, K.E. (1972) [Questions of work hygiene in hydrometallurgy of copper, nickel and cobalt in a sulfide ore.] Gig. Tr. prof. Zabol., 16: 42-45 (in Russian).

SUTHERLAND, R.B. (1959) Respiratory cancer mortality in workers employed in an Ontario nickel refinery covering the period 1939-1957, Ottawa, Ontario Department of Health, Division of Industrial Hygiene (Unpublished report, November, 1959).

SWANN, M. (1984) Malignant soft-tissue tumour at the site of a total hip relacement. J. bone joint Surg., B 66: 629-631.

SWIERENGA, S.H., & BASRUR, P.K. (1968) Effect of nickel on cultured rat embryo muscle cells. Lab. Invest., 19: 663-674.

SWIERENGA, S.H., & MCLEAN, J.R. (1985) Further insights into mechanisms of nickel-induced DNA damage studies with cultured rat liver cells. In: Brown, S.S., & Sunderman, F.W. Jr, ed. Progress in Nickel toxicology. Proceedings of the 3rd

International Conference on Nickel Metabolism and Toxicology, Paris, 4-7 September, 1984, Oxford, Blackwell Scientific Publications, pp. 101-104.

SWIERENGA, S.H.H., GILMAN, J.P.W., & MCCLEAN, J.R. (1987) Cancer risk from inorganics. Cancer Metastasis Rev., 6: 113-154.

SWIERENGA, S.H.H., MARCEAU, N., KATSUMA, Y., FRENCH, S.W., MULLER, R., & LEE, F. (1989) Altered cytokeratin expression and differentiation induction during neoplastic transformation of cultured rat liver cells by nickel subsulfide. Cell Biol. Toxicol., 5(3): 271-286..

SZADKOWSKI, D., SCHULTZE, H., SCHALLER, K.H., & LEHNERT, G. (1969a) [On the significance of heavy metal contents in cigarettes.] Arch. Hyg., 153: 1-8 (in German).

SZADKOWSKI, D., KOHLER, G., & LEHNERT, G. (1969b) [Electrolyte-changes in serum and electrical-mechanical heart action in industrial heat stress.] Aertzl. Forsch., 23(8): 271-284 (in German).

SZATHMARY, S.C., & DALDRUP, T. (1982) [Determination of nickel by GC, GC-MS and AAS in biological materials in a case of lettral poisoning.] Fresenius Z. anal. Chem., 313: 48 (in German).

TABILLION, R., WEBER, F., & KALTWASSER, H. (1980) Nickel requirement for chemolithotropic growth in hydrogen oxidizing bacteria. Arch. Microbiol., 124: 131-136.

TAGAKI, Y., MATSUDA, S., IMAI, S., OHMORI, Y., MASUDA, T., VINSON, J.A., MEHRA, M.C., PURI, B.K., & KANIEWSKI, A. (1986) Trace elements in human air. Bull. environ. Contam. Toxicol., 36: 793-899.

TAKENAKA, S., HOCHRAINER, D., & OLDIGES, H. (1985) Alveolar proteinosis induced in rats by long-term inhalation of nickel oxide. In: Brown, S.S., & Sunderman, F.W. Jr, ed. Progress in nickel toxicology. Proceedings of the 3rd International Conference on Nickel Metabolism and Toxicology, Paris, 4-7 September, 1984, Oxford, Blackwell Scientific Publications, pp. 89-92.

TANAKA, I. (1985) Ion-chromatographic determination of divalent metal ions (Co, Ni, Cu, Zn, Cd) using EDTA as complexing agent. Fresenius Z. anal. Chem., 320(2): 125-127.

TANAKA, I., ISHIMATSU, S., MATSUNO, K., KODAMA, Y., & TSYCHIYA, K. (1985) Biological half time of deposited nickel oxide aerosol in rat lung by inhalation. Biol. Trace Elem. Res., 8: 203-210.

TANAKA, I. ,ISHIMATSU, S., HARATAKE, J., HORIE, A., & KODAMA, Y. (1988) Biological half-time in rats exposed to nickel monosulfide (amorphous) aerosol by inhalation. Biol. Trace Elem. Res., 17: 237-246.

TANDON, S.K. (1982) Disposition of nickel-63 in the rat. Toxicol. Lett., 10(1): 71-73.

TANDON, S.K., MATHUR, A.K., & GAUR, J.S. (1977) Urinary excretion of chromium and nickel among electroplaters and pigment industry workers. Int. Arch. occup. environ. Health, 40(1): 71-76.

TASK GROUP ON LUNG DYNAMICS (1966) Deposition and retention models for internal dosimetry of the human respiratory tract. Health Phys., 12: 173- 207.

TATARSKAYA, A.A. (1960) [Occupational disease of upper respiratory tract in persons employed in electrolytic nickel refining departments.] Gig. Tr. prof. Zabol., 6: 35-38 (in Russian).

TATARSKAYA, A.A. (1965) [On the problem of occupational cancer of the upper respiratory tract in the nickel-refining industry.] Gig. Tr. prof. Zabol., 9(2): 22-25 (in Russian).

TAYLOR, D., MADDOCK, B.G., & MANCE, G. (1985a) The acute toxicity of nine gray list metals (arsenic, boron, chromium, copper, lead, nickel, tin, vanadium, and zinc) to two marine fish species: dab (*Limanda limanda*) and grey mullet (Chelon labrosus). Aquat. Toxicol., 7(3): 135-144.

TAYLOR, P., DAMS, R., & HOSTE, R. (1985b) The determination of cadmium, chromium, copper, nickel and zinc in city waste incinerator ash using inductively coupled plasma-atomic emission spectrometry. J. anal. Let., 18(A19): 2361-2368.

TAYTON, K.J.J. (1980) Ewing's sarcoma at the site of a metal plate. Cancer, 45: 413-415.

TEDESCHI, R.E., & SUNDERMAN, F.W. (1957) Nickel poisoning. V. The metabolism of nickel under normal conditions and after exposure to nickel carbonyl. Arch. ind. Health, 16: 486-488.

TERAKI, Y., & UCHIUMI, A. (1988) Experimental studies on metal carcinogenesis: a comparative assessment of nickel and lead. In: Fourth International Conference on Nickel Metabolism and Toxicology, Abstracts, Espoo, Finland, 5- 9 September, 1988, Helsinki, Institute of Occupational Health, p. 5.

THURSTON, G.D., & SPENGLER, J.D. (1985) A quantitative assessment of source contributions to inhalable particulate matter pollution in metropolitan Boston. Atmos. Environ., 19(1): 9-25.

TIMOURIAN, H., & WATCHMAKER, G. (1972) Nickel uptake by sea urchin embryos and their subsequent development. J. exp. Zool., 182: 379-388.

TOLA, S., KILPIOE, J., & VIRTAMO, M. (1979) Urinary and plasma concentrations of nickel as indicators of exposure to nickel in an electroplating shop. J. occup. Med., 21(3): 184-188.

TOLOT, F., BROUDEUR, P., & NEULAT, G. (1956) Troubles pulmonaires asthmatiformes chez des ouvriers exposés à l'inhalation de chrome, nickel et aniline. Arch. Mal. prof. Méd. Trav. Séc. soc., 18: 291-293.

TORJUSSEN, W., & ANDERSEN, I. (1979) Nickel concentrations in nasal mucosa, plasma, and urine in active and retired nickel workers. Ann. clin. lab. Sci., 9(4): 289-298.

TORJUSSEN, W., SOLBERG, L.A., & HOGETVEIT, A.C. (1979a) Histopathologic changes of nasal mucosa in nickel workers: a pilot study. Cancer, 44(3): 963- 974.

References

TORJUSSEN, W., SOLBERG, L.A., & HOGETVEIT, A.C. (1979b) Histopathological changes of the nasal mucosa in active and retired nickel workers. Br. J. Cancer, 40(4): 568-580.

TOSSAVAINEN, A., NURMINEN, M., MUTANEN, P., & TOLA, S. (1980) Application of mathematical modelling for assessing the biological half-times of chromium and nickel in field studies. Br. J. ind. Med., 37(3): 285-291.

TRAUL, K.A., HINK, R.J., WOLFF, J.S., & WLODYMR, K. (1981) Chemical carcinogesus *in vitro*: an improved method for chemical transformation in Rauscher leukemia virus-infected rat embryo cells. J. appl. Toxicol., 1(1): 32-37.

TREAGAN, L., & FURST, A. (1970) Inhibition of interferon synthesis in mammalian cell cultures after nickel treatment. Res. Commun. chem. Pathol. Pharmacol., 1: 395-401.

TURHAN, U., HAIN, C., WOLLBURG, C., & SZADKOWSKY, D. (1983) [Chromium and nickel content in human lung. Medical aspects.] In: [Proceedings of the 1983 Annual Congress of the German Association for Occupational Medicine, Göttingen], Stuttgart, Gentner, pp. 477-480 (in German).

TURK, J.L., & PARKER, D. (1977) Sensitization with Cr, Ni, and Zn salts and allergic type granuloma formation in the guinea-pig. J. invest. Dermatol., 68: 341-345.

TWEATS, D.J., GREEN, M.H.L., & MURIEL, W.J.A. (1981) A differential killing assay for mutagens and carcinogens based on an improved repair-deficient strain of E. coli. Carcinogenesis, 2: 189-194.

UMEDA, M., & NISHIMURA, M. (1979) Inducibility of chromosomal aberrations by metal compounds in cultured mammalian cells. Mutat. Res., 67: 221-229.

UPITIS, V.V., NOLLENDORF, A.F., & PAKALNE, D.S. (1980) [Relevance of nickel for culturing microalgae.] In: Anke, M., Schneider, H.-J., & Brückner, Chr., ed. [3. Trace Element Symposium Nickel], Jena, German Democratic Republic, 7-11 July, 1980, Jena, Friedrich-Schiller University, pp. 171-183 (in German).

US BUREAU OF MINES (1985) Nickel. In: Minerals yearbook 1984, Washington, DC, Department of the Interior, Vol. I, pp. 665-684.

US BUREAU OF MINES (1986) Nickel. In: Mineral commodity summaries 1986. An up-to-date summary of 87 nonfuel mineral commodities, Wahington, DC, Department of the Interior, pp. 108-109.

US EPA (1986) Health assessment document for nickel and nickel compounds, Washington DC, US Environmental Protection Agency, 438 pp (EPA/600/8-83/012 FF).

US EPA (1987) Acute toxicity handbook of chemicals to estuarine organisms, Gulf Breeze, Florida, US Environmental Protection Agency, 274 pp (EPA/600/8-87/017).

US NIOSH (1977a) Special occupational hazard review and control recommendations for nickel carbonyl. Cincinatti, Ohio, US National Institute for Occupational Safety and Health, 45 pp (DHEW (NIOSH) Publication No. 77-184).

US NIOSH (1977b) Criteria for a recommended standard occupational exposure to inorganic nickel. Cincinatti, Ohio, US National Institute for Occupational Safety and Health, 282 pp (prepared by Stanford Research Institute, Menlo Park) (DHEW (NIOSH) Publication No. 77-164).

VAN SOESTBERGEN, M., & SUNDERMAN, F.W. (1972) ^{63}Ni complexes in rabbit serum and urine after injection of ^{63}NiCl$_2$. Clin. Chem., 18(12): 1478-1484.

VALENTA, P., OSTAPCZUK, P.H., PIKLAR, B., & NURNBERG, H.W. (1981) New applications of voltammetry in the determination of toxic trace metals in food. Proceedings of the 3rd CEC/WHO International Conference on Heavy Metals in the Environment, Amsterdam, 1981. Edinburgh, CEP Consultants Ltd, pp. 619-621.

VALENTINE, R., & FISHER, G.L. (1984) Pulmonary clearance of intratracheally administered ^{63}Ni$_3$S$_2$ in strain A/J mice. Environ. Res., 34: 328-334.

VAN BAALEN, C., & O'DONNELL, R. (1978) Isolation of a nickel-dependent blue-green algae. J. gen. Microbiol., 105: 351-353.

VANDENBERG, J., & EPSTEIN,W. (1963) Experimental skin contact sensitisation in man. J. invest. Dermatol., 41: 413-416.

VAN ROSEN, G. (1954) Breaking of chromosomes by the action of elements of the periodical system and by some other principles. Hereditas, 40: 258-263.

VANDENBERG,J., & EPSTEIN, W. (1963) Experimental skin contact sensitisation in man. J. invest. Dermatol., 41: 413-416.

VERMA, H.S., BEHARI, J.R., & TANDON, S.K. (1980) Pattern of urinary ^{63}Ni excretion in rats. Toxicol. Lett., 5(3-4): 223-226.

VINSON, L.J., & CHOMAN, B.R. (1960) Percutaneous absorption and surface active agents. J. Soc. Cosmet. Chem., 11: 127-137.

VOGEL, E. (1984) The relation between mutational pattern and concentration by chemical mutagens in *Drosophila*. In: Screening tests in chemical carcinogenesis. Proceedings of a workshop organized by IARC, and the Commission of the European Communities, Brussels, 9-12 June, 1975, Lyon, International Agency for Research on Cancer, pp. 117-137 (IARC Scientific Publications No. 12).

VOLLKOPF, U., GROBENSKI, Z., & WELZ, B. (1981) Determination of nickel in serum using graphite furnace atomic absorption. At. Spectrosci., 2: 68-70.

VOLINI, F., DE LA HUERGA, J., & KENT, G. (1968) Trace metal studies in liver disease using atomic absorption spectrometry. In: Sunderman, F.W., & SUNDERMAN, F.W. Jr, ed. Laboratory diagnosis of liver diseases, St. Louis, Missouri, Warren H. Green, pp. 199-206.

VOROSHILIN, B.I., PLOTKO, E.G., NIKIPHOROVA, V.YA., & FINK, T.V. (1977) Cytogenetic effects of some inorganic chemicals on human leukocytes in cultures. In: Problems of genetics and selection in Urals (information data). Sverdlorsk. The Urals Scientific Center of the USSR Academy of Sciences, pp. 46-47.

VOS, L., KOMY, Z., REGGERS, G., ROEKENS, E., & VAN GRIEKEN, R. (1986) Determination of trace metals in rain water by differential-pulse stripping voltammetry. Anal. chim. Acta, 184: 271-280.

VUOPALA, U., HUHTI, E., TAKKUNEN, J., & HUIKKO, M. (1970) Nickel carbonyl poisoning. Report of 25 cases. Ann. clin. Res., 2: 214-222.

References

WACHS, B. (1982) Concentration of heavy metals in fishes from the river Danube. Z. Wasser Abwasser Forsch., 15(2): 43-84.

WAHLBERG, J.E. (1975) Nickel allergy and atopy in hairdressers. Contact Dermatit., 1(3): 161-165.

WAHLBERG, J.E. (1976) Sensitization and testing of guinea-pigs with nickel sulfate. Dermatology, 152: 321-330.

WAHLBERG, J.E., LINDSTEDT, G., & EINARSSON, O. (1977) Chromium, cobalt and nickel in Swedish cement, detergents, mould and cutting oils. Berufsdermatosen, 25(6): 220-228.

WAKSVIK, H., & BOYSEN, M. (1982) Cytogenetic analyses of lymphocytes from workers in a nickel refinery. Mutat. Res., 103: 185-190.

WAKSVIK, H., BOYSEN, M., & HOGETVEIT, A. Chr. (1984) Increased incidence of chromosomal aberrations in peripheral lymphocytes of retired nickel workers. Carcinogenesis, 5(11): 1525-1527.

WALL, L.M., & CALNAN, C.D. (1980) Occupational nickel dermatitis in the electroforming industry. Contact Dermatit., 6(6): 414-420.

WALLACE, A., ROMNEY, E.M., KINNEAR, J., & ALEXANDER, G.U. (1980a) Single and multiple trace metal excess effects on three different plant species. J. Plant Nutr., 2: 11-23.

WALLACE, A., ROMNEY, E.M., MULLER, R.T., & ALEXANDER, G.U. (1980b) Calcium-trace metal interactions in soybean plants. J. Plant Nutr., 2: 79-86.

WALTHARD, B. (1926) [Experimental nickel idiosyncrasy in laboratory animals.] Schweiz. Med. Wochenschr., 7: 603-604 (in German).

WALTSCHEWA, W., SLATEWA, M., & MICHAILOW, I. (1972) [Testicular changes due to long-term administration of nickel sulphate in rats.] Exp. Pathol., 6(3): 116-120 (in German).

WAN, C.C., CHIANG, S., & CORSINI, A. (1985) Two-column method for preconcentration of trace metals in natural waters on acrylate resin. Anal. Chem., 57: 719-723.

WARNER, J.S. (1984) Occupational exposure to airborne nickel in producing and using primary nickel products. In: Nickel in the human environment, Proceedings of a Joint Symposium, Lyon, 8-11 March, 1983, Lyon, International Agency for Research on Cancer, pp. 419-437 (IARC Scientific Publications No. 53).

WARNER, J.S. (1985) Estimating past exposures to airborne nickel compounds in the Copper Cliff Sinter Plant. In: Brown, S.S. & Sunderman, F.W. Jr, ed. Progress in nickel toxicology. Proceedings of the 3rd International Conference on Nickel Metabolism and Toxicology, Paris, 4-7 September, 1984, Oxford, Blackwell Scientific Publications, pp. 203-206.

WASE, A.W., GOSS, D.M., & BOYD, M.J. (1954) The metabolism of nickel. I. Spatial and temporal distribution of Ni^{63} in the mouse. Arch. Biochem. Biophys., 51: 1-4.

WATANABE, H., GOTO, K., TAGUCHI, S., MCLAREN, J.W., BERMAN, S.S., & RUSSELL, O.S. (1981) Preconcentration of trace elements in sea-water by complexation with 8-hydroxyquinoline and adsorption on C_{18}-bonded silica gel. Anal. Chem., 53: 738-739.

WATERMAN, A.H., & SCHRIK, J.J. (1985) Allergy in hip arthroplasty. Contact Dermatit., 13(5): 294-301.

WATERS, M.D., GARDNER, D.E., & COFFIN, D.L. (1975) Toxicity of metallic chlorides and oxides for rabbit alveolar macrophages in vitro. Environ. health Perspect., 10: 267-268.

WATRAS, C.J., MACFARLANE, J., & MOREL, F.M.M. (1985) Nickel accumulation by Scenedesmus and Daphnia: Food-chain transport and geochemical implications. Can. J. Fish. aquat. Sci., 42(4): 724-730.

WEAST, R.C. (1981) CRC Handbook of chemistry and physics, 62nd ed., Boca Raton, Florida, CRC Press, pp. B123-B124.

WEBER, P.C. (1986) Epithelioid sarcoma in association with total knee replacement. J. bone joint Surg., B 68: 824-826.

WEBSTER, J.D., PARKER, T.F., ALFREY, A.C., SMYTHE, W.R., KUBO, H., NEAL, G., & HULL, A.R. (1980) Acute nickel intoxication by dialysis. Ann. intern. Med., 92: 631-633.

WEHNER, A.P., & CRAIG, D.K. (1972) Toxicology of inhaled NiO and CoO in Syrian golden hamsters. Am. Ind. Hyg. Assoc. J., 33: 146-155.

WEHNER, A.P., BUSCH, R.H., OLSON, R.J., & CRAIG, D.K. (1975) Chronic inhalation of nickel oxide and cigarette smoke by hamsters. Am. Ind. Hyg. Assoc. J., 36(11): 801-810.

WEHNER, A.P., MOSS, O.R., MILLIMAN, E.M., DAGLE, G.E., & SCHIRMER, R.E. (1979) Acute and subchronic inhalation exposures of hamsters to nickel-enriched fly ash. Environ. Res., 19(2): 355-370.

WEHNER, A.P., DAGLE, G.E., &. MILLIMAN, E.M. (1981) Chronic inhalation exposure of hamsters to nickel-enriched fly ash. Environ. Res., 26(1): 195- 216.

WEIDENAUER, M., & LIESER, K.H. (1985) [Determination of trace elements in river water by means of voltammetry.] Fresenius Z. anal. Chem., 320: 550-555 (in German).

WEINZIERL, J., & UMLAND, F. (1982) [Determination of Co (and Ni) in the ppb-range by differential pulse polarography.] Fresenius Z. anal. Chem., 312: 608-610 (in German).

WELLENREITER, R.H., ULLREY, D.E., & MILLER, E.R. (1970) Nutritional studies with nickel. In: Mills, C.F., ed. Trace element metabolism in animals. Proceedings of 1st WAAP/IBP International Symposium, Aberdeen, Scotland, July 1969, Edinburgh, London, Livingstone, pp. 52-58.

WELLS, G.C. (1956) Effects of nickel on the skin. Br. J. Dermatol., 68: 237-242.

WEST, B., & SUNDERMAN, F.W. (1958) Nickel poisoning. VII. The therapeutic effectiveness of alkyl dithiocarbamates in experimental animals exposed to nickel carbonyl. Am. J. med. Sci., 236: 15-25.

WHANGER, P.D. (1973) Effects of dietary nickel on enzyme activities and mineral contents in rats. Toxicol. appl. Pharmacol., 25: 323-331.

WHITE, D.H., KING, K.A., MITCHELL, L.A., & MULHERN, B.M. (1986) Trace elements in sediments, water, and american coots (*Fulica* americana) at a coal-fired power plant in Texas. Bull environ. Contam. Toxicol., 36(3): 376-383.

WHO (1984) Guidelines for drinking-water quality. Vol. 1 - Recommendations, Geneva, World Health Organization, p. 57.

WHO (1988) Air quality guidelines for Europe., Copenhagen, World Health Organization, pp. 285-296 (WHO Regional Publications, European Series No 23).

WIERNIK, A., JOHANSSON, A., JARSTRAND, C., & CAMNER, P. (1983) Rabbit lung after inhalation of soluble nickel. I. Effects on alveolar macrophages. Environ. Res., 30: 129-141.

WILLS, M.R., BROWN, C.S., BERTHOLF, R.L., & SAVORY, J. (1985) Serum and lymphocyte aluminium and nickel in chronic renal failure. Clin. chim. Acta, 145: 193-196.

WILSON, J.D., STENZEL, M.R., LOMBARDOZZI, K.L., & NICHOLS, C.L.C. (1981) Monitoring personal exposure to stainless steel welding fumes in confined spaces at a petrochemical plant. Am. Ind. Hyg. Assoc. J., 42: 431-436.

WILSON, L.W., & DINUNZIO, J.E. (1981) Enrichment of nickel and cobalt in natural hard water by Donnan Dialysis. Anal. Chem., 53: 692-695.

WILSON, W.W., & KHOOBYARIAN, N. (1982) Potential identification of chemical carcinogens in a viral transformation system. Chem. biol. Interact., 38: 253-259.

WINDHOLZ, M., BUDAVARI, S., BLUMETTI, R.F., & OTTERBEIN, E.S. (1983) The Merck Index, 10th ed., Rahway, New Jersey, Merck & Co., Inc.

WOLNIK, K.A., FRICKE, F.L., CAPAR, S.C., BRAUDE, G.L., MEYER, M.W., SATZGER, R.D., & KUENNEN, R.W. (1983) Elements in major raw agricultural crops in the United States. 2. Other elements in lettuce, peanuts, potatoes, soybeans, sweet corn and wheat. J. Agric. food Chem., 31: 1244-1249.

WOLNIK, K.A., FRICKE, F.L., CAPAR, S.G., MEYER, M.W., SATZGER, R.D., BONNIN, E., & GASTON, C.M. (1985) Elements in major raw agricultural crops in the United States. 3. Cadmium, lead, and eleven other elements in carrots, field corn, onions, rice, spinach and tomatoes. J. Agric. food. Chem., 33: 807-811.

WULF, H.C. (1980) Sister chromatid exchanges in human lymphocytes exposed to nickel and lead. Danish med. Bull., 27: 40-42.

YAMASHIRO, S., GILMAN, J.P.W., HULLAND, T.J., & ABANDOWITZ, H.M. (1980) Nickel sulphide-induced rhabdomyosarcomata in rats. Acta pathol. Jpn., 30(1): 9-22.

YAMASHIRO, S., BASRUR, P.K., GILMAN, J.P., HULLAND, T.J., & FUJIMOTO, Y. (1983) Ultrastructural study of Ni_3S_2-induced tumors in rats. Acta pathol. Jpn., 33(1): 45-58.

YAN, N.D., & STRUS, R. (1980) Crustacean zooplankton communities of acidic, metal-contaminated lakes near Sudbury, Ontario. Can. J. Fish. aquat. Sci., 37(12): 2282-2294.

YAN, N.D., MILLER, G.E., WILE, I., & HITCHIN, G.G. (1985) Richness of aquatic macrophyte floras of soft water lakes of differing pH and trace metal content in Ontario, Canada. Aquat. Bot., 23(1): 27-40.

YANG, X.H., BROOKS, R.R., JAFFRE, T., & LEE, J. (1985) Elemental levels and relationships in the flacourtiaceae of New Caledonia and their significance for the evaluation of the serpentine problem. Plant Soil, 87(2): 281-292.

YARITA, T., & NETTESHEIM, P. (1978) Carcinogenicity of nickel subsulfide for respiratory tract mucosa. Cancer Res., 38: 3140-3145.

YOSHIMURA, K., THOSHIMITSU, Y., & OHASKI, S. (1980) Ion-exchanger colorimetry - VI micro-determination of nickel in natural water. Talanta, 27: 693-697.

ZAKOUR, R.A., TKESHELASHVILI, L.K., SHERMAN, C.W., KOPLITZ, L.M., & LOEB, L.A. (1981) Metal-induced infidelity of DNA synthesis. J. cancer Res. clin. Oncol., 99: 187-196.

ZISLIN, D.M., GANYUSKINA, S.M., DUBINIA, E.S., & TYUSHNYAKOVA, N.V. (1969) [Residual air in a complex evaluation of the respiratory system function in initial and suspected pneumonconiosis.] Gig. Tr.prof. Zabol., 13: 26-29 (in Russian).

ZOBER, A. (1981a) [Symptoms and findings at the bronchopulmonary system of electric arc welders. II. communication: pulmonary fibrosis.] Zentralbl. Bakteriol. Hyg., I. Abt. Orig. B, 173: 120-148 (in German).

ZOBER, A. (1981b) [Symptoms and findings at the bronchopulmonary system of electric arc welders. I. communication: epidemiology.] Zentralbl. Bakt. Hyg., I. Abt. Orig. B 173: 92-119 (in German).

ZOBER, A. (1982) [Possible dangers to the respiratory tract from welding fumes.] Schweissen Schneiden, 34: 77-81 (in German).

ZOBER, A., KICK, K., SCHALLER, K.H., SCHELLMANN, B., & VALENTIN, H. (1984) [Nickel and chromium content of selected human organs and body fluids.] Zentralbl. Bakteriol. Hyg., I. Abt. Orig. B, 179: 80-95 (in German).

1. RESUME ET CONCLUSIONS

1. Identité, propriétés physiques et chimiques et méthodes d'analyse

Le nickel est un élément métallique qui appartient au groupe VIIIb de la classification périodique. Il résiste aux alcalis mais se dissout en général dans les acides oxydants dilués. Le carbonate, le sulfure et l'oxyde sont insolubles dans l'eau, alors que le chlorure, le sulfate et le nitrate sont solubles. Le nickel carbonyle est un liquide volatil incolore qui se décompose au-dessus de 50 °C. Sous forme ionique, le nickel existe essentiellement à l'état de nickel(II). Dans les systèmes biologiques, le nickel une fois dissous peut former des complexes avec divers ligands et se fixer aux matières organiques.

L'analyse des échantillons biologiques et environnementaux se fait le plus fréquemment par spectroscopie d'absorption atomique et voltammétrie. Pour obtenir des résultats fiables, en particulier dans le domaine des ultratraces (ultramicrotraces), il faut suivre un mode opératoire précis pour réduire au minimum le risque de contamination lors des prélèvements, de la conservation, du traitement, et de l'analyse des échantillons. Selon les traitements préalables subis par l'échantillon et les méthodes d'extraction ou d'enrichissement utilisées, on peut atteindre des limites de détection de 1 à 100 ng/litre dans les produits biologiques et l'eau.

2. Sources d'exposition humaine et environnementale

Le nickel est présent à l'état de traces un peu partout dans le sol, l'eau, l'air et la biosphère. La teneur moyenne de la croûte terrestre en nickel est d'environ 0,008%. Les terrains agricoles en contiennent entre 3 et 1000 mg/kg. Dans les eaux naturelles, les concentrations varient de 2 à 10 µg/litre (eau douce) et de 0,2 à 0,7 µg/litre (eau de mer). Dans l'atmosphère, les concentrations en nickel varient de 0,1 à 3 ng/m^3, au-dessus des zones écartées.

Les gisements de nickel consistent en accumulations de minéraux sulfurés (essentiellement de la pentlandite) et en latérites. Le nickel est extrait de ces minerais par affinage pyro- et hydrométallurgique. Le nickel est principalement utilisé pour la production d'aciers inoxydables qui sont très résistants à la corrosion et aux températures élevées. On l'utilise sous forme d'alliages et de dépôts galvaniques pour la confection de pièces d'automobiles, de machines-outils, d'armements, d'outils, de matériel électrique, d'appareils ménagers, et de pièces de monnaie. Les dérivés du nickel sont également utilisés comme catalyseurs, comme pigments, et dans les batteries d'accumulateurs. En 1985, la production minière mondiale de nickel a été d'environ 67 millions de kg. Les principales émissions de nickel dans l'air ambiant proviennent de la combustion du charbon et du mazout pour le chauffage et la production d'énergie, de l'incinération des déchets et des boues de stations d'épuration, de la production minière et de la métallurgie du nickel, de la fabrication de l'acier, du nickelage, et de diverses autres sources, comme les cimenteries. Dans les polluants atmosphériques, le nickel est présent principalement sous forme de sulfates, d'oxydes, de sulfures, et, dans une moindre mesure, de nickel métallique.

Le nickel rejeté lors de diverses opérations industrielles ou autres finit par aboutir dans les eaux usées. Après traitement de ces dernières, les résidus sont éliminés par injection dans des puits profonds, décharge en mer, enfouissement dans le sol et incinération. Les effluents rejetés par les usines de traitement des eaux usées contiendraient jusqu'à 0,2 mg de nickel par litre.

3. Transport, distribution, et transformation dans l'environnement

Le nickel qui pénètre dans l'environnement peut être d'origine naturelle ou artificielle et il circule à travers tous les compartiments du milieu, soit sous l'effet de processus physicochimiques, soit par transport biologique au sein d'organismes vivants.

On estime que le nickel atmosphérique est essentiellement présent sous forme d'aérosols particulaires dont la teneur en nickel varie en fonction de leur origine. C'est en général les particules les plus petites qui présentent les teneurs les plus élevées en nickel. Le nickel carbonyle est instable à l'air et il se décompose pour former de l'oxyde de nickel.

Résumé et conclusions

Le transport et la distribution des particules de nickel en direction ou à l'intérieur des divers compartiments du milieu dépend beaucoup de la granulométrie des particules et des conditions météorologiques. Cette granulométrie dépend à son tour principalement de la source émettrice. En général, les particules d'origine artificielle sont de dimension plus réduite que celles qui proviennent de la poussière naturelle.

Le nickel peut pénétrer dans l'hydrosphère soit à partir de l'atmosphère, soit par lessivage superficiel, soit encore par décharge de déchets industriels ou municipaux ou érosion naturelle des sols et des roches. Dans les cours d'eau, le nickel est principalement transporté par des particules enrobées de matières précipitées ou en association avec des matières organiques; dans les lacs, le transport est assuré sous forme ionique, essentiellement en association avec des matières organiques. Le nickel peut également s'adsorber sur des particules d'argile ou être fixé par des organismes vivants. A l'inverse, il peut se désorber et s'échapper ainsi des sédiments où il était fixé. Une partie du nickel ainsi transporté par les cours d'eau aboutit à l'océan. On estime que l'apport sous forme de particules en suspension dans les cours d'eau est d'environ 135×10^7 kg/an.

Selon le type de sol, le nickel peut présenter une forte mobilité et parvenir jusqu'aux eaux souterraines, puis aux lacs et aux rivières. C'est principalement par leurs racines que les plantes terrestres fixent le nickel présent dans le sol. La quantité ainsi fixée dépend de divers paramètres géochimiques et physiques, notamment le type de sol, son pH, sa teneur en eau et en matières organiques, ainsi que la concentration en nickel extractible. L'exemple le mieux connu d'accumulation de nickel est fourni par un certain nombre d'espèces végétales "hyperaccumulatrices", qui concentrent le nickel à raison de plus de 1mg/kg de poids sec, et qui poussent sur des sols serpentinifères relativement stériles. A partir de 50 mg/kg de poids sec, le nickel est toxique pour la plupart des végétaux. On a observé une accumulation de nickel et des effets toxiques chez les végétaux cultivés sur sols traités par des boues de stations d'épuration ainsi que dans la végétation située à proximité de sources d'émission. Certaines plantes aquatiques concentrent fortement le nickel. Des études de laboratoire portant sur des poissons ont montré que le nickel n'avait guère tendance à s'accumuler, quelques que soient les espèces étudiées. Dans des eaux non polluées, les concentrations relevées dans le poisson entier (et rapportées au poids de tissus frais), variaient de 0,02 à 2 mg/kg. Ces valeurs pourraient être jusqu'à 10 fois supérieures dans les poissons qui vivent en eau polluée. Toutefois rien n'indique que le nickel subisse une bioamplification tout au long de la chaîne alimentaire.

Les animaux sauvages, par exemple, les herbivores et leurs prédateurs carnivores, absorbent du nickel par voie alimentaire, qui se retrouve ensuite dans de nombreux organes et tissus.

4. Concentrations dans l'environnement et exposition humaine

Dans les organismes terrestres et aquatiques, les concentrations de nickel peuvent varier de plusieurs ordres de grandeur. L'homme est exposé à des concentrations atmosphériques dont les valeurs caractéristiques vont de 5 à 35 ng/m^3 en zones rurales et urbaines, d'où une absorption par inhalation de 0,1 à 0,7 µg de nickel par jour. L'eau de boisson en contient généralement moins de 10 µg/litre mais il arrive que du nickel soit libéré dans l'eau d'adduction par les éléments de la plomberie et atteigne des concentrations de 500 µg/litre.

Dans les produits alimentaires, les concentrations sont généralement inférieures à 0,5 mg/kg en poids de substance fraîche. Le cacao, les graines de soja, certains légumes secs, diverses noix, et la farine d'avoine en contiennent de fortes concentrations. L'ingestion journalière de nickel par la voie alimentaire est très variable du fait de la diversité des habitudes alimentaires: les quantités ingérées peuvent varier de 100 à 800 µg/jour; en moyenne l'apport alimentaire dans la plupart des pays est de 100 à 300 µg/jour. La libération de nickel par les ustensiles de cuisine peut augmenter sensiblement les quantités ingérées. Chez les fumeurs, les quantités absorbées par voie pulmonaire peuvent aller de 2 à 23 µg/jour pour une consommation quotidienne de 40 cigarettes.

L'exposition cutanée dans l'environnement général joue un rôle important dans l'apparition et l'entretien d'une hypersensibilité de contact qui peut être due à un contact quotidien avec des objets nickelés ou en alliage de nickel (par exemple des bijoux, des pièces de monnaie, des agrafes).

Les implants et prothèses en alliage de nickel, les liquides pour injection intraveineuse ou dialyse, et les produits de contraste utilisés en radiographie peuvent entraîner une exposition iatrogène au nickel. On estime qu'un malade sous dialyse reçoit à chaque traitement environ 100 µg de nickel en moyenne par voie intra-veineuse.

Dans les ambiances de travail, les concentrations de nickel en suspension dans l'air peuvent varier de quelques µg jusqu'à, parfois, plusieurs mg/m^3, selon le processus industriel et la teneur en nickel des produits manipulés.

Partout dans le monde, des millions de travailleurs sont exposés à des poussières et des fumées contenant du nickel lors de travaux tels que le soudage, le nickelage, le broyage, l'extraction minière, l'affinage du nickel ainsi que dans les aciéries, fonderies, et autres industries métallurgiques. Une exposition cutanée au nickel peut intervenir à l'occasion des activités professionnelles les plus diverses, qu'il s'agisse d'une exposition directe à du nickel dissous, par exemple, lors de l'affinage, du nickelage ou de l'électroformage ou lors de la manipulation d'outils contenant du nickel. Le nettoyage par voie humide peut comporter une exposition au nickel qui s'est dissous dans les eaux de lavage.

5. Cinétique et métabolisme chez l'homme et l'animal

Chez l'homme et l'animal le nickel peut être absorbé par inhalation, ingestion, ou par voie percutanée. En cas d'exposition professionnelle, la principale voie de pénétration dans l'organisme est l'absorption respiratoire avec résorption gastro-intestinale secondaire (formes solubles et insolubles). Une proportion importante des matières inhalées est avalée après avoir été éliminée des voies respiratoires sous l'action de l'ascenseur muco-ciliaire. Une hygiène personnelle insuffisante et de mauvaises habitudes de travail peuvent accroître l'exposition par voie digestive. L'absorption percutanée est négligeable quantitativement mais joue un rôle important dans la pathogénèse de l'hypersensibilité de contact. La résorption dépend de la solubilité du composé et elle décroît en général dans l'ordre suivant: nickel carbonyle

> composés solubles > composés insolubles. De tous les dérivés du nickel, c'est le nickel carbonyle qui est le plus rapidement et le plus complètement résorbé dans l'organisme de l'homme et des animaux. Les études comportant sur l'administration de nickel par voie respiratoire sont limitées. Des études au cours desquelles on a administré à des hamsters et à des rats de l'oxyde de nickel insoluble ont montré que celui-ci était faiblement résorbé, la plus grande partie du produit étant retenue dans les poumons après plusieurs semaines. En revanche, la résorption du chlorure de nickel soluble et du sulfure de nickel amorphe ont été rapides. Dans le sang, le nickel est principalement transporté par l'albumine.

Le nickel est plus ou moins résorbé au niveau gastrointestinal en fonction de la composition de l'alimentation. Lors d'une étude récente sur des volontaires humains, on a constaté que le nickel était résorbé à hauteur de 27% à partir de l'eau, contre moins de 1% à partir de la nourriture. L'excrétion peut s'effectuer en principe dans toutes les sécrétions et notamment l'urine, la bile, la sueur, les larmes, le lait et le liquide muco-ciliaire. Le nickel qui n'est pas résorbé est éliminé dans les matières fécales. On a mis en évidence l'existence d'un transfert transplacentaire chez des rongeurs. Après administration parentérale de sels de nickel, on a constaté que c'est au niveau des reins, des glandes endocrines, des poumons et du foie que l'accumulation de nickel était la plus importante: des concentrations élevées ont été également observées dans l'encéphale après administration de nickel carbonyle. Les données relatives à l'excrétion correspondent à un modèle bicompartimental. Chez des adultes en bonne santé exposés en dehors de leur activité professionnelle, on a observé des concentrations sériques et urinaires de 0,2 µg/litre (limites: 0,05–1,1 µg/litre) et de 1,5 µg/g de créatinine (limites: 0,5–4,0 µg/g de créatinine), respectivement. Après exposition professionnelle, on a constaté une augmentation des concentrations de nickel dans ces deux liquides. Pour un adulte non exposé de 70 kg, la charge de l'organisme en nickel est de 0,5 µg.

6. Effets sur les êtres vivants dans leur milieu naturel

Chez les microorganismes, on a constaté une inhibition de la croissance lorsque la concentration en nickel dans le milieu était de 1 à 5 mg/litre (cas des actinomycètes, des levures et des eubactéries marines ou non marines) ou de 5 à 1000mg/litre (cas des champignons filamenteux). Dans le cas des algues, on n'a pas observé de croissance à des concentrations d'environ 0,05–5 mg de nickel par litre. La toxicité du nickel dépend également de facteurs abiotiques tels que le pH, la dureté, la température et la salinité du milieu ainsi que de la présence de particules organiques ou minérales.

Chez les invertébrés aquatiques, la toxicité du nickel varie dans de très importantes proportions selon l'espèce et les facteurs abiotiques. Chez la daphnie, on a observé une CL_{50} à 96 h. de 0,5 mg/litre alors que chez trois mollusques, deux gastéropodes d'eau douce et un bivalve, ce paramètre était respectivement égal à 0,2 mg/litre et 1100 mg/litre.

Chez les poissons, les valeurs de la CL_{50} à 96 h. sont généralement comprises entre 4 et 20 mg/litre mais elles peuvent être plus élevées chez certaines espèces. Des études à long terme sur les poissons et leur développement effectuées dans des eaux douces ont montré qu'à des concentrations ne dépassant pas 0,05mg/litre, le nickel exerçait certains effets, en particulier sur la truite arc-en-ciel. Pour les plantes terrestres, des concentrations de nickel dépassant 50 mg/kg de poids à sec sont généralement toxiques. On a observé que le cuivre potentialisait la toxicité du nickel alors que le calcium la réduisait. On ne dispose que de données limitées au sujet des effets du nickel sur les animaux terrestres.

Les lombrics paraissent être relativement insensibles au nickel, du moins si le milieu abonde en microorganismes et matières organiques qui captent une partie du nickel aux dépens des vers. On n'a pas envisagé la possibilité que le nickel constitue un polluant majeur à l'échelon planétaire encore que des modifications écologiques, par exemple une diminution du nombre et de la diversité des espèces ait été observée à proximité de sources de nickel. En outre, des études portant sur les microécosystèmes ont montré qu'un apport de nickel dans le sol perturbait le cycle de l'azote.

7. Effets sur les animaux d'experience et sur les systèmes d'épreuve *in vitro*

Le nickel joue un rôle essentiel dans l'activité catalytique de certaines enzymes végétales et bactériennes. Chez certaines espèces animales on a observé, après administration d'un régime carencé en nickel, un ralentissement du gain de poids, de l'anémie et une moindre viabilité de la progéniture.

Le dérivé du nickel qui présente la plus forte toxicité aiguë est le nickel carbonyle dont le poumon constitue l'organe cible. Ce composé peut provoquer un oedème pulmonaire en l'espace de 4 heures. Les autres dérivés du nickel présentent une faible toxicité aiguë.

Si les allergies de contact sont très fréquentes chez l'homme, on ne parvient à produire une sensibilisation expérimentale chez l'animal que dans certaines conditions particulières. Chez le rat, la souris et le cobaye on a provoqué des lésions des muqueuses et des réactions inflammatoires des voies respiratoires en faisant inhaler pendant une longue période à ces animaux du nickel métallique, de l'oxyde et du sous-sulfure de nickel. On a observé une hyperplasie de l'épithélium chez des rats après inhalation d'aérosols de chlorure et d'oxyde de nickel.

Chez le rat, une exposition intense et prolongée à de l'oxyde de nickel a provoqué une pneumoconiose progressive. Des réactions inflammatoires quelquefois accompagnées d'une légère fibrose ont été observées chez des lapins après exposition intense à la poussière de nickel et de graphite. L'exposition de rats à du sous-sulfure de nickel a provoqué une fibrose pulmonaire.

Des sels de nickel, administrés par voie parentérale à des rats, des lapins et des poulets ont provoqué une hyperglycémie transitoire rapide. Ce phénomène pourrait être lié aux effets qui ont été observés sur les cellules alpha et béta des ilots de Langerhans. Le nickel réduit également la libération de prolactine. L'administration par voie orale ou par inhalation de chlorure de nickel a réduit la fixation d'iode par la thyroïde.

Administrés par voie intraveineuse à des chiens, des sels de nickel ont entraîné une réduction du débit sanguin des artères coronaires; à forte concentration, le nickel diminue la contractilité du myocarde du chien *in vitro*.

Le chlorure de nickel perturbe le système des cellules T et inhibe l'activité des cellules tueuses naturelles (cellules NK). On a observé que l'administration parentérale de chlorure et de sous-sulfure de

Resume et conclusions

nickel entraînait la mort intrautérine des foetus et réduisait le gain de poids chez le rat et la souris. Chez des rats et des hamsters, l'inhalation de nickel carbonyle a provoqué la mort des foetus et une diminution de leur poids, avec en outre des effets tératogènes. Les études en question ne donnaient aucune information sur la toxicité des composés du nickel pour les femelles gestantes. On a fait état de mutations létales dominantes obtenues chez des rats par exposition au nickel carbonyle.

Un certain nombre de dérivés minéraux du nickel ont fait l'objet d'études de mutagénicité dans divers systèmes d'épreuve. En général les dérivés du nickel se révèlent inactifs chez les bactéries sauf dans le cas des tests de fluctuation. On a observé des mutations chez plusieurs types de cellules mamaliennes en culture. Les dérivés du nickel inhibent la synthèse de l'ADN chez un grand nombre d'organismes. En outre, ces dérivés produisent des aberrations chromosomiques et des échanges entre chromatides soeurs tant dans les cellules mamaliennes que dans les cellules humaines en culture. Chez des personnes exposées, de par leur profession, à des composés insolubles ou solubles du nickel on a effectivement observé des aberrations chromosomiques, mais pas d'échange entre chromatides soeurs (sauf dans le cas d'une étude portant sur des ouvriers travaillant dans des ateliers d'électrolyse). *In vitro*, le nickel provoque la transformation des cellules.

Lors d'une étude d'inhalation, on a observé que du sous-sulfure de nickel provoquait l'apparition de tumeurs pulmonaires bénignes et malignes chez le rat. Quelques tumeurs pulmonaires ont également été observées chez des rats soumis à une série d'études où on leur faisait inhaler du nickel carbonyle. On n'a pas noté d'augmentation significative des tumeurs pulmonaires chez des rats soumis à l'inhalation de nickel métallique dans des conditions d'étude satisfaisantes. L'inhalation d'oxyde noir n'a pas provoqué non plus l'apparition de tumeurs pulmonaires chez des hamsters dorés (espèce qui résiste à la cancérogénèse au niveau pulmonaire). On ne dispose pas de bonnes études de cancérogénicité basées sur l'exposition par inhalation à d'autres composés du nickel. Toutefois on a constaté qu'après l'instillation intratrachéenne réitérée de sous-sulfure de nickel, de poudre de nickel métallique et d'un oxyde non précisé, des tumeurs pulmonaires bénignes se formaient chez le rat.

Le nickel carbonyle, le nickelocène, ainsi qu'un grand nombre de composés légèrement solubles ou insolubles comme le sous-sulfure de nickel, le carbonate, le chromate, l'hydroxyde et divers sulfures, des seléniures, des arséniures, des téllurures, des antimoniures, différents oxydes non spécifiés, deux oxydes de nickel et de cuivre, du nickel métallique et divers alliages, ont provoqué l'apparition de tumeurs mésenchymateuses locales chez divers animaux d'expérience après administration intramusculaire, sous-cutanée, intrapéritonéale, intrapleurale, intraoculaire, intraosseuse, intrarénale, intra-articulaire, intratesticulaire et intra-adipeuse. On n'a pas observé d'effet cancérogène local lors d'une étude portant sur des doses uniques de certains alliages de nickel, d'hydroxyde colloïdal de nickel et de deux échantillons d'oxyde, préparés spécialement en vue d'une étude de cancérogénicité par calcination à 735 et 1045 °C.

On a provoqué l'apparition de tumeurs de la cavité péritonéale chez des rats par administration intrapéritonéale réitérée de sulfate et d'acétate de nickel; toutefois aucun effet n'a été obtenu avec le chlorure de nickel.

On a étudié la cancérogénicité du nickel métallique et d'un grand nombre de dérivés du nickel par administration parentérale; à de rares exceptions près, toutes ces substances ont provoqué l'apparition de tumeurs locales.

C'est seulement dans le cas du sous-sulfure de nickel qu'on a des preuves convaincantes d'un pouvoir cancérogène après exposition par voie respiratoire. Toutefois le nombre d'études de ce type qui sont vraiment satisfaisantes reste très limité.

Lors d'études au cours desquelles on a procédé à des instillations intratrachéennes répétées de poudre de nickel, d'oxyde et de sous-sulfure de nickel, on a observé l'apparition de tumeurs pulmonaires.

Trois sels solubles de nature différente qui n'avaient pas produit de tumeurs locales lors d'études antérieures ont été réétudiés en procédant cette fois à une administration intrapéritonéale réitérée: deux de ces sels ont alors manifesté des effets cancérogènes.

8. Effets sur l'homme

C'est le nickel carbonyle qui présente la toxicité aiguë la plus forte pour l'homme. Elle se traduit par des céphalées frontales, des vertiges, de la nausée, des vomissements, de l'insomnie et de l'irritabilité, puis par des symptômes pulmonaires évoquant une pneumonie d'origine virale. Parmi les lésions pulmonaires pathologiques qui ont été observées on peut citer des hémorragies, de l'oedème et une désorganisation cellulaire. Le foie, les reins, les surrénales, la rate et le cerveau sont également affectés. On a aussi fait état d'intoxications par le nickel chez des malades sous dialyse après contamination du liquide de dialyse par des dérivés du nickel ainsi que chez des nickeleurs qui avaient bu accidentellement de l'eau contaminée par du sulfate et du chlorure de nickel.

On a signalé chez des nickeleurs et chez des ouvriers d'ateliers d'affinage du nickel des effets chroniques tels que rhinite, sinusite, perforation de la cloison nasale et asthme. Certains auteurs font état d'anomalies pulmonaires, notamment une fibrose, chez des ouvriers amenés à inhaler de la poussière de nickel. En outre, on a signalé des cas de dysplasie nasale chez des ouvriers travaillant dans des ateliers d'affinage du nickel. L'hypersensibilité de contact est très largement documentée tant dans la population générale que chez un certain nombre d'ouvriers exposés de par leur profession à des dérivés solubles du nickel. Dans plusieurs pays on indique que 10% de la population féminine et 1% de la population masculine présentent une sensibilité au nickel. Parmi ces personnes, 40 à 50% présentent un eczéma vésiculeux des mains qui peut être très grave dans certains cas et entraîner une incapacité. L'ingestion de nickel peut aggraver l'eczéma siégeant au niveau des mains ou d'autres régions du corps protégées de tout contact avec du nickel.

On a également signalé que les prothèses et autres implants chirurgicaux en alliage de nickel pouvaient produire une sensibilisation à ce métal ou aggraver une dermatite préexistante.

Des effets néphrotoxiques, comme un oedème rénal avec hyperémie et dégénérescence du parenchyme, ont été observés dans des cas d'exposition industrielle accidentelle à du nickel carbonyle.

Des effets néphrotoxiques passagers ont été enregistrés après ingestion accidentelle de sels de nickel.

On a fait état d'un risque très élevé de cancers pulmonaire et nasal chez des ouvriers travaillant dans des ateliers d'affinage du nickel où l'on effectue le grillage à mort des minerais sulfurés, opération qui entraîne une exposition importante au sous-sulfure, à l'oxyde et éventuellement au sulfate de nickel. L'exposition à des dérivés solubles du nickel (électrolyse, extraction du sulfate de cuivre, hydrométallurgie), souvent associée à une exposition à l'oxyde de nickel mais sans exposition trop importante au sous-sulfure, comporte des risques analogues. Pour les mineurs et les ouvriers des ateliers d'affinage, le risque est beaucoup plus faible. Chez les soudeurs d'acier inoxydable ou chez ceux qui travaillent dans des industries employant du nickel, les taux de cancer sont généralement proches de la normale, sauf dans le cas où il y a également exposition au chrome, en particulier dans des ateliers de galvanoplastie. Il semblerait cependant que les ouvriers employés à la fabrication des accumulateurs au nickel/cadmium, qui sont exposés à de fortes concentrations de nickel et de cadmium, soient exposés à un risque légèrement plus important de cancer du poumon.

On a signalé occasionnellement un excès de divers cancers autres que ceux qui siègent au niveau des poumons ou du nez, par exemple des cancers rénaux, gastriques ou prostatiques, chez des travailleurs de l'industrie du nickel, mais cette surmorbidité n'a pas été systématiquement constatée.

On peut s'appuyer sur les données épidémiologiques pour répondre à deux questions importantes: i) certains dérivés du nickel sont-ils effectivement cancérogènes, et ii) l'étude de cohortes faiblement exposées permet-elle d'établir la limite supérieure du risque pour des niveaux d'exposition donnés?

a) Dérivés solubles

On a mis en evidence un certain risque de cancer chez des ouvriers exposés à des concentrations en sels solubles de l'ordre de 1 à 2 mg/m^3, qu'il s'agisse d'ouvriers d'ateliers d'électrolyse ou travaillant à la préparation de sels solubles. Ces ouvriers étaient également exposés à d'autres dérivés du nickel mais souvent à des concentrations plus faibles que dans les processus à haut risque. En l'absence d'une évaluation rétrospective de l'exposition, il est impossible de tirer des conclusions définitives de ces observations mais il existe certainement une forte présomption d'une cancérogénicité des dérivés solubles du nickel. Les poussières des ateliers d'affinage contiennent quelques fois une proportion non négligeable de sulfate de nickel à côté du sous-sulfure. Il se pourrait donc que le risque de cancer très important qui a été observé chez les ouvriers qui travaillent au grillage à mort des minerais sulfurés soit dû pour une part aux dérivés solubles du nickel.

b) Sous-sulfure de nickel (sulfure nickeleux)

Dans les ateliers d'affinage où le risque de cancer est élevé, on a constaté que l'exposition au sous-sulfure de nickel était toujours accompagnée d'une exposition à l'oxyde et peut-être même au sulfate (voir plus haut). Il est donc difficile à partir des seules données épidémiologiques d'affirmer que le sous-sulfure est cancérogène, encore que ce soit probable.

c) Oxyde de nickel

Dans tous les cas où l'on a observé un risque élevé de cancer, il y avait présence d'oxyde de nickel accompagné d'une ou de plusieurs autres formes (sous-sulfure, sels solubles, nickel métallique). Comme dans le cas du sous-sulfure, il est difficile d'affirmer ou d'écarter l'existence d'un pouvoir cancérogène sur la base des seules données épidémiologiques.

d) Nickel métallique

On n'a pas constaté d'accroissement du risque de cancer chez des ouvriers exposés uniquement à du nickel métallique. L'ensemble des données relatives aux ouvriers qui travaillaient sur des alliages du nickel ou dans des ateliers de diffusion gazeuse et qui tous étaient exposés à des concentrations moyennes de l'ordre de 0,5 mg/m^3, ne font pas ressortir d'élévation du risque, encore que le nombre total de cancers du poumon parmi ces cohortes ait été trop faible pour qu'on puisse exclure un léger accroissement du risque à cette concentration.

e) Conclusion

Bien que quelques-uns, voire tous les dérivés du nickel puissent être cancérogènes, il n'existe que peu ou pas de risque décelable dans la plupart des secteurs de l'industrie du nickel aux niveaux d'exposition actuels; cette observation vaut aussi pour certains procédés dont on estimait jusqu'ici qu'ils comportaient un risque très élevé de cancers pulmonaire et nasal. Une exposition prolongée à des dérivés solubles du nickel à des concentrations de l'ordre de 1 mg/m^3 peut entraîner une augmentation sensible du risque relatif de cancer pulmonaire mais ce risque est à peu près égal à l'unité chez les ouvriers exposés à des concentrations en nickel métallique de l'ordre de 0,5 mg/m^3 en moyenne. Pour un niveau d'exposition donné, le risque de cancer est plus important dans le cas des composés solubles que dans le cas du nickel métallique et peut-être des autres dérivés du nickel. L'absence d'un risque notable de cancer pulmonaire chez les nickeleurs n'est pas surprenante étant donné qu'en moyenne l'exposition aux dérivés solubles est beaucoup plus faible dans les ateliers de galvanoplastie que dans les ateliers d'affinage électrolytique ou de traitement des sels de nickel.

RESUMEN Y CONCLUSIONES

1. **Identidad, propiedades físicas y qumícas y métodos analíticos**

   El níquel es un elemento metálico perteneciente al grupo VIIIb de la tabla periódica. Es resistente a los álcalis, pero en general se disuelve bien en ácidos oxidantes diluidos. El carbonato de níquel, el sulfuro de níquel y el óxido de níquel no son solubles en agua, mientras que sí lo son el cloruro de níquel, el sulfato de níquel y el nitrato de níquel. El carbonilo de níquel es un líquido incoloro y volátil que se descompone por encima de los 50 °C. La forma iónica predominante es la de níquel (II). En los sistemas biológicos, el níquel disuelto puede formar componentes complejos con diversos ligandos y unirse a materiales orgánicos.

   Los métodos que más se utilizan en el análisis de materiales de origen biológico y medio ambiental son la espectroscopía de absorción atómica y la voltamperimetría. A fin de obtener resultados fidedignos, sobre todo en el orden de las ultratrazas, hay que adoptar procedimientos específicos para reducir al mínimo el riesgo de contaminación durante la recogida, el almacenamiento, el tratamiento y el análisis de muestras. En los materiales de origen biológico y en el agua se puede llegar límites de detección de 1-100 ng/litro, atendiendo a los procedimientos de tratamiento previo, extracción y enriquecimiento de la muestra.

2. **Fuentes de exposición humana y ambiental**

   El níquel es un metal traza muy ampliamente distribuido, que se encuentra en el suelo, el agua, el aire y la biosfera. En la corteza terrestre hay un porcentaje medio del 0,008%. Los terrenos cultivables contienen entre 3 y 1000 mg de níquel/kg. En el agua natural los niveles oscilan entre 2 y 10 µg/litro (agua dulce) y de 0,2 a 0,7 µg/litro (agua de mar). Las concentraciones atmosféricas de níquel en zonas remotas son de 0,1 a 3 ng/m^3.

Los yacimientos metalíferos de níquel están formados por acumulaciones de minerales de sulfuro de níquel (principalmente pentlandita) y lateritas. El níquel se extrae de la mena mediante procesos de afinado piro e hidrometalúrgico. La mayor parte del níquel obtenido se destina a la producción de acero inoxidable y a otras aleaciones muy resistentes a la corrosión y a la temperatura. Las aleaciones de níquel y el niquelado se utilizan en vehículos, maquinaria industrial, armamento, herramientas, equipo eléctrico, utensilios domésticos y la acuñación de monedas. También se utilizan compuestos de níquel como catalizadores, pigmentos y en las baterías. En 1985, la producción minera mundial de níquel fue de unos 67 millones de kg. Las principales fuentes de emisión de níquel a la atmósfera son la combustión de carbón y petróleo para la obtención de calor o energía, la incineración de desechos y fangos cloacales, la extracción minera y la producción primaria de níquel, la fabricación de acero, la galvanoplastia y otras fuentes, como la producción de cemento. En el aire contaminado predominan el sulfato, los óxidos y los sulfuros de níquel, y en menor cantidad el níquel metálico.

El níquel procedente de los distintos procesos industriales y de otras fuentes termina llegando a las aguas residuales. Los restos que se obtienen del tratamiento de esas aguas se eliminan mediante inyección en pozos profundos, vertido en los océanos, tratamiento en tierra e incineración. Los efluentes de las instalaciones de tratamiento de aguas residuales contienen, según los datos disponibles, hasta 0,2 mg de níquel/litro.

3. Transporte, distribución y transformación en el medio ambiente

El níquel se introduce en el medio ambiente a partir de fuentes tanto naturales como de origen humano y circula por todos sus compartimentos mediante procesos físicos y químicos, así como por el transporte biológico que efectúan los organismos vivos.

Se considera que en la atmósfera el níquel está principalmente en forma de aerosoles de partículas con distintas concentraciones del metal, en función de su procedencia. Las concentraciones más altas de níquel en el aire suelen aparecer en las partículas más pequeñas.

El carbonilo de níquel es inestable en el aire y se descompone formando óxido de níquel.

El transporte y la distribución de las partículas de níquel a los diferentes compartimentos del medio ambiente o entre ellos depende en gran medida del tamaño de las partículas y de las condiciones meteorológicas. La distribución por tamaños de las partículas depende sobre todo de las fuentes de emisión. En general, las partículas de fuentes artificiales son más pequeñas que las que se encuentran en el polvo natural.

El níquel ingresa en la hidrosfera por eliminación desde la atmósfera, por la escorrentía superficial, por la descarga de residuos industriales y municipales, y también tras la erosión natural de los suelos y de las rocas. En los ríos, este elemento se transporta principalmente en forma de revestimiento precipitado sobre partículas y unido a materia orgánica; en los lagos, se transporta en forma iónica y de manera predominante unido a materia orgánica. La partículas de arcilla también lo pueden absorber, y a la biota pasa por ingestión o absorción. El proceso de absorción se puede invertir, lo que se traduce en una liberación del níquel del sedimento. Los ríos y arroyos transportan una parte hasta el mar. El aporte fluvial de partículas suspendidas se estima en 135×10^7 kg/año.

En función del tipo de suelo, el níquel puede presentar una gran movilidad dentro del perfil edáfico hasta alcanzar las aguas subterráneas y, con ello, los ríos y lagos. La lluvia ácida tiene una marcada tendencia a movilizar el níquel del suelo. Las plantas terrestres lo absorben principalmente por las raíces. La cantidad absorbida depende de diversos parámetros geoquímicos y físicos, entre los que figuran el tipo de suelo, su pH y humedad, su contenido en materia orgánica y la concentración de níquel extraíble. El caso más conocido de acumulación de níquel es el de algunas especies de plantas ("hiperacumuladoras"), que crecen en suelos serpentinosos relativamente poco fértiles y en las que se han encontrado niveles superiores a 1 mg/kg de peso seco. Los niveles de níquel superiores a 50 mg/kg de peso seco son tóxicos para la mayoría de las plantas. Se han observado casos de acumulación y efectos tóxicos en hortalizas cultivadas en suelos tratados con fangos residuales y en la vegetación próxima a las fuentes de emisión del metal. En plantas acuáticas se han detectado factores de concentración elevados. Los estudios de laboratorio han puesto de manifiesto que la capacidad de acumulación de níquel en los peces examinados es escasa. En aguas no contaminadas, las concentraciones halladas en los peces enteros (sobre la base del peso fresco) oscilaron entre 0,02 y 2 mg/kg. Estos valores pueden ser hasta diez veces superiores en los peces procedentes de aguas contaminadas. Sin embargo, no hay pruebas de que se produzca bioamplificación del níquel en la cadena de los alimentos.

El níquel se encuentra en muchos órganos y tejidos de animales silvestres, debido a su ingestión con los alimentos por los animales herbívoros y sus predadores carnívoros.

4. Niveles medioambientales y exposición humana

Los niveles de níquel en los organismos terrestres y acuáticos pueden diferir en varios ordenes de magnitud. La exposición humana a los valores habituales en la atmósfera, que oscilan entre alrededor de 5 y 35 ng/m3 en zonas rurales y urbanas, hace que la inhalación de este elemento sea de 0,1–0,7 µg/día. Aunque el agua potable contiene menos de 10 µg/litro, en ocasiones los empalmes de las tuberías pueden liberarlo, dando lugar a concentraciones de hasta 500 µg/litro.

Su concentración en los alimentos suele estar por debajo de 0,5 mg/kg de peso fresco. El cacao, la soja, algunas legumbres secas, diversos frutos secos y la harina de avena contienen elevadas concentraciónes de níquel. Su ingestión diaria con los alimentos varía ampliamente en función de los diferentes hábitos alimentarios, y puede oscilar entre 100 y 800 µg/día; la ingestión media de níquel en la dieta es, en la mayoría de los países, de 100–300 µg/día. Los utensilios de cocina liberan níquel, lo que contribuye de manera considerable a su ingestión. Fumando 40 cigarrillos diarios pueden absorberse a través de los pulmones 2–23 µg/día.

La exposición cutánea en el medio ambiente general es importante, porque induce y mantiene la hipersensibilidad por contacto que produce en la piel el uso diario de objetos niquelados o con aleaciones de este metal (por ejemplo, joyería, monedas, "clips").

Puede haber exposición iatrogénica al níquel debida a las implantaciones y las prótesis hechas con aleaciones que lo contienen, a fluidos intravenosos o de diálisis y a los medios de contraste radiográfico. Se estima que la absorción media de níquel por vía intravenosa a partir de los fluidos de diálisis es de 100 µg por tratamiento.

En el ámbito laboral, la concentración en el aire puede variar desde varios g/m3 a, en ocasiones, unos mg/m3, en función del proceso de que se trate y del contenido en níquel del material que se maneja.

Hay millones de trabajadores de todo el mundo expuestos a polvo y humo que contienen níquel durante las operaciones de soldadura, niquelado y pulimentado, la extracción minera, el afinado del níquel, y en las acerías, fundiciones y otras industrias metalúrgicas.

Hay una gran variedad de empleos en los que puede haber exposición cutánea al níquel, sea por contacto directo con el metal disuelto, como por ejemplo en las industrias de afinado, galvanoplastia y electroconformación, o bien por el manejo de herramientas que lo contienen. En los trabajos de limpieza con agua puede haber una exposición a este metal, a causa de las cantidades que lleva disueltas el agua de lavado.

5. Cinética y metabolismo en el hombre y en los animales

El hombre y los animales pueden absorber níquel por inhalación, por ingestión o por vía cutánea. La absorción respiratoria, con asimilación gastrointestinal secundaria de níquel (soluble e insoluble), es la principal vía de penetración durante la exposición en el trabajo. Una cantidad importante del material inhalado se traga con el fluido mucociliar procedente del tracto respiratorio. La mala higiene personal y las prácticas laborales deficientes pueden contribuir a la exposición gastrointestinal. La absorción percutánea es cuantitativamente insignificante, pero es importante en la patogenia de la hipersensibilidad por contacto. La absorción está relacionada con la solubilidad del compuesto, y sigue la relación general carbonilo de níquel > compuestos de níquel solubles > compuestos de níquel insolubles. El carbonilo de níquel es el compuesto de este metal que absorben de forma más rápida y completa el hombre y los animales. Hay pocos estudios en los que se haya administrado níquel por inhalación. En experimentos con hámsters y ratas tratados con óxido de níquel insoluble se observó una escasa absorción, con retención pulmonar de gran parte del material después de varias semanas. En cambio, la absorción de cloruro de níquel soluble o de sulfuro de níquel amorfo fue rápida. El níquel es transportado en la sangre, principalmente ligado a la albúmina.

Su absorción gastrointestinal es variable y depende de la composición de la dieta. En un estudio reciente con un grupo de voluntarios se encontró que la absorción de níquel a partir del agua era del 27%, mientras que a partir de los alimentos era inferior al 1%. Todas las secreciones orgánicas, como la orina, la bilis, el sudor, las lágrimas, la leche y el fluido mucociliar, son posibles vías de eliminación. El níquel que no se absorbe se elimina con las heces. En roedores se ha demostrado que atraviesa la placenta. Tras la administración parenteral de sales de níquel, los niveles más altos de acumulación se producen en el riñón, las glándulas endocrinas y el hígado; si lo que se administra es carbonilo de níquel, también se observa una alta concentración en el cerebro. Los datos sobre excreción del níquel sugieren un modelo bicompartimental. Su concentración en el suero y la orina de adultos sanos no expuestos en el trabajo es de 0,2 µg/litro (oscilación: 0,05–1,1 µg/litro) y 1,5 µg/g de creatinina (oscilación: 0,5–4,0 µg/g de creatinina), respectivamente. Se ha observado que después de la exposición profesional, la concentración aumenta en ambos fluidos. La carga corporal de níquel en un adulto de 70 kg de peso no expuesto es de 0,5 µg.

6. Efectos en los seres vivos del medio ambiente

En los microorganismos, con concentraciones de 1–5 mg/litro de níquel en el medio se inhibía, en general, el crecimiento en actinomicetos, levaduras y eubacterias marinas y no marinas, y con niveles de 5–1000 mg/litro, en hongos filamentosos. En el caso de las algas, con una concentración aproximada de 0,05–5 mg de níquel/litro no se observó crecimiento. Ciertos factores abióticos, como el pH, la dureza, la temperatura y la salinidad del medio, así como la presencia de partículas orgánicas e inorgánicas, influyen en la toxicidad del níquel.

La toxicidad del níquel para los invertebrados acuáticos varía considerablemente en función de las especies y de los factores abióticos. En el género *Daphnia* se ha encontrado una CL_{50} a las 96 horas de 0,5 mg de níquel/litro, mientras que en moluscos los valores de la CL_{50} a las 96 horas fueron de unos 0,2 mg/litro en dos especies de caracol de agua dulce y de 1100 mg/litro en un bivalvo.

En los peces, los valores de la CL_{50} a las 96 horas suelen oscilar entre 4 y 20 mg de níquel/litro, pero pueden ser más altos en algunas especies. Los estudios a largo plazo realizados sobre los peces y su crecimiento en aguas blandas pusieron de manifiesto que niveles tan bajos como 0,05 mg de níquel/litro tenían algunos efectos en la trucha arco iris. En plantas terrestres se ha encontrado que los niveles superiores a 50 mg/kg de peso seco suelen ser tóxicos. Se ha observado que desde el punto de vista toxicologica el cobre tiene un efecto sinérgico con el níquel, mientras que el calcio reduce su toxicidad. Hay pocos datos acerca de los efectos del níquel en animales terrestres.

La lombriz de tierra parece ser relativamente insensible a este metal si el medio es rico en microorganismos y materia orgánica, que dejan menos níquel disponible para aquella. Aunque el níquel no se ha considerado un contaminante mundial en gran escala, en las cercanías de fuentes emisoras se han observado cambios ecológicos, como la disminución del número y la diversidad de especies. En estudios de microecosistemas se ha puesto de manifiesto que la adición de este elemento al suelo altera el ciclo del nitrógeno.

7. **Efectos en los animales de experimentación y en sistemas de prueba** *in vitro*

El níquel es indispensable para la actividad catalítica de algunas enzimas vegetales y bacterianas. En algunas especies animales se ha señalado que una dieta carente de níquel provoca lentitud en la ganacia de peso, anemia y una disminución de la viabilidad de la descendencia.

El compuesto de toxicidad más aguda es el carbonilo de níquel, cuyo órgano destinatario es el pulmón; dentro de las cuatro horas siguientes a la exposición puede aparecer edema pulmonar. La toxicidad aguda de los demás compuestos del níquel es baja.

Aunque la alergia de contacto causada por este metal es muy frecuente en el hombre, en animales sólo se consigue inducir sensibilización experimental en determinadas condiciones. La exposición a un largo período de inhalación de níquel metálico, óxido de níquel o subsulfuro de níquel causó lesiones en las mucosas y una reacción inflamatoria del tracto respiratorio en ratas, ratones y cobayos. En ratas expuestas a la inhalación de aerosoles de cloruro de níquel u óxido de níquel se observó hiperplasia epitelial.

La exposición a altos niveles de óxido de níquel durante un período prolongado produjo una neumoconiosis progresiva en ratas. En conejos expuestos a un nivel elevado de polvo de níquel-grafito se observó una reacción inflamatoria, a veces acompañada de una ligera fibrosis. Las ratas expuestas a subsulfuro de níquel presentaron fibrosis pulmonar.

La administración de sales de níquel a ratas, conejos y pollos por vía parenteral indujo una hiperglucemia rápida transitoria. Estos cambios pueden estar relacionados con los efectos en las células alfa y beta de los islotes de Langerhans. El níquel también disminuyo la liberación de prolactina. Se ha informado de que la administración por vía oral o respiratoria de cloruro de níquel reduce la absorción de yodo por el tiroides.

La administración intravenosa de sales de níquel a perros hizo disminuir el flujo sanguíneo en las arterias coronarias; la concentración elevada de este metal redujo la contractibilidad del miocardio de perro *in vitro*.

El cloruro de níquel afecta al sistema de células T y suprime la actividad de las células destructoras naturales. Se ha observado que la administración parenteral de cloruro de níquel y de subsulfuro de níquel a ratas y ratones causa mortalidad intrauterina y una disminución de la ganancia de peso. En ratas y hámsters expuestos a inhalaciones de carbonilo de níquel se produjo la muerte de fetos y una disminución de la ganancia de peso, y tuvo también un efecto teratógenico. En ninguno de esos estudios se informa acerca de la toxicidad materna. Se ha señalado que el carbonilo de níquel ocasiona mutaciones letales dominantes en la rata.

Resumen y conclusiones

Se han hecho pruebas con distintos sistemas para determinar la mutagenicidad de varios compuestos inorgánicos del níquel. Estos, en general, se mostraron inactivos en los ensayos de mutagénesis efectuados con bacterias, excepto en los casos en que se utilizaron pruebas de fluctuación. Se observaron mutaciones en varios tipos de células de mamíferos cultivadas. Los compuestos del níquel inhibieron la síntesis de ADN en una gran variedad de organismos. Además, produjeron aberraciones cromosómicas e intercambio de cromátidas hermanas en cultivos de células tanto de mamíferos como humanas. En personas profesionalmente expuestas a compuestos de níquel insolubles o solubles se observaron aberraciones cromosómicas, pero no intercambio de cromátidas hermanas (excepto en un estudio sobre personas que trabajan en electrólisis). El níquel indujo transformación celular *in vitro*.

En un estudio de inhalación en ratas, el subsulfuro de níquel provocó la aparición de tumores pulmonares benignos y malignos. En una serie de estudios de inhalación de carbonilo de níquel en la rata, se observaron algunos tumores pulmonares. En un estudio de inhalación de níquel metálico efectuado en ratas no se apreció un aumento importante de tales tumores. La inhalación de óxido de níquel negro no indujo la aparición de tumores pulmonares en hámsters dorados sirios (especie resistente a la carcinogénesis pulmonar). No se conocen estudios adecuados sobre la carcinogénesis por inhalación de otros compuestos del níquel. Sin embargo, la instilación intratraqueal repetida de subsulfuro de níquel, de polvo de níquel metálico y de un óxido de níquel no especificado produjo en ratas tumores pulmonares benignos y malignos.

El carbonilo de níquel, el niqueloceno y gran número de compuestos del níquel ligeramente solubles o insolubles, entre ellos el subsulfuro, el carbonato, el cromato, el hidróxido, los sulfuros, los seleniuros, los arseniuros, el telururo, el antimoniuro, diversas preparaciones de óxidos no identificadas, dos óxidos de níquel y cobre, el níquel metálico y diversas aleaciones de níquel provocaron tumores mesenquimáticos locales en distintos animales de experimentación tras la administración por vía intramuscular, subcutánea, intraperitoneal, intrapleural, intraocular, intraósea, intrarrenal, intraarticular, intratesticular o intraadiposa. No se observó respuesta carcinogénica local en los estudios de dosis única con algunas aleaciones de níquel, hidróxido de níquel coloidal o dos muestras de óxido de níquel, especialmente preparados para pruebas de carcinogénesis mediante calcinación a 735 °C o bien a 1045 °C.

La administración repetida de sulfato y acetato de níquel a ratas por vía intraperitoneal indujo tumores en la cavidad peritoneal, pero no la de cloruro de níquel.

Se han realizado pruebas de carcinogénesis mediante la administración por vía parenteral de níquel metálico y de una gran variedad de compuestos; salvo

algunas excepciones, se produjeron tumores locales.

Sólo en el caso del subsulfuro de níquel se ha demostrado de manera convincente que la inhalación produce cáncer. Sin embargo, hay muy pocos estudios adecuados sobre la exposición por vía respiratoria.

En los estudios de aplicación de instilaciones intratraqueales repetidas de polvo de níquel, óxido de níquel y subsulfuro de níquel aparecieron tumores pulmonares.

Cuando se realizaron pruebas con tres sales solubles de níquel distintas que en estudios anteriores no habían provocado la aparición de tumores locales, utilizando la administración intraperitoneal repetida, dos de las sales indujeron una respuesta carcinogénica.

8. Efectos en la especie humana

En relación con la salud del ser humano, el carbonilo de níquel es el compuesto del metal con toxicidad más aguda. Entre los efectos de la intoxicación aguda que causa están dolor de cabeza frontal, vértigo, náuseas, vómitos, insomnio e irritabilidad, seguidos de trastornos pulmonares análogos a los producidos por una neumonía vírica. Entre las lesiones patológicas de los pulmones figuran hemorragias, edema y desorganización celular. También se ven afectados el hígado, los riñones, las cápsulas suprarrenales, el bazo y el cerebro. Ha habido asimismo casos de intoxicación en pacientes sometidos a diálisis con contaminación de níquel y en galvanizadores que accidentalmente han ingerido agua contaminada por sulfato y cloruro de níquel.

Resumen y conclusiones

Se han notificado casos de efectos crónicos, como rinitis, sinusitis, perforación del tabique nasal y asma, entre trabajadores del afinado del níquel y del niquelado. Algunos autores han descrito alteraciones pulmonares con fibrosis en personas que en el trabajo respiraban polvo de níquel. Además, se ha informado de casos de displasia nasal en trabajadores del afinado del níquel. Hay abundante documentación acerca de la hipersensibilidad por contacto, tanto en la población en general como en algunos trabajadores profesionalmente expuestos a compuestos solubles de níquel. En varios países se ha comunicado que el 10% de la población femenina y el 1% de la masculina son sensibles al níquel. De ellos, el 40-50% tienen eccema vesicular en las manos; en algunos casos puede ser muy grave y causa de incapacidad laboral. La ingestión de níquel puede agravar la eccema vesicular de las manos y, posiblemente, inducir también su aparición en otras partes del cuerpo que no han tenido contacto cutáneo con este metal.

Se sabe que las prótesis y otras implantaciones quirúrgicas de aleaciones con níquel causan sensibilización a este elemento o agravan la dermatitis existente.

Se han descrito efectos nefrotóxicos, como edema renal con hiperemia y degeneración parenquimatosa, en casos de exposición industrial accidental a carbonilo de níquel. También se han registrado efectos nefrotóxicos pasajeros por ingestión accidental de sales de níquel.

Se ha informado de un elevadísimo riesgo de cáncer del pulmón y la nariz en los trabajadores del afinado del níquel que se ocupan de tostar a alta temperatura los minerales de sulfuro, lo que supone una gran exposición al subsulfuro, el óxido y quizás el sulfato de níquel. Se han señalado riesgos análogos en los procesos que entrañan exposición al níquel soluble (electrólisis, extracción de sulfato de cobre, hidrometalurgia), a menudo combinados con alguna exposición al óxido de níquel pero muy poca al subsulfuro. El riesgo de los mineros y otros trabajadores del afinado parece ser mucho menor. El índice de cáncer entre los soldadores de acero inoxidable y los trabajadores de industrias que usan níquel ha sido, en general, casi normal, salvo cuando hay exposición al cromo, particularmente en la galvanoplastia. Sin embargo, los trabajadores que hacen baterías de níquel/cadmio y están expuestos a altos niveles de ambos metales corren un riesgo ligeramente mayor de contraer cáncer de pulmón.

Ocasionalmente se ha notificado en trabajadores del níquel una frecuencia excesiva de otros tipos de cáncer distintos del de pulmón y el nasal, como por ejemplo renal, gástrico o de próstata, pero ninguno se ha encontrado de manera constante.

Se pueden utilizar los datos epidemiológicos para plantear dos importantes cuestiones: i) si se ha demostrado que ciertos compuestos de níquel que son carcinógenos; y ii) si las cohortes sometidas a bajas exposiciónes presentan límites superiores de riesgo a determinados niveles de exposición.

a) Níquel soluble

Se obtuiseron pruebas de que los trabajadores expuestos a concentraciones de níquel soluble del orden de 1–2 mg/m^3, tanto en el sector de la electrólisis como en el de la preparación de sales solubles, corrían riesgo de cáncer. Esos trabajadores estaban expuestos también a otros compuestos de níquel, pero con frecuencia a niveles más bajos que en otros procesos de alto riesgo. En ausencia de mediciones anteriores de la exposición, es imposible llegar a conclusiones definitivas, pero las pruebas de que el níquel soluble es carcinógeno son ciertamente sólidas. El polvo del afinado contiene a veces una proporción importante de sulfato de níquel, además de subsulfuro. Esto plantea la posibilidad de que el altísimo riesgo de cáncer observado en los trabajadores que se ocupan de la oxidación de subsulfuro de níquel a alta temperatura pueda deberse en parte al níquel soluble.

b) Subsulfuro de níquel

En las zonas de afinado donde el riesgo de contraer cáncer era alto, la exposición al subsulfuro de níquel casi siempre iba acompañada de la exposición al óxido y, quizás, al sulfato (véase más arriba). Así pues, basándose solamente en los datos epidemiológicos, es difícil demostrar que el subsulfuro de níquele sea carcinógeno, aunque parece probable.

c) Oxido de níquel

El óxido de níquel estaba presente en casi todas las circunstancias en que el riesgo de cáncer era elevado, acompañado de una o más formas del elemento (subsulfuro de níquel, níquel soluble, níquel metálico). Como en el caso del subsulfuro de níquel, es difícil afirmar o negar su carácter carcinogénico basándose exclusivamente en los datos epidemiológicos.

d) *Níquel metálico*

No se ha demostrado que los trabajadores expuestos exclusivamente al níquel metálico corran mayor riesgo de cáncer. Los datos combinados relativos a trabajadores que se ocupan de aleaciones de níquel y los dedicados a difusión gaseosa, todos ellos expuestos a concentraciones medias del orden de 0,5 mg de níquel/m^3, no muestran un riesgo excesivo, aunque el número total de casos de cáncer de pulmón en estas cohortes era demasiado pequeño para excluir un pequeño aumento del riesgo a este nivel.

e) *Conclusión*

Aunque algunas, y quizás todas, las formas del níquel pueden ser carcinógenas, con los niveles normales de exposición en muchos sectores de la industria de este metal hay un riesgo pequeño o no detectable; entre ellos se encuentran algunos procesos que en el pasado estaban asociados con un riesgo muy alto de cáncer pulmonar y nasal. La exposición prolongada a concentraciones del orden de 1 mg/m^3 de níquel soluble puede ocasionar un importante aumento del riesgo relativo de cáncer de pulmón, pero este riesgo entre los trabajadores expuestos a un nivel medio de níquel metálico de alrededor de 0,5 mg/m^3 es aproximadamente 1. El riesgo de cáncer con un determinado nivel de exposición puede ser más alto para los compuestos de níquel solubles que para el níquel metálico, y posiblemente, que para otras formas. No es extraña la ausencia de cualquier riesgo pronunciado de cáncer pulmonar entre los niqueladores, puesto que el promedio de exposición al níquel soluble es mucho más bajo que el que se produce en el afinado electrolítico o en el tratamiento de sales de níquel.

www.ingramcontent.com/pod-product-compliance
Ingram Content Group UK Ltd.
Pitfield, Milton Keynes, MK11 3LW, UK
UKHW021315180426
11947UKWH00015B/1236